Stefano Della Torre · Massimiliano Bocciarelli ·
Laura Daglio · Raffaella Neri
Editors

Buildings for Education

A Multidisciplinary Overview of The Design of School Buildings

Editors
Stefano Della Torre
Architecture, Built Environment
and Construction Engineering—ABC
Department
Politecnico di Milano
Milan, Italy

Massimiliano Bocciarelli
Architecture, Built Environment
and Construction Engineering—ABC
Department
Politecnico di Milano
Milan, Italy

Laura Daglio
Architecture, Built Environment
and Construction Engineering—ABC
Department
Politecnico di Milano
Milan, Italy

Raffaella Neri
Architecture, Built Environment
and Construction Engineering—ABC
Department
Politecnico di Milano
Milan, Italy

https://doi.org/10.1007/978-3-030-33687-5

Preface

This book belongs to a series, which aims at emphasizing the impact of the multidisciplinary approach practiced by ABC Department scientists to face timely challenges in the industry of the built environment. Following the concept that innovation happens as different researches stimulate each other, skills and integrated disciplines are brought together within the department, generating a diversity of theoretical and applied studies.

Therefore, the books present a structured vision of the many possible approaches—within the field of architecture and civil engineering—to the development of researches dealing with the processes of planning, design, construction, management, and transformation of the built environment. Each book contains a selection of essays reporting researches and projects, developed during the last six years within the ABC Department (Architecture, Built Environment, and Construction Engineering) of Politecnico di Milano, concerning a cutting-edge field in the international scenario of the construction sector. The design of schools has been recognized as one of the hottest topics in architectural research, also for the criticalities detected in the current conditions of Italian school buildings.

The papers have been chosen on the basis of their capability to describe the outputs and the potentialities of researches and projects, giving a report on experiences well rooted in the reality and at the same time introducing innovative perspectives for the future.

With the aim of exploring the evolutionary scenario of school design as an architectural topic, the collected papers were selected according to a comprehensive and multidisciplinary overview. Researches on typology and spatial organization are enriched through the contribution of a historical and social perspective to enlarge the focus on the urban role of the school buildings. Moreover, innovative approaches and tools have been highlighted both in the design process and in the education techniques. The presented experiences include best practices of consistent and coordinated contributions of the several disciplines involved in the design of school buildings, also implementing digital tools. Finally, the issues related to the challenges of the existing built stock triggered the development of more technical and specialized, albeit multidisciplinary, investigations and case studies' reports.

Stefano Della Torre
Head of the Department Architecture
Built Environment and Construction Engineering
Politecnico di Milano
Milan, Italy
e-mail: stefano.dellatorre@polimi.it

Introduction

Background

The design of educational spaces dedicated to school is a rather recent topic in Italy, since until the end of the nineteenth century and the unification of the country,[1] children were educated exclusively in private or ecclesiastical environments; and only later, the school education was recognized for its significant role in the teaching and learning processes (Pennisi 2012). The evolution of the architectural school typology and of the primary school in particular, can be analyzed as a complex combination of political, cultural, social and urban planning issues and as a reflection of the historical situation. Through the analysis of the educational buildings erected in the different periods, it is possible in fact to detect the evolution of the legislative framework, aimed at defining hygienic and comfort requirements, and of the organization of spaces required by the different pedagogical approaches. The study of the architecture of existing schools reveals a sequence of construction systems, both traditional and innovative, from masonry walls to reinforced concrete frames and to prefabricated solutions, which were employed to better respond to changing needs (in particular, low construction and maintenance cost and construction time reduction). Finally, and with a strict connection with the above considerations, the role of the school building in the city is remarkable at the urban level also, for its ability to promote the development of entire neighborhoods of a city or for the ability to revitalize an existing portion of a city in relation to other public services and open spaces.

[1]The compulsory education was introduced in Italy with the Casati Law, issued by the Minister of Public Education Gabrio Casati in 1860. This law entrusted the central government the obligation to enact laws in relation to school education and the management of public schools and gave private individuals the possibility of founding and managing institutions, but without the right to confer educational qualifications. In this period, elementary education became free, compulsory only for the first two out of four years (i.e., for pupils aged 6–7 years) but only present in cities with over 4000 inhabitants or in secondary education institutions (Laurenti and Dal Passo 2018).

The Current Situation

The results of a more than a centenary process of school buildings' construction are significant from a quantitative point of view. The whole stock of educational buildings of all levels and dimensions amounts to 42,408 units, hosting 7,816,408 students in 370,597 classes (Miur 2017), distributed all over the national territory (see Fig. 1). However, this is an extremely heterogeneous heritage,[2] because of the aging, the functional and often physical obsolescence, which ultimately does not respond to the current demands in terms of teaching and learning methodologies, but also because of the low comfort and safety performances and of fruition and accessibility problems (lack of compliance with "Universal Design" goals).

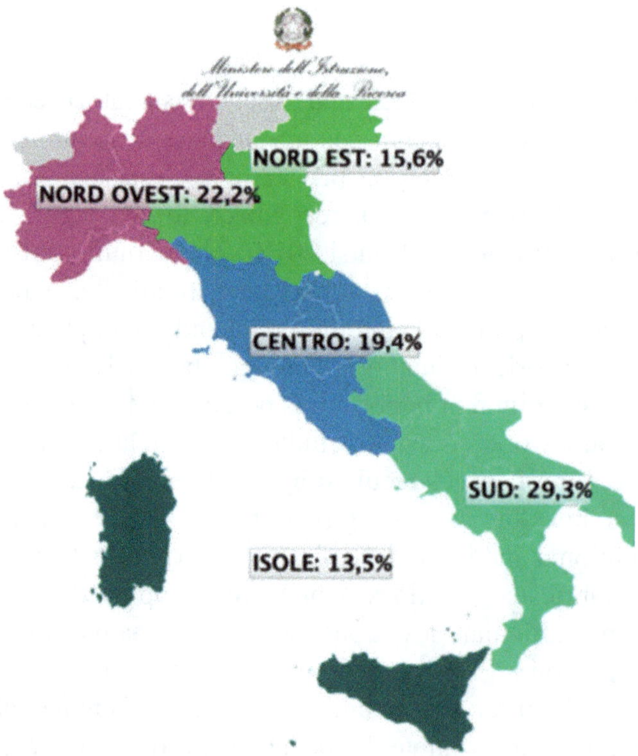

Fig. 1 Distribution of the educational buildings on the Italian territory (Source: MIUR—*Portale unico dei dati della scuola, Anagrafe scuola*)

[2] Thirty-two percent of the schools was built after 1976, 27% between 1961 and 1975, 12% between 1946 and 1960, 8% between 1921 and 1945, 4% between 1900 and 1920, 3% in the nineteenth century, and 1% before 1800. There is no information for the remaining 13% (Miur 2017).

In addition to the hydrogeological hazard that can affect some schools positioned in risk areas, one of the most urgent issues is related to the high seismic vulnerability characterizing most of the existing schools, which indeed were designed with respect to gravity loading only.

The identification of the seismic areas in Italy started at the beginning of the twentieth century, through the instrument of the royal decree, issued after the destructive earthquakes of Reggio Calabria and Messina on December 28, 1908. Since 1927, the areas hit by earthquakes have been divided into two categories, in relation to their degree of seismicity and their geological constitution. Therefore, the seismic map in Italy was nothing but the map of the territories affected by the strong earthquakes after 1908, while all the territories struck before that date (most of the seismic areas of Italy) were not classified as seismic and, consequently, there was no obligation to build in compliance with anti-seismic regulations. Only in 1974, through the law of February 2, 1974, n. 64, a new national seismic regulation was established which defined the reference framework for the seismic classification methods of the entire national territory, as well as for the drafting of technical standards. Immediately after the earthquake of October 31, 2002, that hit the territories on the border between Molise and Puglia, the Civil Protection adopted the ordinance of March 20, 2003, n. 3274, in order to provide an immediate response to the need to update the seismic classification and seismic regulations. According to the ordinance n. 3274, and unlike the provisions of the previous regulations, the entire national territory was classified as seismic and divided into four zones, characterized by different seismic hazard.

This brief history demonstrates that seismic regulations in Italy are quite recent. Indeed, according to the new registry launched by the Ministry of Education University and Research (Miur 2017), only 8% of the schools was designed in compliance with seismic regulations, 54% is in a vulnerable zone, and around 19,000 buildings are situated in high-risk seismic areas. The collapse of educational buildings in the 2009 and 2016 earthquakes in central Italy and the tragedy of San Giuliano di Puglia (2002), where 27 children died in the primary school building collapse, represent a clear symbol of the gravity of this problem.

A second major issue is related to the inadequate energy performance of the educational buildings, again due to the old construction date and to the evolution of the regulations on the energy performance of the buildings, the first being enacted only in 1976, but with very low requirements in comparison with the current situation. Although the European Energy Performance of Buildings Directive (EPBD) requires that "*the public sector in each Member State should lead the way in the field of energy performance of buildings*" and "*buildings occupied by public authorities and buildings frequently visited by the public should set an example*," almost 85% of the school buildings in Italy belongs to the bottom classes of the energy performance ranking. Only 5% (Legambiente 2018) of the stock can be classified among the first three classes, a percentage corresponding to the constructions completed after the 2001, when the first regulations requiring a high standard of energy efficiency were enacted. Hence, if the lack of sufficient structural safety can appear as a real threat, the inadequate energy performance is certainly a waste of resources and a lost chance as well. Energy retrofit programs in fact can become lighthouse projects not only because schools are public buildings visited by pupils, their parents, and the staff, but also because the direct understanding of the behavior of the building envelope and technical systems can help children learn how to support energy savings as responsible users and transfer the knowledge to their families. A further issue to add to the serious situation of the national heritage, related to both structural safety and energy poor performance, is the significant gap between northern and southern regions; an imbalance which characterizes also the funding for ordinary repairs, let aside renovation interventions.

Furthermore, health and indoor comfort requirements should be addressed, especially when considering that almost 10% (Legambiente 2018) of the existing complexes should be cleaned from asbestos.

Finally, the shift toward a knowledge society where information and knowledge are expanding in quantity and accessibility is introducing major changes in teaching and learning models. The information revolution has changed the way we interact with people and things. We live in a society where information is spread out in a large-scale dimension, and new technologies become new tools to change the relationship between time and space. Learning happens everywhere. The new generation of net-native pupils, with an increasingly different set of expectations about space and time, will require constant access to learning materials and resources to share within and beyond the school. Inter-disciplinary learning and collaborative peer-to-peer learning will become increasingly common. New educational models and approaches will be required to help multiple generations, belonging to diversified cultures and in different fields. This will require a general rethinking of the school layouts to overcome the actual strict zoning of the functions and to respond with a higher flexibility to the rapidly changing demand.

The barriers toward the starting of a concrete policy for the renovation or the replacement of the existing stock are varied. It is not just a problem of economic resources but also of a complex set of different issues related to both the diversity of the heritage and the heterogeneous set of institutions responsible for the construction/renovation process. The schools in fact are managed by municipalities as well as by provinces and also directly by the central state. The interventions, considering the major presence of public buildings, are very often subjected to the national public works legislation, requiring a significant effort in planning and organization. One of the challenges is thus how to support municipalities or institutions, especially the smallest ones, in the process from the design activity, to the tendering, to the site inspections and co-ordination during execution, until the final acceptance testing.

The decision for the construction or the retrofit of the school building should consider the relationship with the urban context and the possible potentials that the public building and its annexes can add to the community, for example, in terms of quality of the public spaces, additional resilience in case of emergency[3] and of lifelong learning[4] or integration with other public facilities. A new construction or a requalification can also trigger the regeneration of the surrounding neighborhoods.

The Challenge of Renovation and New Buildings Design

From 2014, in Italy a vast program[5] of construction of new schools and requalification of existing educational buildings that affect, in different ways, every level of education, from primary schools to universities, have been public financed. Different architectural design competitions were also proposed, beyond the attribution of the design task, to collect innovative proposals able to explore new solutions and approaches for the renovation of the educational facilities. Many examples and competition applications are collected in this book.

This program concerned the transformation of educational and pedagogical approaches, aimed at improving the effectiveness of learning models, as well as the requalification of the existing buildings from an energy-saving and structural safety point of view, the latter with particular regard to seismic vulnerability of the existing buildings.

These themes have long been a field of great interest, experimentation, and research, aimed at developing projects, models, and intervention strategies where different disciplines and skills are involved. The possibility of giving old places a new identity, to update buildings according to the new educational and teaching models, to develop projects that take into account the actual needs of energy savings and structural safety is deeply investigated in the following chapters.

On a broader scale, all these needs offer the possibility of redesigning complex existing buildings and developing projects that play an important role also at the urban level, by becoming reference places, opportunities for redevelopment of degraded parts of a city, new cultural, and civic centers.

This book describes the results of some of the research and consulting works, carried out at the Department of Architecture, Built Environment and Construction engineering (Politecnico di Milano), related to the design of new schools and to the

[3] A structural safe school building in seismic areas can be used, for example, as a possible emergency center or temporary accommodation in case of necessity.

[4] The often-unused spaces of a school building during the evening or weekends can host courses for adults or other continuous learning programs or different activities for the whole community.

[5] Of the ten billion euros invested, five have been spent by municipalities, provinces, and metropolitan cities to construct 300 new buildings and start 12,000 renovation projects. ItaliaSicura, the Council of Ministers authority created to lead and manage the renovation programme, was closed in July 2018 (https://www.corriere.it/scuola/primaria/18_luglio_05/edilizia-scolastica-ambiente-governo-chiude-italiasicura-adef7264-8017-11e8-841c-47290107a48c.shtml).

requalification of existing ones. The description of these activities has been organized into three sections, where particular emphasis is given to the effective collaboration with institutions at various levels and the synergetic combination of the different disciplines involved, needed to respond to their requests through applied and basic theoretical research works.

The chapters, organized into the three different sections, investigate central themes about the buildings for education, focusing, in particular, on the definition of multidisciplinary approaches for the design of new schools and for the upgrading of existing ones. Among the main topics highlighted, the first section focuses on the relationship between the city and the school as a civic building with a public role for the community also to possibly host different functions. Accordingly, some recent concept designs are featured, carried out within national and international competitions, and analytical and historical studies on the theme of schools and on their typology, as well as on the role of these buildings at the urban level, are reported. In the second section, innovative solutions for both the design and the construction process are analyzed, and in some applications, particular relevance is given to the building information modeling (BIM) strategy as an optimal tool to achieve a synergetic combination of the different disciplines involved. Finally, the third section focuses on the built heritage, particularly: (i) on the tools, technologies, and approaches required to upgrade the existing buildings, in order to comply with the new regulations (in terms of seismic resistance and energy performance); (ii) on the possible transformation of unused constructions into buildings for education, and (iii) on the management of the existing stock. Theoretical as well as applied research paths are reported to illustrate the topic both from the methodological point of view and through real case studies.

Massimiliano Bocciarelli
Laura Daglio
Raffaella Neri

References

Laurenti A, Dal Passo F (2018) La scuola italiana. Le riforme del sistema scolastico dal 1848 ad oggi, Novalogos

Legambiente (2018) Rapporto Ecosistema scuola. Retrieved from https://www.legambiente.it/wp-content/uploads/ecosistema_scuola_2018.pdf. visited 1st Aug 2019

MIUR (2017) Portale unico dei dati della scuola. Anagrafe scuola

Pennisi S (2012) L'edilizia scolastica: evoluzione di una tipologia attraverso un secolo di storia. In: Storia dell'Ingegneria. Atti del 4° Convegno Nazionale, pp 785–798

Contents

About the Editors

Stefano Della Torre Graduated in Civil Engineering and in Architecture, he is a full professor in restoration at the Politecnico di Milano. He is the director of the ABC Department - Architecture, Built environment and Construction engineering. He is the author of more than 250 publications. He serves as an advisor for CARIPLO Foundation (Cultural districts) Province of Como and Lombardy Region (policies of programmed conservation of historical-architectural heritage). He is the president of Building SMART Italia - national chapter of association Building SMART international.

Massimiliano Bocciarelli is an associate professor at the Politecnico of Milan, he has been lecturing in the areas of structural and solid mechanics at the School of Industrial Engineering and of steel and concrete structures within the School of Architecture. He graduated at the Politecnico of Milan, completed a Master of Science in Structural Engineering at Chalmers University of Technology in Sweden and a Ph.D. in Structural Engineering at the Politecnico di Milano. His research interests have been primarily focused on numerical methods for the modeling of the service and ultimate behavior of materials and structures with particular regard to the diagnosis of masonry historical structures. He is author of more than 40 papers on international journals and two book chapters.

Laura Daglio, Ph.D. is a registered architect and an associate professor of Architectural Technology at the Politecnico di Milano (Department of Architecture, the Built Environment and Construction Engineering) where she works on research issues concerning building and construction design with a special interest in social housing, environmental design and sustainability in architecture for new construction as well as renovation. She is involved in research programs funded by Ministries and Public Bodies at different levels and in international projects. She is the author of books, essays, articles in reviews and of academic papers included in international conference proceedings, on topics related to sustainability in architecture and environmental comfort at different scales. She has been in charge of various projects for public and private buildings' new construction and refurbishment and achieved mentions and awards in design competitions.

Raffaella Neri Graduated in Architecture in 1986, she is a full professor at the Politecnico di Milano. In 1993 she gained her Ph.D. in Architectural Composition from IUAV, Venice, with a dissertation entitled, "Essay on construction. Research into the role of construction and architectural design in relation to type and decoration". In 1994–1995 she worked on organizing the exhibition entitled, "The Center elsewhere" (coordinated by A. Monestiroli), La Triennale di Milano. Since 2003 she has been a member of the teaching staff body for the Ph.D. in Architectural Composition at IUAV, Venice. Her research activities include the theory of architecture, urban design and the role of construction in design. In recent years she has studied compositional principles for residential developments and the issue of redevelopment of brownfield sites and former military zones. She participates in design contests, winning the Luigi Cosenza National Architecture Award in 1996.

Urban and Social Role of School Buildings

Massimiliano Bocciarelli, Laura Daglio and Raffaella Neri

The school system has a fundamental role in the construction of the urban collective places.

Settlement's principles and architectural typologies are closely related to the educational models: the idea of the city, the principles of the construction of the public spaces and school systems are intertwined in order to establish the relevance of the formal education and of the corresponding places in the development of a civil awareness and the identity of places.

The different orders of schools—from those for children to universities—strongly diversify their urban role and their ability to establish new centers: they define a hierarchy of urban places that represent, from time to time, benchmarks for the districts, the city or a vast territory.

In this section, some significant case studies are explored: they especially show an important and evident relationship among educational models, architectural and typological choices and the urban role of the school complexes. The role and the capacity of the new settlements to adhere to the recent educational models and to give adequate response to the construction of the modern city have been tested through architectural designs and competitions.

The first contribution offers a general overview of the buildings for education in different historical and cultural contexts. Two important pedagogical models are compared that have strongly influenced the design of two ancient complexes in Milan: the Rinnovata Pizzigoni (1911) and the School of the Trotter Park (1918). Inspired by northern European experiences, these open-air schools highlight the role of green spaces as the center of a new educational approach based on low-rise pavilions organized inside a park and thus complying with the hygiene and health ideal, which originated them.

Another reform that expanded the education models is then investigated through the work of outstanding Milanese architects: Ignazio Gardella, Arrigo Arrighetti and Roberto Menghi who contributed to develop the civil role of schools during the social and urban reconstruction period after the Second World War. A significant research area analyses prefabricated construction systems typology and layout—to respond to the fast increase of the students and the need of new educational spaces —and the school autonomous isolated new settlements as a model for urban growth.

The third contribution explores the function of educational facilities as community outposts for the construction of new city expansions, aimed at triggering its growth and development. As a case study, the research and design activity carried out by Guido Canella on Milanese outskirts and suburbs are analyzed.

A second group of contributions provides a historical retrospective on the relationship between schools and their cultural context. In the Chapter "*Imagining the School of the Future*", the connection between education and the school as an institution is analyzed in its legislative, social, organizational and spatial dimensions starting from the work of Ciro Cicconcelli, Director of the "*Centro Studi per l'edilizia scolastica*" (a national think tank for school buildings), who recognized in 1958 the classroom as the minimum unit of the school organization. Twelve architects were invited to imagine and represent their idea of a classroom for the future to be shown in a public exhibition.

The historical case study of the agricultural reclamation and planning of the Agro Pontino in the fascist period illustrates how the new schools had a pivotal role as urban public buildings and new civic centers, with the task of determining the identity of the new settlements as well as an important educational character.

A comparable example, although in a different context, is represented by the schools in Northern Greece in the period between the two World Wars. In this case, school buildings represent the effort to reinterpret, in the light of a contemporary perspective, the Greek rural culture between tradition and modernity, to provide an identity not only to the places but also to the newborn nation.

A further case study belongs to a different context, the city of Buenos Aires in Argentina and investigates the role of schools as a mean of social and cultural integration. This chapter analyzes a massive construction plan implemented between the nineteenth and the twentieth century to build up the identity of the growing capital city.

Another group of contributions introduces the topic of higher education and of campuses: this is a specific topic involving advanced teaching and learning models and, therefore, a different set of principles concerning spatial organization.

A special study, reporting the results of a historical research, analyzes the many influences that led Thomas Jefferson, the third President of the United States, to conceive and implement the innovative project for American Universities.

The two following contributions collect some of the proposals submitted for a competition launched by Milan Municipality and Politecnico di Milano and inviting the academic staff to design a masterplan for the new campus in Milano Bovisa, in compliance to the new urban planning regulations. The projects presented interpret the topic integrating the new university settlement and its urban surroundings with the research, didactic and residential (student and conventional housing) facilities together with public services in order to regenerate the suburban neighborhood.

The contribution by Castaldo et al. illustrates the winning entry design for "*the Collegio di Milano*" extension, offering a rethinking of the student housing typology with a special focus on the issue of design quality and its development up to the detailing phase.

An urgent topic in Italy is represented by the challenge of reconstructing the school buildings and the towns after disastrous natural events. The case of the 2016 earthquake in Central Italy is presented here, with the analysis of the role of the educational facilities in the regeneration of the areas hit by disasters and in the re-creation of their identity. The methodologies of intervention are discussed whether to adopt a philological approach, a modern language or a typological reinterpretation of the destroyed buildings.

The last contribution is a collection of new school design proposals submitted for the national competition ("*Scuole innovative*") recently launched in 2015 together with a massive investment program for the renovation and upgrading of the existing stock. In spite of the different levels (nursery, primary and secondary schools) the projects offer an overview of the current approaches investigating the spatial and layout changes, generated by the current evolution of the educational models and the relationship between the school complex and the neighborhood, as a collective space able to interpret the identity of the urban places.

The Open-Air School Typology in the Milanese Experience: The Trotter and the Rinnovata Pizzigoni

Enrico Bordogna

Abstract This text investigates the particular typology of open-air schools in Milanese architecture in relation to similar experiences in modern European architecture from Unification until the 1930s.

Keywords Open-air school · Typology · Trotter · Rinnovata Pizzigoni

The first three decades of the twentieth century in Milan saw the appearance of a typological alternative to traditional school building of great importance, which was the fruit of two converging trends: on the one hand, the strict adherence of the building structure to an innovative pedagogical concept, specifically the one trialled since 1911 by the teacher Giuseppina Pizzigoni; on the other hand, a reference to the experience, in many ways similar, of open-air schools in the Central-European and North American traditions. This was the Renewed Elementary School at Ghisolfa—the Rinnovata Pizzigoni—and the open-air Trotter School at Turro, both designed and built during the First World War and both located, perhaps not by chance, in two very characteristic areas at the outskirts of old Milan, and which became constitutive elements of their respective cultural and settlement identities.

If in the Rinnovata at Ghisolfa what prevailed was certainly the pedagogical and educational spirit, in the Trotter school at Turro what was more marked was the hygienic-health aspect accompanied by an anti-urban reaction against the ills of the contemporary city. In both designs, however, there is an active educational imprint, founded on direct experience and observation, on the relationship with nature, on the value of socialization, manual activities and work, marked, albeit with some significant architectural and compositional differences, by an analogous typological system, with only one floor above ground, filled with collective environments and classrooms for special activities, and the regular classrooms opening directly onto the green spaces, which are in turn equipped with workshops, animal pens, orchards, glasshouses, and a swimming pool.

E. Bordogna (✉)
Architecture, Built Environment and Construction Engineering—ABC Department,
Politecnico di Milano, Milan, Italy
e-mail: enrico.bordogna@polimi.it

S. Della Torre et al. (eds.), *Buildings for Education*, Research for Development,
https://doi.org/10.1007/978-3-030-33687-5_1

The origin of the school at Ghisolfa dates back to 1911, while the Trotter School—into the bargain directly influenced by the Rinnovata—was a project of the 1918 Giunta Caldara; however, both acquired their definitive typological and constructive form between 1925 and 1927.

The Rinnovata at Ghisolfa represents a wholly original case in the panorama of the Milanese school typology, the first, and perhaps only example, in which an innovative pedagogical programme, that of Giuseppina Pizzigoni, became the direct matrix of its typological conformation.

In 1911, a manifesto was published by the Municipality of Milan notifying the citizenship that "in two sections of the first class at the Ghisolfa school the experimental method proposed by the Committee for the Rinnovata will be applied, with the approval and backing of the school authorities, the City Hall, and the Government. There are 60 places, 30 for girls and 30 for boys. In these two sections, the timetable will run from 9 am to 5 pm, with a two-hour break spent at school to eat and rest. The curriculum will be the same as at other schools but taught using a different method".[1] This different method was the experimental one advocated by Pizzigoni, whose fundamental elements can be summarized as follows: full-time education with lunch taken at school; classes with a maximum of 30 pupils against the 50–70 permitted by the regulations; ample room allocated to physical education, outdoor activities, and extra-school excursions; "coeducation", that is, the introduction of mixed-sex classes; faithfulness to the ministerial programmes, but taught according to a method which gave prime importance to observation and the children' direct experience.

When it opened, the Rinnovata did not have its own premises but was housed in four Döcker pavilions (a sort of prefabricated hut of the time) at the municipal primary school of Ghisolfa, situated "on the northern boundary of Milan between Bovisa, Cagnola and Ghisolfa [where] there are the council housing estates of the Ente Autonomo and [where] vast meadows and fields and orchards still lie".[2] Pizzigoni's reference is to the famous IACP neighbourhoods "Mac Mahon", "Cialdini", "Villapizzone", and "Campo dei Fiori", built between 1909 and 1919 and still existing (except for "Campo dei Fiori", which was demolished in the 1960s).

Of the various components which made up the Pizzigoni method, manual work was the fundamental characterizing element. This was "a tool of the method, a coefficient of education [...] The manual work that we introduce in school will give the pupils dexterity and will contribute powerfully to the wealth of ideas around things, to the richness of individual vocabulary, and instil the concept that study, that life itself, is work".[3]

This method fitted into the more general movement which in the second half of the nineteenth century fought for a radical reform of teaching and learning, and which,

---

[1]In P. F. Nicoli, *Storia della Scuola Rinnovata secondo il metodo sperimentale*, Ufficio Propaganda Opera Pizzigoni, Milan 1947, p. 32.

[2]In G. Pizzigoni, *La Scuola Rinnovata, Conferenza tenuta alla Regia Società Italiana di Igiene il 6 marzo 1914*, in Nicoli, op. cit., pp. 41–42.

[3]In G. Pizzigoni, *Conferenza tenuta nell'Aula Magna del Ginnasio Beccaria, marzo 1911*, in Nicoli, op. cit., pp. 26–28.

on the threshold of the new century, came out of the closet of theoretical debate in order to compete in a series of concrete experiments in some countries of Central Europe and North America: from the *Landerziehungsheime* (houses of education in the country) and the *Waldschulen* (schools in the forest) in Germany, the open-air schools in Britain, the *Écoles des Roches* (schools of the rocks) in France, and the numerous active schools inspired by the teachings of J. Dewey in America.

In the city of Milan itself in those years initiatives sprang up that were close to Pizzigoni's ideas, such as the *Per la Scuola* association, founded in 1907 by a group of scholars and benefactors and, especially, the *Società Umanitaria* (Humane Society), whose many activities concerning community education greatly aroused Pizzigoni's interest.[4] However, her own programme differed from these Milanese initiatives and from most of the international experiments in so far as it was not aimed at special categories (frail children and those with TB at the open-air schools rather than the workers at the *Umanitaria* trade schools) but presented itself as a model for teaching healthy children from state primary schools, which would have needed radically reforming to fit with the experimental method.

Naturally, her scheme also required a profound transformation of the traditional school building. In fact, right from the first draft of her programme in 1907, Pizzigoni envisaged a typology commensurate with the needs of the new method, until, in 1914, with the help of the engineer Erminio Valverti, a member of the Rinnovata School Committee, she herself published a methodical project for a school of 400 pupils.[5]

"By an environment suitable for a new school," commented Pizzigoni, "I mean firstly a simple block whose architectural lines and furnishings apply a severe and serene concept to the house for study, which will serve to facilitate each scholastic task and educate the aesthetic sense. It must be provided with capacious classrooms, cheered by large French windows through which the light enters in torrents and the children can rapidly and frequently exit from; a well-furnished kitchen, a refectory, lavatories, showers, and decent healthcare services. There must also be a room for projections, one for music, for work, for a museum […]. The building is located in the middle of grounds that offer a playing field, a garden and a kitchen garden […]. The garden should have a henhouse, a rabbit hutch, a pool with fish, an apiary, and a cage with birds".[6]

The project designed by Valverti is an almost diagrammatic version of this description of Pizzigoni. When, in 1924, the Administration guided by Luigi Mangiagalli allocated a new seat to the Rinnovata between the Northern Railway and Via Artieri, Via Monte Ceneri, Via General Govone (today Castellino da Castello), Amerigo Belloni of the Municipal Technical Office merely copied Valverti's project, adapting

[4]The relationship of mutual interest between the Società Umanitaria and Pizzigoni was confirmed by the conference she held at a major convention on popular education, organized by the Umanitaria at its headquarters in October 1916, on the invitation of its Director, Augusto Osimo. See Nicoli, op. cit., p. 92.

[5]See G. Pizzigoni, *La Scuola Elementare Rinnovata secondo il metodo sperimentale*, Milan, Paravia, undated (but 1914), pp. 22–23, and p. 39.

[6]*Ibid*, pp. 36–37.

it to the selected lot of approximately 20,000 m^2. The result was a simple building with only one floor above ground (except for the entranceway) with an L-shaped plan.

On the ground floor, besides the offices and classrooms (29) for regular teaching, each equipped with changing rooms and with direct access to the garden, we find "the workshops (for printers, engravers, blacksmiths, mechanics, shoemakers, and carpenters, all grouped together at the northern end of the building so as not to cause disturbance to the teaching that will take place simultaneously in the classrooms), the gym, the theatre, the room for projections, the refectory, and the classrooms for drawing and sculpting [...]. On the first floor (above the entranceway) are arranged the school museum and the boys' and girls' work rooms, for music, typing [...]. In part of the area lying along Via Artieri, there is the school farm (with two pavilions: one for agriculture lessons and the other for the animal pens and the homesteader's dwelling), playing fields for the pupils' recreation and the indoor swimming pool".[7] The architectural forms, as an attentive observer of the time wrote, are "those characteristic of the Lombard farmstead"[8]; or, more accurately, those copied from mediaeval Lombard architecture, as Boito might have said, selected by the architects of Eclecticism for buildings to meet the new social needs (Figs. 1–2–3–4–5).

If educational and pedagogical reforms fundamentally underlay the Renewed Elementary School at Ghisolfa, the intent and criteria that govern the idea of an open-air school in general are predominantly healthcare and hygiene issues. In fact, the idea of the open-air school also brings together a widespread sensitivity, of a generically humanitarian and progressive stamp, identifying in the rejection of the big city and the return to nature the antidote to the evils of contemporary urban planning, generically identified as the ultimate cause of the violent conflicts induced by industrialization (slums, tuberculosis, infantile frailty, alcoholism, etc.).

Already by 1910, in the second issue of the magazine *Le case popolari e le città giardino* an editorial appeared in favour of open-air schools with documentation on the one created in 1907 in Letchworth, the first garden city in the UK, and with the news that also in Milan a public announcement had been made of the project "to establish for frail children an open-air school similar to the German *Waldeschulen* and open-air British schools".[9]

And in 1912, in a pamphlet entitled *La Scuola all'aperto*, written by Prof. Carlo A. Mor, one of the main leaders of the *Per la Scuola* association, after a series of considerations on the "social scourge, the fearsome plague" of tuberculosis, we read: "So it is not about correcting and modifying all the current trends of the city's municipal schools where smart Administrations—as in Milan—have become zealous and strict interpreters of hygiene in school, it is, instead, about creating and bringing

---

[7]In A. Belloni, *Relazione tecnica al Progetto della Nuova Sede della Scuola Elementare Rinnovata da erigersi in* Via *General Govone, 3 giugno 1924*, Archivio Storico Comunale, Fondo Finanze-Beni Comunali, cart. 206, manuscript.

[8]In L. L. Secchi, *Edifici scolastici italiani primari e secondari*, Hoepli, Milan 1927, p. 216.

[9]In the editorial, *Le scuole all'aperto*, in "Le Case Popolari e le Città-Giardino", year 1, no. 2, 1910, pp. 59–60.

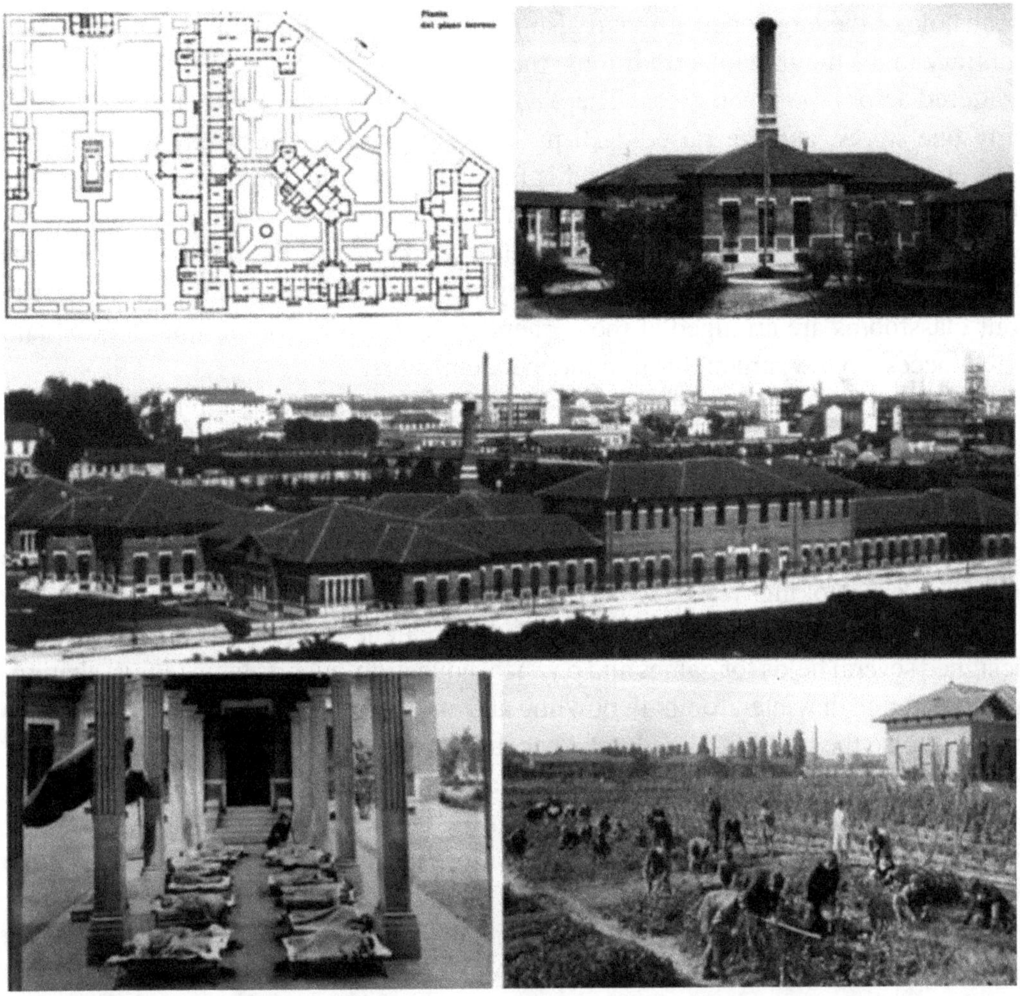

Figs. 1–2–3–4–5 A. Belloni (Municipal Technical Office), Pizzigoni Renovated Elementary School in Via General Govone (now Via Castellino da Castello), Milan, 1924–1927: ground floor plan and views; educational and recreational activities in some images of the mid-thirties

order to what, methodically, does not exist. It is about establishing special schools for frail children".[10]

Among these initiatives, the "Umberto di Savoia" open-air school at the Trotter Park at Turro represented the most important achievement, according to contemporaries, the greatest in Italy and among the greatest in Europe. It sits in the old harness-racing field between Via Monza and Via Padova; an area of approximately 128,000 m^2 (partly occupied by the racing tracks, the stand, the stables, and the services, but otherwise arranged inside large, densely wooded meadows) which the Municipality of the Giunta Caldara acquired in 1919 from the Società del Trotter.

The project, designed by the engineer G. Folli of the Municipal Technical Office in 1919, is of great interest not only due to the scope of the intervention, which occupies

[10] In C. A. Mor, *La Scuola all'aperto e i criteri informativi di assistenza educativa*, Tip. A. Antonini & C., Milan 1912, p. 12.

the whole of the large enclosure but also for the typological layout adopted. In fact, contrary to all the previous traditions, the choice was that of a school *in pavilions* scattered across the green areas. The pavilions for the classrooms number 12, have only one storey, and are raised 1.20 m above the ground. Each pavilion includes four classrooms to take 35 pupils and is formed by two double blocks joined at the rear by a veranda which functions as an indoor recreation area and refectory; at the ends of the veranda are rooms for the doctor, teachers, and the janitor. In a smaller pavilion at the back, connected by a canopy, are the lavatories and the showers. The four classrooms are arranged at the corners so as to enjoy double illumination and direct access to the garden through independent stairs.

Given the high number of pupils (about 1700), the school is divided into three sections with four pavilions each. Each section is equipped with a building in a central position to house the general services of the kitchen, head office, healthcare management, teaching museum, library, and so on. Of the three buildings for general services, two would be new-builds and the third a conversion of the old Trotter buildings overlooking Via Padova. Completing the school's amenities are a wide range of facilities: "In the central part of the area enclosed by the track, we have designed several fields for games and physical education, such as lawn tennis, skating, and ice skating in winter, football, bowling and croquet and finally a large swimming-pool".[11] Also the racetrack was kept, "because it will be an excellent asset for sports since it can serve for exercises of walking or running and cycling [...]. It will also serve as a place for heliotherapeutic cures during the summer months, as was already practised last year for the summer camp established for frail children with a positive outcome and that we believe will continue operating every year during the school summer holidays".[12] Then, finally two cinemas, one in a new pavilion, the other obtained from the old Trotter buildings on Via Padova, complete the complex.

The architectural appearance of the school is deliberately low-key, recalling "the type of constructions of a Swiss chalet, with pediments, gutters and wooden cladding and with a sober painted decoration and friezes along the walls". A deliberate sobriety, it is stated shortly afterwards, dictated by the consideration that "the character of these buildings distributed in the midst of green fields and heavily wooded is not such as to require decorative displays".[13]

The project for the large Trotter school was an initiative of the Socialist Giunta Caldara, which by 1920 had already begun the works. These were protracted in subsequent years under the so-called *national block* Giunta, which governed between 1922 and 1926 (fascists, liberals, and populists in which the councillor for urban planning was the liberal Cesare Chiodi).

In the course of the work the original project underwent some marginal changes, but above all it was enriched with new facilities: the *Casa dei Bambini* for children of pre-school age (designed by the engineer L. Beretta in keeping with the school's

[11] In G. Folli, *Progetto di una scuola all'aperto nel recinto del Trotter nel Riparto di Turro. Relazione tecnica, 30 luglio 1919*, Archivio Storico Comunale, Fondo Finanze - Beni Comunali, Cartella 208, typed manuscript, p. 2.

[12] *Ibid.*

[13] *Idem*, p. 4.

pavilions), the *Casa del Sole* boarding school, housing 160 beds to accommodate the children of parents suffering from TB, two gyms, and a small church.

By 1928, the works were almost complete, so that in the August issue of the magazine "Milano", which punctually documented the doings of the city administration, a richly illustrated and legitimately eulogistic special was included on the new school which rises "on the old field for harness racing, salvaged providentially from the Municipality by private speculation and designed to accommodate the largest and perhaps most beautiful open-air school in Europe".[14] The new complex, whose many features were listed (the school itself, of an agrarian type and therefore with flowerbeds, gardens, orchards, animal pens, chicken runs, etc.; the *Casa Montessoriana* for young children; the *Casa del Sole* boarding school; the summer camp, attended by 2400 frail children selected from all the city's schools) was presented as an alternative to the traditional school system. In fact, "our primary schools, even the most beautiful and the most modern, are inevitable 'school barracks', as crowded and buzzing as hives. The interminable corridors, the classrooms that do not always see sunlight, the closed courtyards […]. Hence, in reaction to these indoor environments, came the need to create an antidote to what might be defined as the painful school phenomenon of urbanization".[15]

Thus the Trotter school at Turro and the Rinnovata Pizzigoni elementary school at Bovisa, even with the differences observed earlier, represented a fundamental innovation in the panorama of Milanese schools in the first decades of the century, a turning point among schools created by positivist Eclecticism and those that were the fruit of subsequent rationalist research. Indeed, together, they are respectably on par with the best-known examples produced a few years later, between the end of the Twenties and the Thirties, by modern European architecture, whether the so-called "redbrick schools", such as those of Wilhelm Dudok and Fritz Schumacher in Hilversum and Hamburg, respectively, which still interpreted the school as a monumental element in the construction of a city, or the so-called "experimental schools", like those created by Ernst May and Martin Elsässer in Frankfurt, or by Otto Haesler in Celle, or again by Bruno and Max Taut in Berlin, which represented the most limpid outcome of a long in-depth analysis of the school organism, broken down into the individual functions and elements of which the educational mechanism is made. All these schools, according to the interpretation of Julius Posener,[16] represent a divergence in the European panorama of the Thirties between a "pole of tradition", still attentive to the urban value of the school building as an element in the construction of a city, and a "pole of radical reformers", which, although welcoming even extremely diversified positions among its members, identifies itself with the common commitment to functionally remove the inner device.

---

[14]In L. Clerle, *La scuola all'aperto "Umberto di Savoia", la "Casa del Sole"*, in "Milano", year XLIV, no. 8, August 1928, p. 33.

[15]*Ibid.* On the open-air Trotter school complex, see also L. L. Secchi, op. cit., pp. 132–133, pp. 155–158, pp. 165–166.

[16]See J. Posener, *Écoles allemandes*, in *Les écoles à l'étranger*, special issue of "L'Architecture d'Aujourd'hui", no. 2, March 1933.

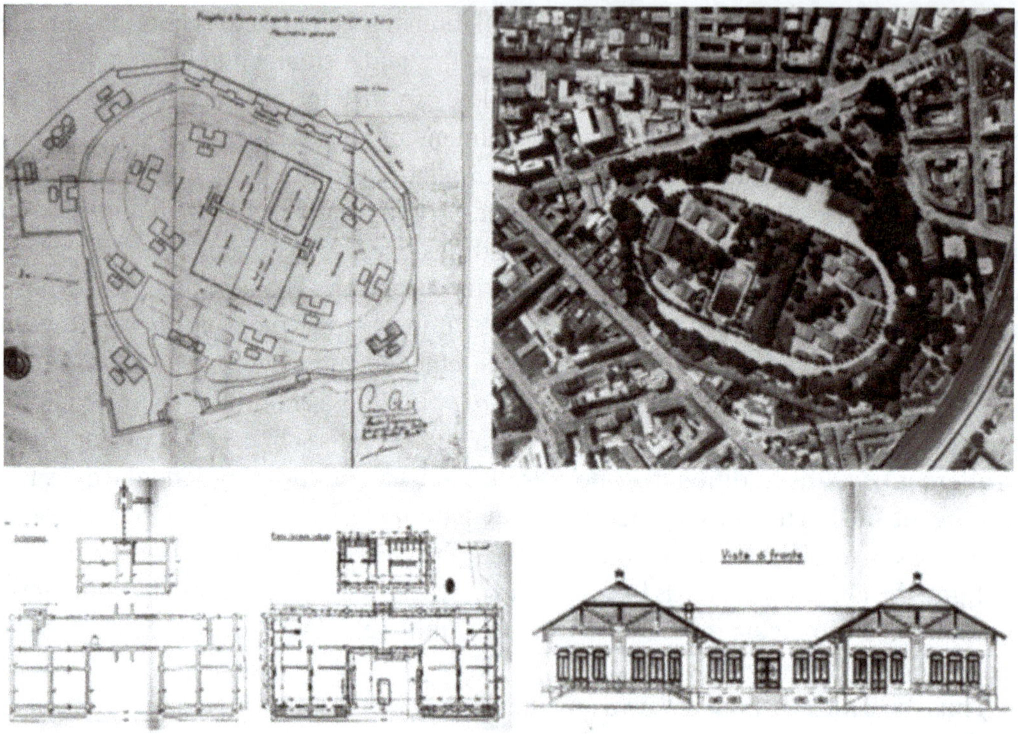

Figs. 6–7 Ing. G. Folli (Municipal Technical Office), Open-air school "Umberto di Savoia" at Trotter in Turro, Milan, 1918–1927: general plan, zenithal view of 1972, plans of the basement and of the mezzanine of a pavilion-type of four classrooms and front and side elevation

Equally relevant would then be the comparison with the open-air schools created by Eugène Beaudouin and Marcel Lods, with the decisive technological contribution of Jean Prouvé, in the Parisian suburb of Suresnes, or the one, singular in its being developed height-wise, by Johannes Duiker in the courtyard of an urban block on the Cliostraat of South Amsterdam, or even that by Neutra for the Rush City Reformed in the USA.

Still today, albeit in changed contextual conditions, the Trotter and Rinnovata Pizzigoni schools actively fulfil their function, as a testimony to the value of typological research carried out almost a century ago, and to their congruency with the cultural and settlement identity of the neighbourhoods in which they were developed and have operated (Figs. 6–7, 8–9–10, 11–12 and 13).[17]

[17]On the history of the scholastic typology in Milanese architecture cfr. also: E. Bordogna, *Radici tipiche dell'architettura scolastica a Milano*, in «Hinterland», n. 17, marzo 1981, pp. 68–79; Idem, *Sperimentazione didattica e innovazione tipologica in un quartiere industriale: la Bovisa e la Rinnovata Pizzigoni*, in G. Fiorese, M. Deimichei (a cura di), *Milano Zona Sette: Bovisa Dergano*, Comune di Milano-Clup, Milano 1984, pp. 61–66; Idem, *Alle origini dell'architettura scolastica milanese*, in G. Fiorese (a cura di), *Milano Zona Due. Centro Direzionale, Greco, Zara*, Comune di Milano, 1987, pp. 115–119; Idem, *La scuola all'aperto del Trotter a Turro, Milano: innovazione tipologica e dettato pedagogico*, in «Edilizia scolastica e culturale», Le Monnier, Firenze, n. 5, maggio-agosto 1987, pp. 85–94.

Figs. 8–9–10 Ing. G. Folli (Municipal Technical Office), "Umberto di Savoia" open-air school at Trotter in Turro, Milan, 1918–1927: view of a type pavilion of four classrooms, view of the gym and view of the pool

Figs. 11–12 Ing. G. Folli (Municipal Technical Office), "Umberto di Savoia" open-air school at Trotter in Turro, Milan, 1918–1927: educational and recreational activities in some images from the 1920s and 1930s

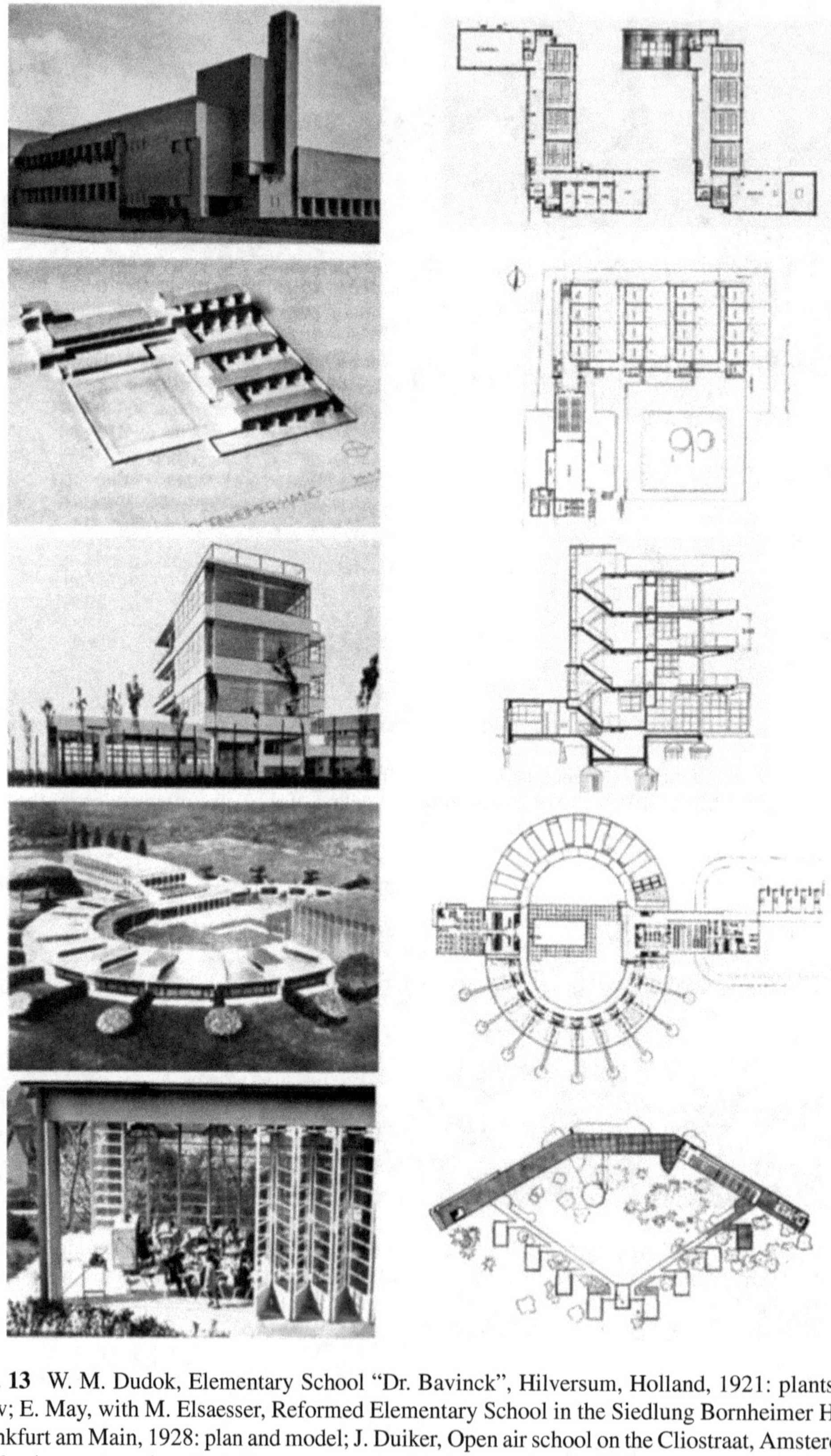

Fig. 13 W. M. Dudok, Elementary School "Dr. Bavinck", Hilversum, Holland, 1921: plants and view; E. May, with M. Elsaesser, Reformed Elementary School in the Siedlung Bornheimer Hang, Frankfurt am Main, 1928: plan and model; J. Duiker, Open air school on the Cliostraat, Amsterdam, 1932: view and section; R. J. Neutra, open-air circular school for the Rush City Reformed, 1928: plan and model; E. Beaudouin and M. Lods, with J. Prouvé, *Open air school in Suresnes*, Paris, 1934: plan and view of a classroom

The Topic of the School Building in the *Milanese* Professionalism

Michele Caja, Martina Landsberger and Angelo Lorenzi

Abstract The question of the relationship between modern architecture and civil themes and of technological experimentation takes on a central role in the school building projects in which, between the 1950s and 1960s, Arrigo Arrighetti, Roberto Menghi and Ignazio Gardella, among others, are involved. The Milanese architects often addressed the issue in the absence of a definite place of intervention and work mainly on flexible systems, adopting the pavilion as the main element of the construction, that allow adaptations, expansions and transformations of the functional program. They experienced innovative technological solutions, such as the prevalent use of prefabricated elements, light metal structures, technological details derived from industrial plants.

Keywords School · Typology · Construction · Roberto Menghi · Ignazio Gardella

In the 1950s, though still struggling with the aftermath of World War II (WWII), Milan was a very lively, creative, open and cosmopolitan city, willing to regain the European and international role it had lost during the fascist era. Nevertheless, in those years, Milan was inhabited by an extraordinary generation of intellectuals, artists and architects: the BBPR group, Franco Albini, Luigi Caccia Dominioni, Mario Asnago, Claudio Vender, Roberto Menghi and Ignazio Gardella, just to mention the best known, animated the Milanese architectural scene with punctual interventions, that were able to define new themes and urban intervention principles, even if they did not change the overall urban structure. These themes will have a great fortune in the following years and they are still at the very core of Italian and European architectural design researches (Bottoni 1954).

Housing and dwelling for the urban bourgeoisie were the main topic with which this generation of architects, attentive to European experiences, dealt. However, also the design of public buildings has a great importance within their researches. Through

M. Caja (✉) · M. Landsberger · A. Lorenzi
Architecture, Built Environment and Construction Engineering—ABC Department, Politecnico di Milano, Milan, Italy
e-mail: michele.caja@polimi.it

S. Della Torre et al. (eds.), *Buildings for Education*, Research for Development,
https://doi.org/10.1007/978-3-030-33687-5_2

his many articles and essays written in the aftermath of WWII, Giulio Carlo Argan, one of the most important Italian architectural historians of the twentieth century, pointed out the museum as a central theme to work on in order to restore dignity and urban quality to those cities that, along with their built heritage, risked losing their identity and historical memory (Argan 1949). In those years, the museum became one of the privileged places for rebirth, a gymnasium for up-to-date museological installations, a challenge to the communicative possibilities of art, a testimony of wide-ranging cultural projects. Moreover the school as a design research theme had a different and more controversial role in Italian culture after World War II. Previously, teaching institutions had assumed a central importance in the context of the fascist reforms. The Gentile Reform of 1923 and the Reform of 1926, through the establishment of the Opera Nazionale Balilla, opened a broad program of civic and physical education for young people. The centrality of classical culture and its contradictory identification with the fascist ideals on the one hand, and the importance of physical education as an element that introduces military discipline on the other hand, are the background of a project of control, propaganda and overall orientation of the young people, future Italian citizens. Obviously, this program included extensive investments also for the construction of schools and scholastic institutions, which represented the testing ground for the Italian architects of the time.

Republican Italy, asserted in June 1946, will struggle for a long time before defining a clear project for the reform of educational institutions. The Gentile Reform remains for a long time one of the benchmarks in the field of education to which the 1947 Constitution adds principles of spreading culture and opening opportunities to the weakest social classes. The main intervention in the teaching sector will take place with the reform of 1962 which introduces the extension to a minimum of 8 years of school education and, above all, establishes a unique system of middle school, eliminating the distinction between professional schools and secondary education, to respond to the needs of a democratic society that wants to offer training opportunities to all citizens. With the birth of the new middle school, education started to be a mass phenomenon, with important and problematic consequences that concern the school buildings in the country. This less known aspect of the research of Milanese professionalism seems to be important to investigate because, in the absence of a clear design on the part of the institutions, architectural culture takes charge of proposing innovative settlement, up-to-date typological and constructive solutions. The question of the relationship between modern architecture and civil themes and of technological experimentation take on a central role in the school building projects in which, between the 1950s and 1960s, Arrigo Arrighetti, Roberto Menghi and Ignazio Gardella, among others, are involved (Menghi 2000; Porta 1985; Samonà 1981). The Milanese architects often addressed the issue in the absence of a definite place of intervention and work mainly on flexible systems, adopting the pavilion as the main element of the construction, that allow adaptations, expansions and transformations of the functional program. They experienced innovative technological solutions, such as the prevalent use of prefabricated elements, light metal structures, technological details derived from industrial plants. They often adopted indirect lighting systems that were not yet used in Italy but were being exploited in Northern Europe.

The text we present is part of a broader research project started in 2016 in collaboration with some professors, researchers and PhD students of the Politecnico di Milano—in particular the ABC Department and the Design Department—and the Università di Parma—Department of Engineering and Architecture—with the CSAC/Centro Studi e Archivio delle Comunicazioni of the Università di Parma. The work involves a group of young researchers and doctoral students coming from different disciplinary fields, and it is based on the extraordinary archives of drawings, photographs and documents on twentieth-century Italian architecture, conserved at the CSAC. The CSAC, founded by Arturo Carlo Quintavalle, preserves original materials related to the field of visual communication, artistic research and Italian design culture and, today, is one of the most important Italian archives on modern architecture.

Recently, the research has focused on Roberto Menghi, Ignazio Gardella and Luigi Vietti, with the aim of reopening, starting from the reading of the works to archive documents, the critical debate on the work of these masters of Italian architecture by suggesting new and different points of view, and, at the same time, in order to investigate the less known and often unpublished works by these architects. During its development, the research has also been carried out in strict relationship with other institutions, museums and research organizations.

1 Roberto Menghi: Pirelli Foundation and French High School

The research of Roberto Menghi's school building reflects typical aspects of architectural and urban culture practiced since the post-war period, although it still dates back to the principles and ideas of cities developed by the Modern Movement. Both projects analyzed here are buildings that have a weak, if not non-existent, relationship with the city: on the one hand, precisely because of their location within non-consolidated contexts, in areas of new expansion or with a thin and not compact urban fabric; on the other hand, for the way in which the buildings are conceived as autonomous objects, with an inner articulation, but without a precise relationship with the surrounding streets. In both cases these are buildings organized in single pavilions placed inside the block and set back from the lot's boundaries. This solution, typical of many schools built between the 1950s and 1970s, not only in Milan, stems from the idea of considering the school as a special building, collective equipment, linked to an idea of the city subdivided into functional sectors—such as the term often still used today "school complex" means—to create autonomous mono-functional parts, often unrelated to the surrounding existing urban fabric. This idea of the plexus, which on a larger scale becomes a campus, remains a subject to be addressed critically today when we are faced with the theme of the school building, to avoid understanding it as if it were an independent organism, foreign to the surrounding context. Something that did not happen in the nineteenth-century school typology—from the Schools of Camillo Boito (1836–1914) to those of the first decades of the twentieth century—in which the building was always included within the morphological constraints posed by the city, such as the relationship with the street, the inclusion in a building curtain and an urban block, the adjacency with residential houses and buildings for other uses, the entrance hall and the courtyard, open to the road or closed, as a collective element, the multi-storey development as a height articulation system.

The choice, precisely, to untie the school building from any urban constraint, the often isolated location within a lot, the decomposition of the organism into a system of isolated pavilions, the introduction of a reduced height of maximum 2–3 floors, the disappearance mostly of the central aggregative space of the court are all aspects that distinguish, mostly, the research on the school building intended as an autonomous specialized typology, up to the present day.

In the architectural production of Roberto Menghi the school building represents a little-frequented theme, which he tackled essentially on two occasions. The first is the project for a training centre for the Pirelli Foundation in Viale Fulvio Testi in Milan (1957–1958), which still exists today despite of variations within it. The second is a more complex project, for the headquarters of the French Lyceum of the Chamber of Commerce, which is developed on two different sites between the years 1958 and 1960–1961. The first project is located at Villa Simonetta in via Stilicone in Milan, one of the most interesting examples of Renaissance architecture in Lombardy, and the second one on an urban block along via Laveno in the outer district of San Siro, where it was partly built (now demolished and replaced by another building housing the Lycée Stendhal).

Despite the diversity of the functional program of the two projects, the theme is developed according to the type of the isolated pavilion, connected by low connecting bodies. In the Pirelli Foundation the two parallel pavilions of different section and heights, respectively, contain the large floor-to-ceiling space of the workshops and the volume of the classrooms, with the adjoining canteen. Connected by a transverse entrance body, with wardrobe, they build an open C-system on Viale Fulvio Testi. The brick construction is completed by a roof structure defined by an articulated section with large metal trusses and windows. If the industrial character and the constructive choices are the most interesting aspects of this building, the project for the French Lyceum touches on different aspects, concerning the urban scale, the research on the scholastic typology, the relationship with the pre-existences, and the different variants of the project.

There are several variants for the proposal on the site of Villa Simonetta; of the first version proposed by Menghi (January 1958) emerges in particular its radical nature—still indebted to the logic of the tabula rasa of the Modern Movement, but also strangely deaf to the topic of environmental pre-existences already theorized at the time by Ernesto N. Rogers—with which he proposes to completely replace Villa Simonetta, except for the porch, which is moved and reassembled along the street. In this case the building has a close relationship with the street, but at the expense

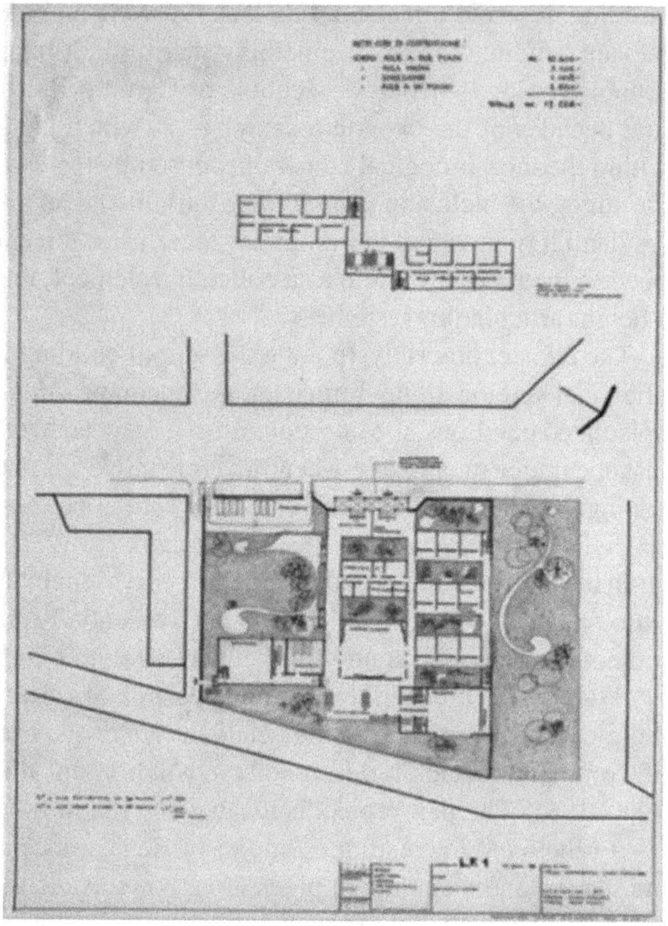

Fig. 1 *French High School, Milan: proposals for Villa Simonetta site*, Milano, 1958, Layout study for the French High School, Mezzanine and first floor plan, January 1958, ink and pencil on tracing paper, mm. 624 X 440, 1:500. (CSAC, Sezione Progetto, Fondo R. Menghi B034577S)

of the overall removal of the historic building. From the relocated portico, a double corridor distribution system develops around a central courtyard, partially occupied by the transverse bodies of the management offices and the aula magna.

On the external sides of this double distribution system, on the left side, is the access to the three bodies of the classrooms, based on a comb-structure, and on the right side of a single terminal body, containing the library and the refectory. That this circulatory system split around a central inner space, with lateral combed bodies (which indirectly refers to the layout of the Elementary School of Fagnano Olona built by Aldo Rossi 15 years later) was congenial to Menghi, it will also be seen in the solutions designed for the second site. A particular solution when compared with the other elaborate variants, which not only preserve the layout of the villa but assume it as a compositional and ordering center for the various additional extension bodies. Instead of replacing the ancient monument, two undated solutions widen the villa toward the street side, extending the main building on the right side or the left side, respectively, and re-proposing the C-shaped portico on the inner side of the court. This hypothesis for the courtyard portico will be resumed in another solution (March 1958), where the fourth side will be occupied by two symmetrical buildings, with a central access passage to the rear garden, containing the refectory and the

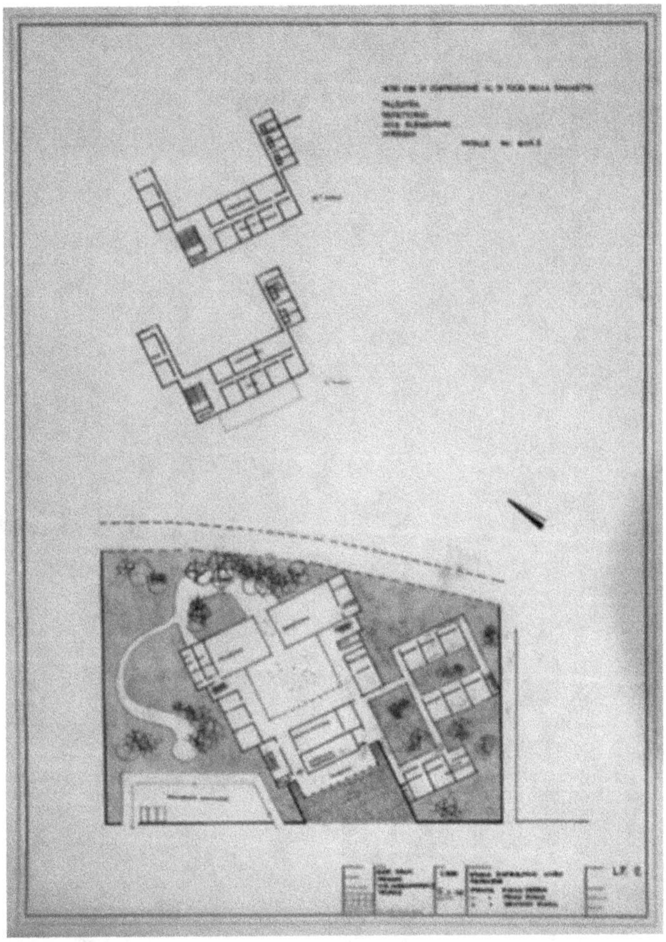

Fig. 2 *French High School, Milan: proposals for Villa Simonetta site*, Milano, 1958, Layout study for the French High School, ground floor, first floor and second floor plans 7.3.58, ink and pencil on tracing paper, mm.618X447, 1:500. (CSAC, Sezione Progetto, Fondo R. Menghi B034576S)

library, while on the right side a corridor system is developed around another court to distribute the one-storey bodies of the classrooms (Figs. 1 and 2).

The layout of the first solution reappears later in the subsequent location, developed on an elongated rectangular lot along the via Laveno in San Siro. There are also several versions of this project, starting from the common layout, founded on a system of pavilions parallel to the road and connected to each other by one-storey distribution bodies, which form a single connecting plate (Fig. 3).

2 Ignazio Gardella: Prefabricated School Projects

As we have seen earlier, the topic of the school offers the designer a double opportunity: first of all, an interpretative possibility in which the typological aspect becomes a fundamental element in the development of the character of the project. A second opportunity, instead, refers to the question of experimentation, to the possibility of working, for example, with prefabricated elements useful to speed up the construction and, above all, to reduce costs. In the post-war period in which Italy must

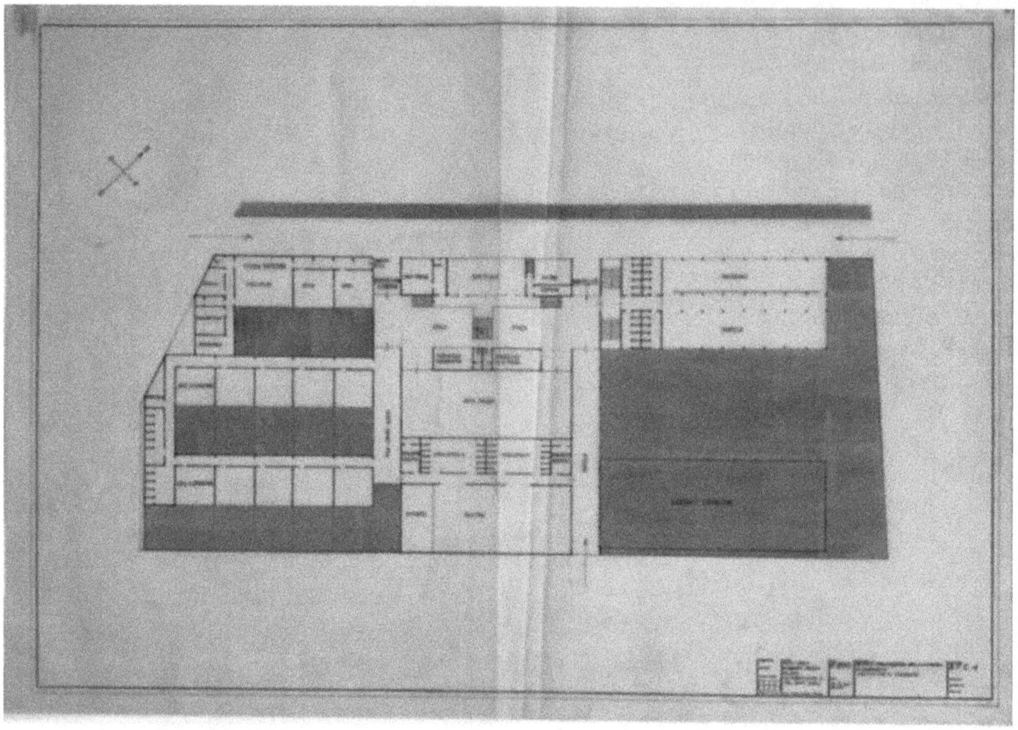

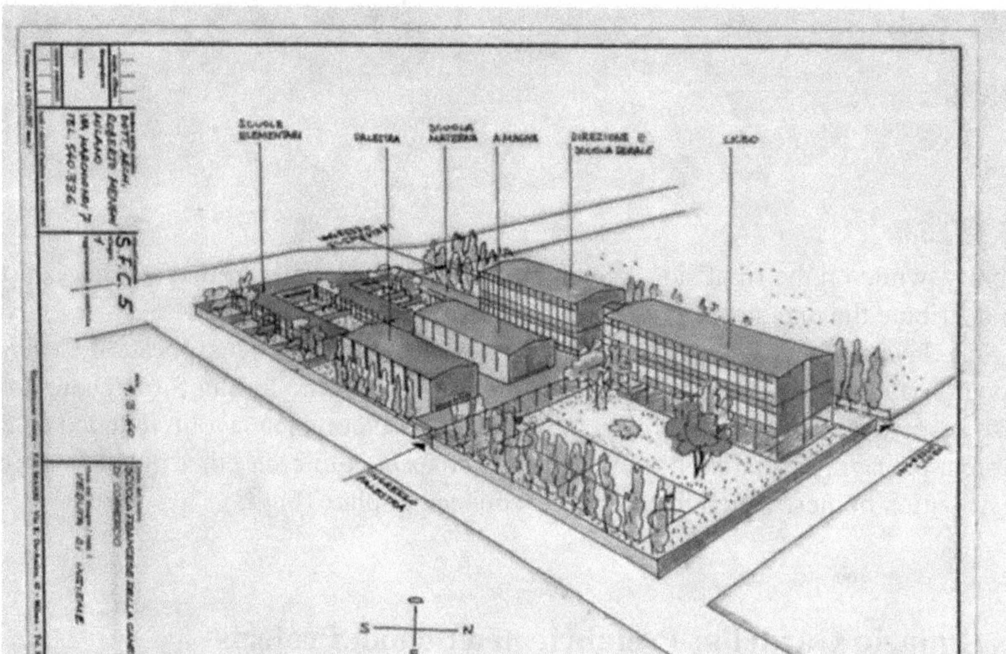

Fig. 3 *French High School, Milan: proposals for Villa Simonetta site*, Milano, 1960–61, Overall view, Axonometry, marker and ink on tracing paper, mm. 240X325. (CSAC, Sezione Progetto, Fondo R. Menghi, B034580S); ground floor plan, 9.03.1960, ink and halftone screen on tracing paper, 1:200. (CSAC, Sezione Progetto, Fondo R. Menghi, B034581S)

substantially rebuild itself "in toto", both from the material point of view—such as buildings—and the more properly "spiritual" one providing a new education to the population, the issue of the school becomes one of the fields on which architects' work converges.

In Milan, for example, from 1950s until 1970, Arrigo Arrighetti (Bodino 1990),[1] director of the Technical Office first, and later of the Urban Planning Office, represents an interesting figure of designer engaged in the search for a new form of public buildings conceived as decisive elements for reconstructing the identity of the city and its representation and recognition in the society. In this sense, the projects of Arrighetti's schools, in their desire to build the relationship with the places in which they are placed, offer a different point of view from that of Roberto Menghi and Ignazio Gardella.

The CSAC archive, which brings together the entire work of Gardella,[2] preserves three school's projects: the design for the elementary school of Nerviano (1946), a design-competition for the construction of a school with the AUCTOR system (1965) and the one for a nursery school in Milan (1964). Only the oldest project for Nerviano presents an urban dimension. The other two projects seem so detached from the city that Gardella himself, almost as if he felt the need to verify its urban role, draws a perspective of the building inserting it in a totally "invented" context to evoke the idea of the city.

While the city does not seem to exist, so that the project is almost totally self-referential, it is instead true that all the three Gardella's projects provide evidence to a careful research on the topic of school typology. The school for Gardella, therefore, is a place built on the relationship that the collective space where all the students can meet together (garden, courtyard, gymnasium, etc.) establishes with the system of classrooms. It is interesting, for example, to look at the project dating back to 1965 in its various stages of study. The starting point is a simple system in which the two buildings of the classrooms are arranged in line, parallel to each other, connected to each other by the gym. In this way a C-shaped composition is defined with a central collective space. This first solution—documented by a series of sketches and distribution schemes—is almost immediately abandoned in favor of a composition that, as already seen in Roberto Menghi's projects, tries to break down the building into its parts determining a composition in which separated pavilions enter into relationship through a system of paths. In this solution, which involves moving the gym to one side and therefore a first disarticulation of the initial composition, the theme of the C of the classrooms in relation to the central space remains stable (Fig. 4).

The drawings dated November 1965,[3] instead present a very different solution from the typological point of view; the desire to identify the idea of the school

---

[1] Arrigo Arrighetti (1922–1989. The book edited by C. Bodino represents the only document that witnesses his work. Few are indeed the critical essays that have taken his figure into consideration.

[2] For this occasion the project for the University of Genoa was deliberately omitted. We believe that it represents a different case because of its scale and for its diversity in approaching to the issue.

[3] It is assumed that this is the final solution. The previous solutions do not report any dating.

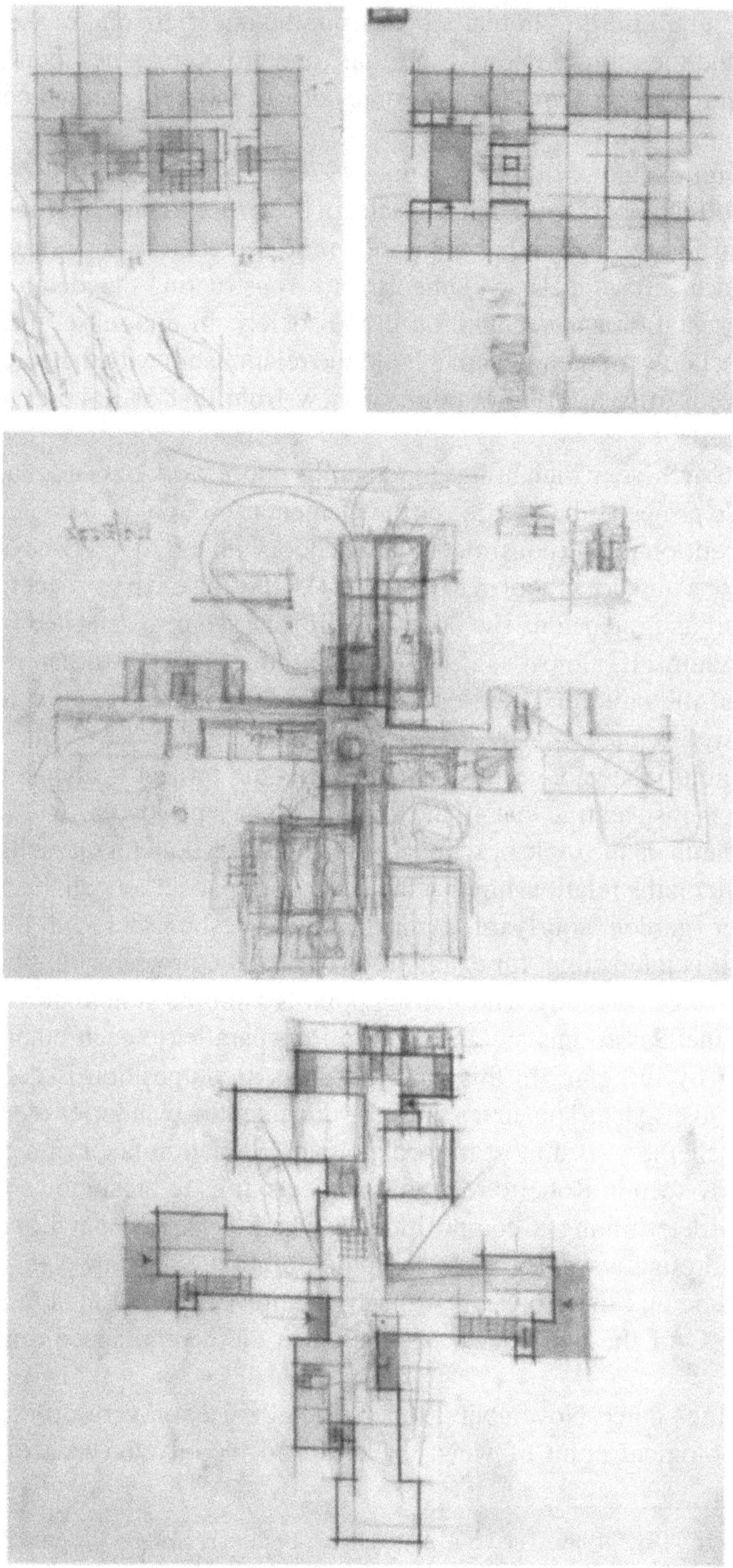

Fig. 4 *Prefabricated Schools for the Municipality (Competition Contract),* via Manin, Varese, 1965. Study Sketches, pencil on tracing paper. (CSAC, Sezione Progetto, Fondo I. Gardella, B001612P)

through the definition of a space able to represent the idea of the community persists. In this solution Gardella moves from a first cross-shaped scheme with a large common space placed in the intersection of the four arms of the cross. Subsequently, Gardella draws a more complex system in which the cross remains losing an arm. This is the solution that Gardella tries to verify in the perspective mentioned above.

The school is conceived as a large building articulated on three floors and built back from the street line.

On the ground floor, in the three arms of the cross, the collective spaces such as special classrooms and laboratories, and the gymnasium are overlooking the equipped back garden or the green space that separates the school from the road. A large atrium, at the intersection of the three arms, marks the entrance to the school and contains the vertical distribution system. The classrooms are located on the upper floors (Fig. 5).

Similar characteristics can be found in the project for a nursery school in Milan. In this case the type is represented through the construction of a large central space useful to collect all the children and in the definition of a sort of enclosure consisting of the blocks of the classrooms.

In this project each volume of the classroom contains in itself also a space destined to "refectory". Between the volumes there is a glazed partition system creating the relationship with the outside and defining the lighting system of the large central courtyard (Fig. 6).

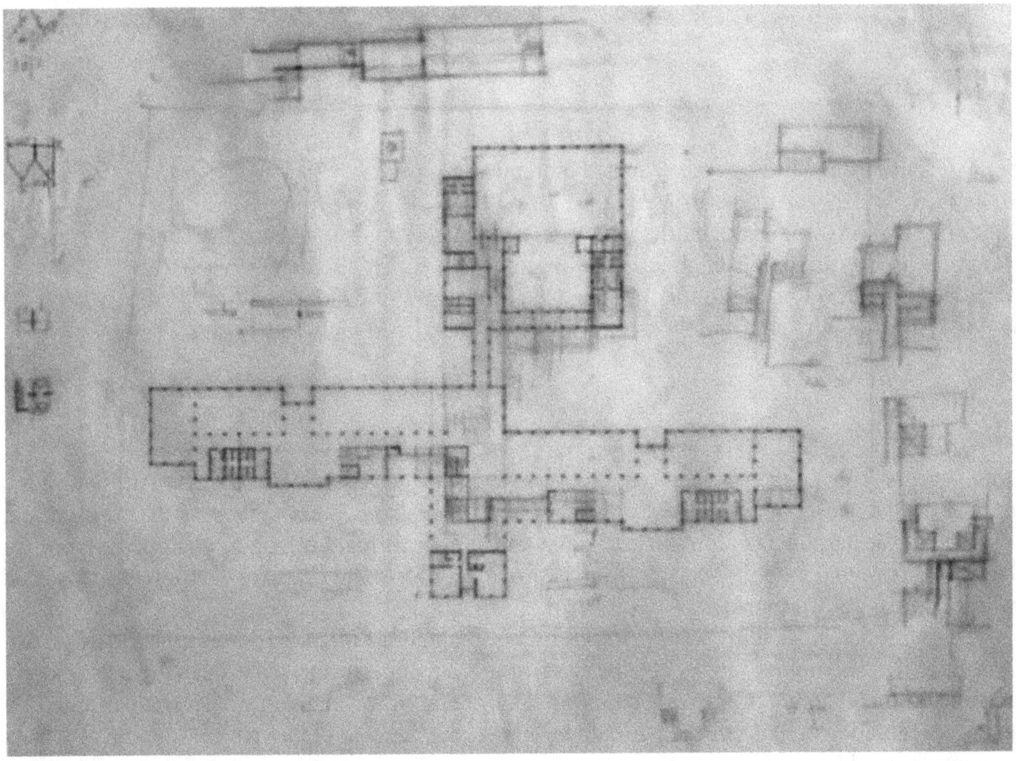

Fig. 5 *Prefabricated Schools for the Municipality (Competition Contract),* via Manin, Varese, 1965, Fondo Gardella, Coll. 68/3. Study Drawing, November 1965, Pencil on tracing paper. (CSAC, Sezione Progetto, Fondo I. Gardella, B001612P)

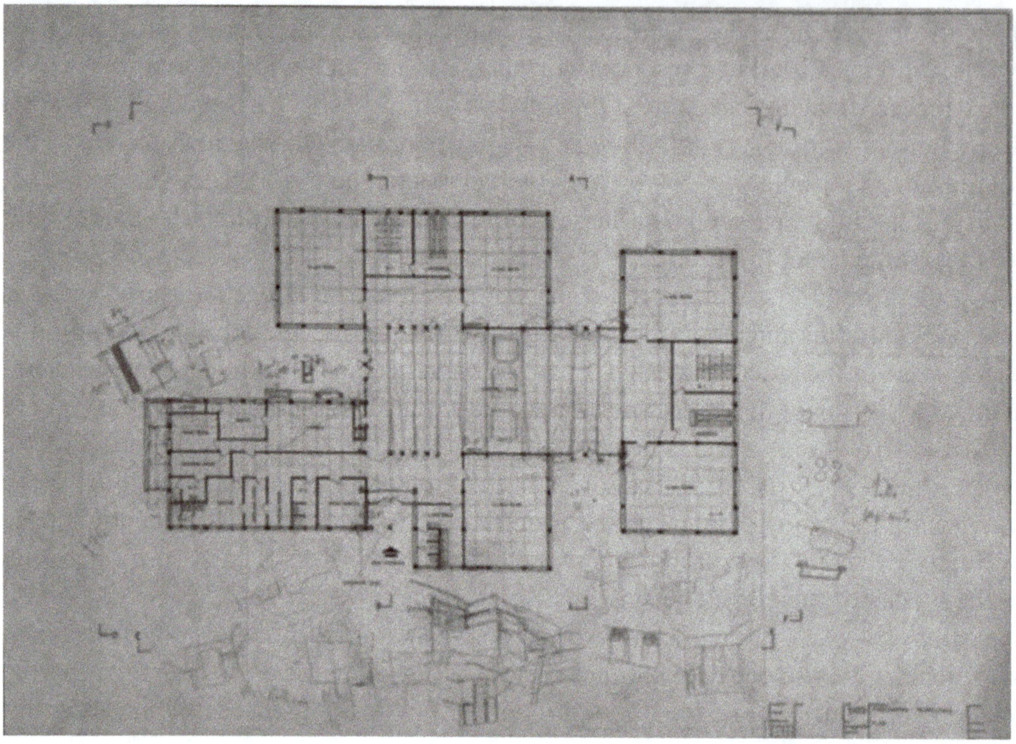

Fig. 6 *Nursery School in Milan*, Project, 1964, Study Drawing, Copy. (CSAC, Sezione Progetto, Fondo I. Gardella, B001601P)

References

Argan GC (1949) Il Museo come scuola. Comunità 3:64–66
Bodino C (ed) (1990) Arrigo Arrighetti
Bottoni P (1954) Antologia di edifici moderni in Milano. Editoriale Domus, Milano
Menghi R (2000) Roberto Menghi. Electa, Milano
Porta M (ed) (1985) L'architettura di Ignazio Gardella. Etas libri, Milano
Samonà A (1981) Ignazio Gardella e il professionismo italiano. Officina, Roma

Space and Figuration of the School Building in the Construction of the Metropolitan Periphery: The School as a Social Emancipation Workshop

Domenico Chizzoniti

Abstract The research concerns the study of settlement forms for school buildings and their susceptibility to becoming public places and centres for aggregation. In particular, schools are analysed here as the set-up of a larger project of social redemption in the metropolitan suburbs. The paper analyses a possible approach to the problem in a situation in which the architectural design of space for education does not renounce the covering of a polygenetic role in the layout of the contemporary city.

Keywords Suburbs · Schools · Typological and figurative aspects · Education · Prototype · Marginality

1 Space and Figuration

According to the particular point of view of the art historian Hans Sedlmayr, a phenomenon has been underway since the middle of the eighteenth century, due to which there is a growing loss of symbolic primacy and mobilising effect, two unitary issues which were until then dominant in the landscape and in the structure of the city: the cathedral and castle-building. To these are added themes which are new, so to speak, such as rental houses, town halls, theatres, the stock exchange, parks, monuments, museums, exhibitions, schools, factories and so on. In fact, as a result of this functional and themed proliferation, as well as a certain tendency for typological perfection and, in any case, beyond Sedlmayr's spiritualistic vision—who, in this phenomenon, interpreted the beginning of the modern city's identity crisis—design

D. Chizzoniti (✉)
Architecture, Built Environment and Construction Engineering—ABC Department, Politecnico di Milano, Milan, Italy
e-mail: domenico.chizzoniti@polimi.it

S. Della Torre et al. (eds.), *Buildings for Education*, Research for Development,
https://doi.org/10.1007/978-3-030-33687-5_3

manuals take on a certain importance and also an operational consistency, tending to differentiate building types for specific requirements. Recently, specialized manuals tend to focus on the supposed infallibility of functionalist particularism, more than leveraging the constantly invoked typological flexibility. Thus, one wonders whether the "functional issue" still exists today, understood in the civil sense, as one which in place of the specialised user's intended use, is still able to bring about instances of the continuing need for a representative characterisation of architecture (Ader 1977). The question put here forces us to consider how the propensity of contemporary architecture to adhere to the most disparate thematic opportunities, "from the spoon to the city" as would have been said in earlier times, pushes the project to be replicated at different scales, sizes and in the most diverse application contexts according to an attitude that is anything but specialised (Paoli 1960; Carbonara 1976). Yet, on closer inspection, it appears quite evident in recent times that even in architecture, themes considered reliable in the making emerge, perhaps in an allegorical sense being the most significant part of an author's poetry: amongst these, for example, school buildings.

Thus, this text relates to unique research which, in the wake of Italian architectural tradition, has contributed to not only building an aesthetic figure but a genuine operational research on the school building facility. It deals with, for example, the extraordinary experience of reorganising educational activities which, including through university research, was exploring didactic prototypes (Petrangeli 1989) on the basis of the work tradition for which both Camillo Boito and Mario Ridolfi highlighted some resistance in comparison with the conventional adaptation to the canons, recognised in those years, of Central and Northern European experimental models (Ward 1976; Perkins 2001). Several architects have moved along these lines such as Carlo Aymonino with his experiences in Pesaro, Aldo Rossi in Broni and Fagnano Olona and Guido Canella with several experimental projects in the Milan Hinterland.[1]

More specifically at this stage, and also due to time constraints, it is the work of Guido Canella which is to be analysed, in particular, the issue surrounding the school facility, its typological experimentation and therefore, also the potential for the user's behavioural induction to the school building. Certain necessary clarifications should be brought forward to clear up the misunderstandings concerning the above-mentioned functionalist particularism which in this case would suggest, given the specificity of the theme, a propensity to technical specialisation deployed to support the different conditions of the project in each case (Panizza 1989). The first issue concerns the role of physical centrality so that the work of architecture, rather than mimetically adapting to it, assumes full economic, productive, social and even representative responsibility, presenting itself as dominant and not only monumental (Pizza 2007). See, in this case, the university settlement's role in the *Competition Project for the University of Calabria in Montalto Uffugo*, Cosenza, 1973 (Fig. 1).

---

[1]Cf. A. Christofellis, *Nel gran teatro dell'Hinterland milanese: scuole materne come case del popolo*, in "*L'architettura. Cronache e storia*", 1976, n. 252, pp. 294–307.

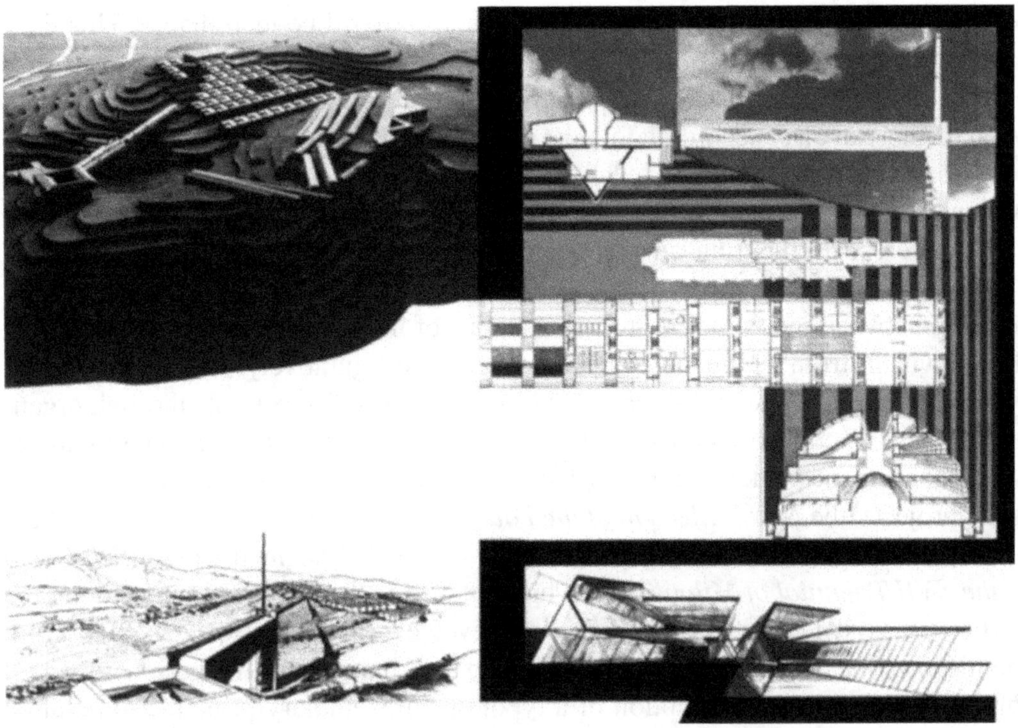

Fig. 1 Guido Canella, *Competition for the University of Calabria*, Cosenza 1973

As a countertrend to a practice that was by now not only prevalent in Italian architecture and to globalism as an aesthetic figure already decreed by the most steadfast supporters of International Style, the research by Canella and his team was exploring two priority areas: first, the aesthetic and compositional values that the considered modern monuments can exercise in the functional and representative revitalisation of entire central parts of the city; secondly, the strategic and structural value that both the conversion of large industrial urban areas and the reorganisation of the urban and regional mobility's large infrastructures (rail and road networks), rather than the reorganisation and strategic reconnection of large metropolitan functions, could have played in reconfiguring the role and the overall fate, for example, of an economically depressed area like Calabria.

These assumptions had already been validated some years before in the construction of a didactic prototype system for the city of Milan and its Hinterland[2] which, in different research occasions, including university research, affirmed a completely overturned perspective in comparison with some trends and a homogenous and repetitive intervention practice, founded on the presumption of the possibility of an improper comparison between different urban areas in terms of nature and culture,

[2]See Guido Canella on this issue: *L'utopia della realtà. Un esperimento didattico sulla tipologia della scuola primaria*, De Donato, Bari 1965; *Il sistema teatrale a Milano*, Dedalo libri, Bari 1966; *Università: ragione, contesto, tipo*, Dedalo libri, Bari, 1975; *Introduzione alla cultura della città*, Clup, Milano 1981; *L'edificio pubblico per la città*, Marsilio, Venezia 1982; *Per un'idea di città*, Cluva, Venezia 1984; *L'architettura italiana oggi. Racconto di una generazione*, Laterza, Bari 1989.

for which a common illness and a unique therapy would be identifiable. This therapy, in turn, emerges in conventional intervention models, deemed to be defined once and for all and universally valid (standard, decentralisation, regeneration, settlement rebalancing, but also, according to most recent models, area projects, pedestrianised historic centres, scientific and technological research parks, turreted and reflective business centres, museum conversions of each historical building and a green belt).

Therefore, strictly from an architectural point of view, what should be noted about this Canella's campaign on the role of the building is the established relationship between induction from the behaviour of human activities and typological design which, from its unique observatory, leads to figuratively (and therefore also typologically) innovate traditional education structures, for example through organisations able to condense collective functions through a programme that anticipates new management models for the hosted activities (Noal 2001). Amongst the unfinished projects, the *New settlement of the Politecnico in Bovisa, Milan,* is highlighted in this regard, presented at the "*Le città immaginate*" (Imagined Cities) *Exhibition of the XVII Triennial of Milan, 1987* which, placed in the historic suburbs along the north-west axis of the city of Milan, tends to reverse the tendency to detachment from the historical density of the productive context of that strip of the periphery, albeit through the figurative redemption of a typological exemplary principle to regain a now lost representative prestige in the university institution (Ader 1977). It is an operation that brings together authoritative typical factors such as that of the Filarete cross of the Ospedale Maggiore in Milan, or the prototype of San Matteo in Pavia, intersecting it with the most compelling Leonardesque allegories in the recovery of existing gasometers which, by virtue of an almost exemplary method of contextualisation, tended to contain architectural representation in the wake of history and in the context of tradition to loom over the city landscape.

Moving on from projects to achievements, it is then worth highlighting the contribution of Canella's aesthetics to school buildings. A poetry that was first engaged with obstinate consistency to subvert certain standards and some architectural trends prevalent in those years which, surpassing the borders of context, function and type, was venturing onto the international scene, spreading from Europe to North America and even to countries with a strong cultural tradition, such as England, Germany and Spain, an emblematic hagiographic calibre, almost extending to the autonomy of sculptural works, when it exhausts its meaning in mere appearance (Bohigas 1997; Curtis 2002; Dudek 2000). Thus, faced with this condition, which is necessarily schematic here, what inspired the work of Canella is the line of "resistance" to globalised internationalism which had revived in certain masters of the Modern Movement, from Dudok to Duiker, Neutra to Lescaze, Sert to Lurcat, from Beaudouin and Lods to Gatepac and so on, a line of aesthetic and moral reliability to be uniquely followed in designing the architecture of the school (Boesiger 1966; Dudek 2008; Cohen 1998).

Meanwhile in Italy, in the years of reconstruction, architectural research moved from the analysis of the conditions in which they debated some of the most engaging and influential current trends—the controversial relationship with the historical and geographical context, the uncertain internal functional requirement, the ambiguous

Fig. 2 Guido Canella, *Infant School at Novegro di Segrate, Milan 1966*

external representative result, contention between suggestions of maximum domesticity and maximum elegance—to finally recognise the most casual criticism of the adoption of atopic internationalism as frontier aesthetics, and how the works increasingly entrusted to the subjective architect–artist inspiration were the most significant. Conversely, for example, the open-air relationship experienced at the school in Suresnes was for Canella not just a technical sagacity with regard to the particular users but rather a training option, able to undermine the conventional educational experience setting to explore experimental and innovative educational models: for example, in the relationship between open and closed spaces, between light and twilight, transparency and opacity, of which the configuration of the architectural space becomes a satisfied accomplice. His fondness of "shielded" projections would perhaps be clearer: these, in fact, from the first proposals, such as the *Infant School in Novegro di Segrate, Milan 1966* (Fig. 2) imposed a special relationship between interior protected space and exterior covered space as if there were a physical extension of the classroom interacting with the extension of the Lombardy plain's landscape. The open-air "theatres" carved as loops to resume the overhang of the outer cover through circular steps, which connected the countryside level to the school, were the custodians of this interaction between the possible attitudes of space to correspond to the needs of institutional and experimental learning.

The same applies to the "*Don Zeno Santini*" *Infant School in the Service Centre at the Incis Village, Pieve Emanuele, Milan, 1968* (Fig. 3) where, for example, the search for that fading effect which he himself would have credited to some of his masters, Terragni and Gardella in particular, looking "against the light" and what filters into the contrast between opacity and transparency, and mass and brightness. In this regard the role of the texture of the glass blocks must be considered, which, like a drapery, buffers the doorways of the reinforced concrete structure, laying bare

Fig. 3 Guido Canella, *"Don Zeno Santini" Infant School in the Services Centre at Incis Village, Pieve Emanuele*, Milan 1968

each tectonic element in its absolute conclusive expressiveness,[3] as well as the effect of the *tholos* coverage created for the *Nursery in Gennara di Abbiategrasso, Milan 1972*. However, it is the prototype created at Opera, *Infant School with Nursery in Zerbo di Opera, Milan, 1972* that the architectural facility for infant schools finds its greatest degree of functional, expressive, typological and figurative exaltation.

Thus, for example, the theatrical dominant is not only taken allegorically, in the symbolic metaphor of the gradually descending trend to the central body, but also functionally engaging a real theatre hall within the architectural body: each of Canella's school buildings has a theatre. The "spiral" structure of architecture takes up a theme on the central plan dear to the author. The spiral is wound around a cylinder in an intermediate position which hosts the theatre on the upper level, those of the free activities directly related to the theatre and with a ramp connected to the countryside level and to the underlying refectory. Just by looking at the creations intended for infant schools, it is possible to isolate some recurring themes which, with different accents and in each individual case, progressively from an almost embryonic state of the mass, architecture "becomes an articulated body" for a typological overlap, figurative contamination, and for linguistic separation.

These themes could be summarised in the physical centrality of the architectural body, in such a way that the work, rather than mimetically adapting to its surroundings, assumes full representative responsibility and serves as a monumental dominant; in the emblematic denotation of its figuration, which adapts to the functional versatility giving rise to two distinct landscapes, exterior and interior; in the allegorical method as a reference alluding to that identity denoting the public and collective character of the "school home".

[3]Guido Canella, *Gardella in controluce,* published in F. Buzzi Ceriani (edited by), Ignazio Gardella. *Progetti e architetture 1933–1990*, Marsilio, Venice 1992, pp. 15–17.

Between research and project, Canella operates in the Milanese and Lombard context, aiming to verify, in line with strategic structural intervention programmes, how a propulsive action of architectural representation on the building is capable of generating a settlement, or at least it can guarantee a physiological balance of the new Hinterland communities, decisive, therefore, in the polygenetic dualism city—countryside, or even more, the centre—periphery in the Milan case. And in this context of Milan and Lombardy, Canella explored and brought to the fore certain distinctive constitutive traits, identifiable in the special polycentric reinforcement which distinguishes them, in which original metropolitan traits were determined at an earlier stage, attempting to isolate certain values which, in their historical persistence, constitute genuine "invariant" settlements, employed operationally as an added value to the project. Thus, in an era of globalised modelling with respect to the school building example, Canella rediscovers some prototypes against which to orientate each intervention proposal.

This is, for instance, the example of the basilica plan which, when initially tested on the *Elementary School in the Services Centre at Incis Village, Pieve Emanuele, Milan 1968–1973* (Fig. 3), assumes the entire flexibility of the typological principle underpinning the three main bodies, the transversal one as a transept and the longitudinal ones as two large lateral halls, to respectively accommodate the atrium, opening onto the large gym lowered by a staircase, the refectory, the secretariat and all the facilities; while the actual classrooms were on two different levels on the longitudinal bodies. The case of the *"Fratelli Cervi" Elementary School and sports field in the Mirasole Village of Noverasco di Opera* (Fig. 4) was resolved a few years later in 1974 with an analogous plan and a similar typological peremptoriness. Here, the three halls, one central and two lateral, are intended for different activities: the central chamber for the gym in continuity with the auditorium, with spaces for free activities all connected at different heights but in visual continuity with each other through the large tiered hall which, with a theatrical layout, looks onto the entire central space; the two-side chambers to the classrooms for the elementary school on both upper levels and the infant school at the bottom level, with the services, the entrance hall, the refectory and the kitchen premises. The large transept houses the

Fig. 4 Guido Canella, *"Fratelli Cervi" Elementary School and the sports field in Mirasole Village of Noverasco di Opera*, Milan 1974

vertical distribution system and brings together the horizontal connections, which open with a pair of large tympana and contain the special rooms that open onto the surrounding countryside.

Finally, some reflections are presented on one last aspect which should be considered trying to illustrate how Canella sought to define an authentic interpretation of those characteristics that are consistently comparable, the reasons of civil functionality and expressive-formal reasons that are evident in some projects and works that his research takes as "certain" references. These are, for example, those experiences in the context of the construction of the modern city, which can be taken as a cognitive reference relative to the problem and to the role of the school building. Suffice it to recall, limiting this only to some of the most significant: the initiatives of industrial paternalism and municipal providence in Milan and in the Lombardy area, where services in early industrialisation (schools, hospitals, boarding schools and economical kitchens), originally using the new types of the first industrial takeoff, are presented as free cornerstones on four fronts, at the same time generating an urbanisation "for centres"; Moscovite construction workers' clubs in the twenties, authentic "social condensers" and monuments with new forms designed to compensate for the precarious urban and housing conditions of post-revolutionary Moscow through functional and behavioural wealth and figurative representativeness; the interventions of modern architects in the municipalities of Parisian suburbs, before and during the Popular Front, where modern architectural forms become an advanced management symbol for the city, and so on. For example, the monumental isolation should be observed to which the building for the *Middle School in Monaca di Cesano Boscone in 1975* was destined, located as the fulcrum of a contentious urban condition between the original core and progressive expansion zones kept isolated from one another by social situation, ethnic background and the urban fabric's historical development. The opportunity to place complementary activities to the school building follows a typological principle. Therefore gym and locker rooms, special classrooms, auditorium-theatre with a stage and dressing rooms and public library were located in a cylindrical building with a central floor. This body was separated from the parallel bodies of ordinary classrooms, four per each floor for a total of 24 classrooms, with stairs and corridors set in an intermediate position to connect the classrooms on all three heights to the central cylindrical body. Such a facility corresponded to involving the entire community after school hours, for example. Not surprisingly, the programme resumed some advanced hypotheses that had already been explored for the *Civic Centre with a Town Hall, School and Sports Field in Pieve Emanuele in 1971* (Fig. 5), where original typological-functional mechanisms creatively combine new patterns of behaviour and figurative innovation in an attempt to counteract the incipient territorial and cultural standardisation processes with architecture in the post-reconstruction and post-economic boom years.[4]

---

[4]L. Fiori, S. Boidi (edited by), Guido Canella. *Centro Civico di Pieve Emanuele*, Editrice Abitare Segesta 1984. E. Bordogna, *Meditazioni gaddiane*, "L'architettura cronache e storia", no. 1st January 1986, pp. 6–47.

Fig. 5 Guido Canella, Civic Centre with a Town Hall, Middle School and Sports Field in Pieve Emanuele, Milan 1971

I believe that this concise and schematic overview of some of Guido Canella's school architecture can perhaps contribute towards rethinking the public building's role for education in the current changed structural conditions and within the same new guidelines of international architectural culture. It seems that architecture which is claimed to be quality architecture increasingly tends to disregard its structural essence and the contextual horizon it is intended for, as well as a functional term apparently considered as increasingly inert and from which the latest design culture seems to consider itself fully liberated (Fig. 6).

Fig. 6 Guido Canella, Infant School with Nursery at Zerbo di Opera, Milan 1972

Moreover, this work on school buildings, which had proposed the goal of a design intended for the hinterland in the Milan metropolitan area, was ultimately able to adequately contend both with the promotion of new collective behaviours and with the authenticity of a figuration rooted in the metropolitan landscape.

References

Ader J (1977) Costruzioni scolastiche. Obiettivi e progetti di scuola secondaria opzionale. A. Armando, Rome

Boesiger W (1966) Richard Neutra. 1961–66 buildings and projects. Artemis, Zurich, pp 78–83, 114–121, 122–129

Bohigas O (1997) Arquitectura y pedagogia. La tradicion escolar catalana. In: Architectural viva, 56, pp 17–25

Carbonara P (1976) Architettura pratica, vol 3, tome 2, Composizione degli edifici, section 7, Gli edifici per l'istruzione e la cultura. UTET, Turin

Cohen JL (1998) André Lurçat (1894–1970). Self-criticism of a modern master. Electa, Milan, pp 149–187

Curtis E (2002) School builders. Wiley-Academy, New York

Dudek M (2000) Architecture of schools. The new learning environments. Architectural Press, Princeton

Dudek M (2008) Schools and kindergarten architecture. Birkhauser, Basilea

Noal S (2001–02) Educational spaces, vols 1/2001, 2/2002, 3/2002. Images Publishing, Mulgrave

Panizza M (1989) Scuole materne, elementari e secondarie, in Architettura Pratica, Aggiornamenti, sezione settima, parte prima - Gli edifici per l'istruzione e la cultura. UTET, Turin, pp 107–274

Paoli E (1960) Gli edifici scolastici. Dalla scuola materna all'università. Cisav, Milan

Perkins B (2001) Building type basics for elementary and secondary schools. Wiley, New York

Petrangeli M (1989) Scuole contemporanee. Dibattito Progetti Realizzazioni 1970–1989. Florence Le Monnier

Pizza A (2007) Gatepac. Scuole per la democrazia. L'istruzione come redenzione sociale. In: Casabella (7–8)757, pp 48–53

Ward C (1976) British school buildings. Designs and appraisals 1964–74. The Architectural Press, London

Imagining the School of the Future

Massimo Ferrari, Claudia Tinazzi and Annalucia D'Erchia

Abstract The idea of school has always merged architecture and pedagogy into a unique body, and its existence is characterised by the close relationship between the definition of an appropriate space for those who inhabit the places of education on a daily basis, and a precise educational model suitable for contemporary society and capable of inventing educational spaces for the present and the near future through a consistent transcription of knowledge modes. In school, individuality and universality become one thing and find the balance required for identifying and understanding diversities within the common needs; a community of original *objects turned out by hand*, who are never the same even if they all are human beings. School architecture represents the concrete opportunity to long for shapes capable to reflect a precise teaching model. In this way, it provides an honest interpretation of all the needs at the basis of a multifaceted theme, with all the peculiarities, the individual accents and the controversies that accompany major transformations occurred over a limited of time. Today's definition of school buildings confirms the uncontrolled frailty and the contradictory and fragmented meanings that characterises contemporary architecture as a whole, no matter what the specific function. In this new transition season, we don't see any consistent attempt to reconsider the principles of a branch of knowledge which seems consumed the speed at which figurative possibilities arise. Forty years ago, typological research was abandoned in favour of partial experiments on management and energy issues.

Keywords School · Architecture applied to school · Ciro Cicconcelli · Maria Montessori · One room school · Typological research

M. Ferrari (✉) · A. D'Erchia
Architecture, Built Environment and Construction Engineering—ABC Department, Politecnico di Milano, Milan, Italy
e-mail: massimo.ferrari@polimi.it

C. Tinazzi
Milan, Italy

S. Della Torre et al. (eds.), *Buildings for Education*, Research for Development,
https://doi.org/10.1007/978-3-030-33687-5_4

1 Imagining the School of the Future

Perhaps one of the most effective means that can help to achieve correct future projections in the field of school architecture might be to share the numerous interdisciplinary components that contribute to defining the complexity of a civil theme translated into architecture; various features that make the topic to be interpreted as somewhat complex; in our case, education. Herbert Read wrote in 1954: "*In a rational society there is only one priority; and no service, other than those referring to nutrition and the protection of human life, must take priority over education*" (Read 1954). A priority, education, has always solicited a reflection on the architectures dedicated to it. The revolution of the learning space, understood as an evolution of the concept of education in its design meaning, at least in its ideal lines, begins in Italy with Ciro Cicconcelli, a Roman architect, winner of the 1949 competition for outdoor schools and nominated in 1958 as director of the Study Centre for school buildings established by the Ministry of Education. This working group—that of the Study Centre—shared by pedagogists architects, doctors, administrators, was created courageously to rewrite the regulations referring to school buildings, firmly guided by Cicconcelli (cf. Fig. 1), with the aim to, above all, reflect on the fundamental passage from the concept of "*instruction-teaching*" to that of "*education*".

"*The design of a modern school*" writes Cicconcelli "*must arise, above all, from the search for a psychologically and functionally suitable space to deal with educational problems. It is therefore necessary to grasp and create spaces capable of favouring the child's tendencies while making them effective; it is necessary to create spaces that accompany children in their biological and psychic growth; children must be at the centre of the search for a school space of our time*" (Cicconcelli 1952).

The careful and passionate look at the experience of Darmstadt and at the school model proposed by Hans Scharoun in 1951 creates the background, for the Study Centre and its director, for re-examining the concept of the *classroom*, up until then

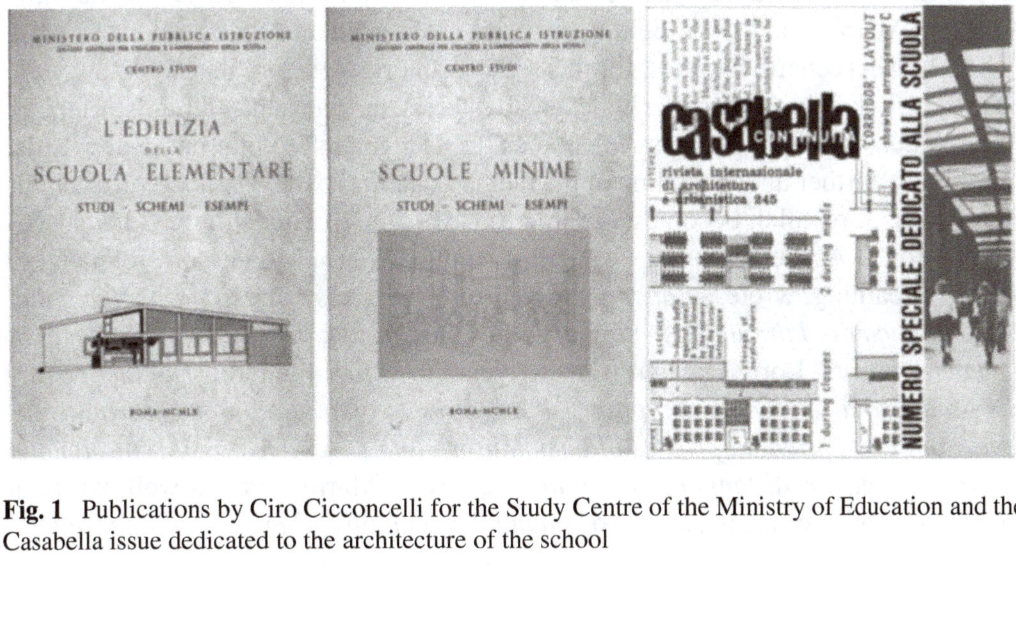

Fig. 1 Publications by Ciro Cicconcelli for the Study Centre of the Ministry of Education and the Casabella issue dedicated to the architecture of the school

habitual, imagining and experimenting the composition of the learning space starting from the capacity for action within the community of children and teachers.

"*Classrooms, for there to be an osmosis process that is established not only between teachers and pupils yet also between the pupils themselves, when they meet in a similar pedagogical function, should be in a position to be coupled and easily transformed; organic transformations, even total, using the same furniture made up of separable and transportable materials*" (Cicconcelli 1952). These words, still today, seem visionary in the extreme contemporaneity of principles; reflections that have accompanied a slow transformation, often simply left on paper or materialized only in very few virtuous examples.

On a smaller scale, the school classroom, a *space for learning* in the most contemporary interpretation of a concept that is both labile and deeply rooted in the idea of education, brings the discussion back to the original dimension of the problem which actually sees children as its main actors, and their ability to share the idea of community for the first time.

The theme of education, in its architectural translation represented by scholastic institutes *of every type and level*, is one of the central topics of our contemporaneity, both political and civil; up to the present day, architecture for schools has anticipated, followed and has sometimes been chased by social transformations, ministerial reforms, educational proposals and has always only quietly articulated the history of our country. At the same time, the numerous projects for school facilities, from kindergartens to universities, have been able to write important passages in the history of architecture, not just national, revolutionizing established principles in the name of a true idea of teaching, free of any pretextual bonds.

Classrooms and corridors, in the words of Aldo Rossi, rooted in the previous avant-gardes, places for education, where the relationship between collective spaces and singular spaces, the ideal form of teaching places, the most adequate didactic idea, make up the backbone of a possible continuous comparison between the teaching methods and the examples in architecture that best interpreted them, starting from the substantial pedagogical readings of the twentieth century, from Rosa and Carolina Agazzi to Maria Montessori, right up to Loris Malaguzzi and Mario Lodi.

The Italian scenario narrates, for this reason, an original story that we know well. The strong link or in any case the widespread interest, the constant attention to applied research deriving from pedagogical studies, to its teachers, to international awards, enriches the further understanding of the themes specific to the composition of buildings which, in parallel, becomes a search for the value of education and the spaces suitable for it. At the beginning of the last century, Maria Montessori, referring to places of learning, wrote: "*Education is a natural process carried out by the child, and is not acquired through listening to words, but through the child's experiences in the environment.*" Loris Malaguzzi, many years later highlighted: "*The atelier (...) has produced a subversive irruption, an additional complication and instrumentation, capable of providing riches of combinatorial and creative possibilities among children's non-verbal languages and intelligences.*" Mario Lodi, as well, wrote in the mid-seventies of the necessity to "*create a community where children feel equal, like companions, like brothers.*" In architecture the environment, ateliers and communities mark numerous possibilities of research still in progress, physical spaces or figurative forms which, with centrifugal force, are able to generate, from the inside, school buildings in their own complexity.

It is therefore *the classroom*, the minimal module (cf. Fig. 2), in the simplification of the text, that represents this generative force directed towards the exterior, the more domestic and at the same time authentic character of the building for education, the deepest seed of possible change in the way to learn and teach in the future.

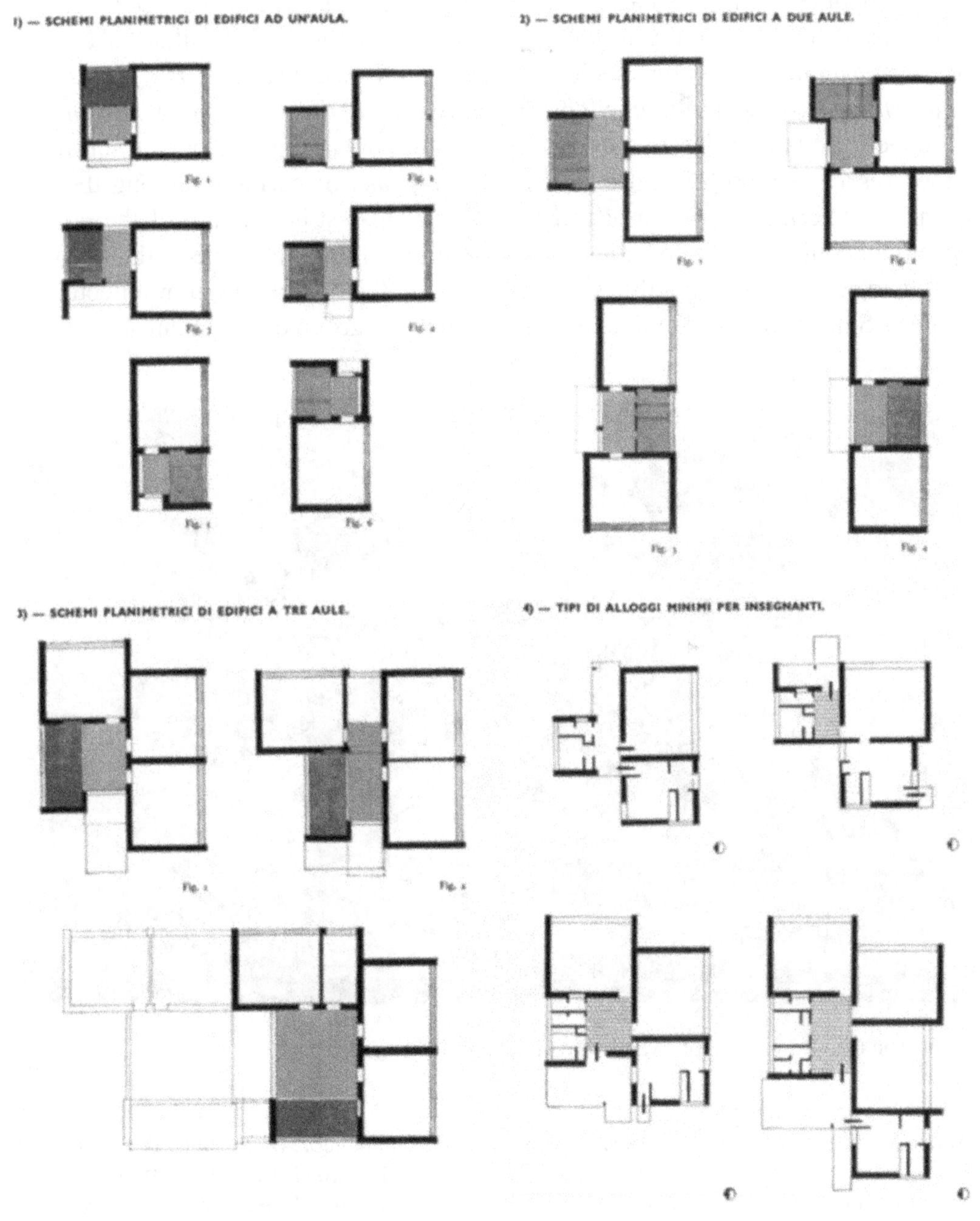

Fig. 2 Ciro Cicconcelli, planimetric studies for schools, 1957

Another ancient example guides us. The American model of the single-class school—*one room school* (cf. Fig. 3)—a model exported to many other countries including Austria, Germany, Australia and Ireland, represents a figurative horizon as opposed to an education goal, a singular suggestion as its uniqueness denounces. These small rural schools were built at the end of the nineteenth century and were surrounded by nature. They were made up of a single space and the few places needed for school life, compressed to their minimum extent: a staircase, an entrance, the teaching room, the bathrooms; a single class for children of different ages, a single teacher to learn to read, write, count, history and geography, a large window to the east to welcome the light. Small buildings with elementary forms that often became the centre of the community in the collective imagination; places that have often represented an idea of a future society, as Abraham Lincoln stated, "*The philosophy of the school in one generation will be the philosophy of the government in the next.*"

Reflecting on the school of the future is therefore not a slogan rather a re-proposal in the present of those central examples, those peaks of harmony among the disciplines that form part of the last century. It means—still—certain of the critical capacity of confrontation, to believe in a generation of Italian architects that are well aware and capable of facing the issue and allowing the quality of our architecture to progress. Some time ago, for this reason, we invited a group of 12 Italian architects,[1]

Fig. 3 The One-Room School: Watson Road School, USA, about 1900

[1] Walter Angonese (*Accademia di Architettura*), Riccardo Campagnola (*Politecnico di Milano*) and Maria Grazia Eccheli (*Università di Firenze*), Armando Dal Fabbro (*Università IUAV di Venezia*), Alberto Ferlenga (*Università IUAV di Venezia*), Luigi Franciosini (*Università degli studi Roma Tre*), Stefano Guidarini (*Politecnico di Milano*), Eleonora Mantese (*Università IUAV di Venezia*), Bruno Messina (*Università degli studi di Catania*), Carlo Moccia (*Politecnico di Bari*), Renato Rizzi

Fig. 4 V Triennale di Milano. The entrance to the pavilion dedicated to the school

during an exhibition at the Triennale di Milano (Ferrari 2015). The twelve architects engaged in research, teaching and criticism of our discipline, belonging to the same generation, were encouraged to imagine and represent their idea of a *classroom* for the future; 12 spaces that differ in shape, character, colour, relationship with light or nature, proportions and flexibility, orientation and possibility of different uses, overlapping and decomposing of places which in general compositional principles refer to as ideas of school. A concrete and proactive attempt, in the variety of the proposed projects, is to imagine, through open confrontation, the various suggestions and declinations of a common goal. The criticism that follows the impossibility, in this essay, of referring to punctual and descriptive images, allows the reader—within the story—to hypothesize, starting from the principles highlighted, a personal figurative interpretation of the exhibited projects as in the picture of the pavilion built in 1933 for the Triennale di Milano (Fig. 4).

The ten most realistic solutions (cf. Fig. 5), exclude, due to lesser concreteness, two equally interesting examples characterised by a more abstract reflection on the quality of the place of teaching and the relationship between teacher and student (Renato Rizzi) and on the infinity of the possibilities of linked spaces for teaching (Paolo Zermani); ten effective suggestions to define more constructively the theme of the space for teaching that we consider as a choral and shared contribution for the places of learning of the future.

The great attention towards new technologies is one of the first interests addressed to the innovative space imagined for students; a digital system capable of educating in the contemporary world through total immersion in the planned places. Regular volumes characterized by simultaneous digital projection on three sides of the room

---

(*Università IUAV di Venezia*), Andrea Sciascia (*Università di Palermo*), Paolo Zermani (*Università di Firenze*).

Fig. 5 Ten future class projects (models by BRG studio)

directly facing the outside through the fourth fragment. An agora engraved on the ground where one can concentrate on the complete involvement of the proposed contents; the only hope where to face the external reality (Walter Angonese).

The sum of different spaces for the lives of children, coupled and equipped to obtain semi-independence with respect to the services offered by the school—the kitchen for example—or subdivided into four units starting from a large collective space, coincides with the research of another two different scenarios proposed. The first, in addition to coupling two contiguous sections, bound by a border of books, envisages both external and internal accessory spaces, private for the two sections, yet integrated into a more complex system (Alberto Ferlenga). The second looks to transparency and filtered light at the perimeter, for the greater character of a collective place that can be divided into four parts, defined in the first instance by an evident metallic cover (Armando Dal Fabbro). The theme of natural light once again defines other hypotheses which, in the rarefied illumination to the zenith and in the complete openness and transparency on the ground, find the most convincing answer in the project of a large covered campus where the space for learning exists through a natural development towards a protected exterior (Luigi Franciosini). Simplicity and ease of composition are the distinguishing marks of numerous offered projects which, particularly in one proposal, prove to be effective in defining the uniqueness of the study space. Solid walls, composed in the shape of a court and covered by the evident recognisability of a pitched roof, make both the unity of the minimum module and the domesticity of the recognizable place (Stefano Guidarini). The symbolic re-proposal of historical spaces, recognised as examples for their quality of inhabited life, distinguishes a different direction that does not envisage any distinction between the spaces of learning within the schools regarding "all types and levels". Only the cultural influence of each level strongly characterizes, in this case, the various identities of the education through the iconicity of furnishings, colours, works of art—created ad hoc for each section—that complete the articulated predefined spaces (Eleonora Mantese).

The theme of one's territory and of the necessary services distinguishes a further proposal that gathers the place of learning in an exalted centrality and distributes to the surroundings the abundant accessories. The proportion in height of this chosen volume makes the main space recognisable; transferring the vertical light to the ground, even from the outside. A tight chequered composition makes up the school (Bruno Messina). Two examples on a minor scale address the issue of the place of early childhood following different experiences. The nursery school in the recognisability of its elementary forms marks an initial hypothesis which, starting from an inhabited perimeter, where educational spaces are distributed, defines a centre conceived as a collective place in the middle of the planimetric geometry. A place, like a covered courtyard, entrusts to a large pillar/tree the role of supporting the flat roof. A primitive social terrain can be imagined under that tree (Carlo Moccia). The second path concretises the close relationship between the idea of home and a schoolroom similar to that, on a major scale, between school and village. The environment created, set up on two floors under a domestic pitched roof, recalls an idea of iconic continuity with the image of the house in the smallest degrees of learning; regarding the school, the grouping of individual houses is arranged to form a small urbanity (Andrea Sciascia). Lastly, a research experience was gained following the executive possibility of a realised project, a reflection on the theme which, starting from the need to expand an existing school, translates the reasoning on the relationship between community and singularity of the school space into a regulating principle. The central manifested place, external fulcrum of the original school, becomes, in the proposed project, the effective centre of a community that finds itself in the theatrical space, the focal point of the two connected interventions (Riccardo Campagnola, Mariagrazia Eccheli).

2 Research in Progress

The need to start from what in the last 100 years has been done, and in particular has been built, with regard to Italian school construction, is a must today for several reasons, including the high number of school buildings distributed more or less homogeneously in our country and above all because the Italian reality requires particular attention to the recovery, restoration and consolidation of existing buildings. In this sense, the recent presentation by the Scholastic Building Registry, after 20 years from its establishment, is undoubtedly a fundamental tool for any advancement in study and knowledge, as well as design, regarding the architecture of the school.

The presence, by the will of the Ministry, of an operational tool that manages to monitor and classify all the Italian heritage related to the theme of school construction is a positive sign, while the research project aims to increase and improve this instrument by comparing it with other interpretative parameters more properly referable to

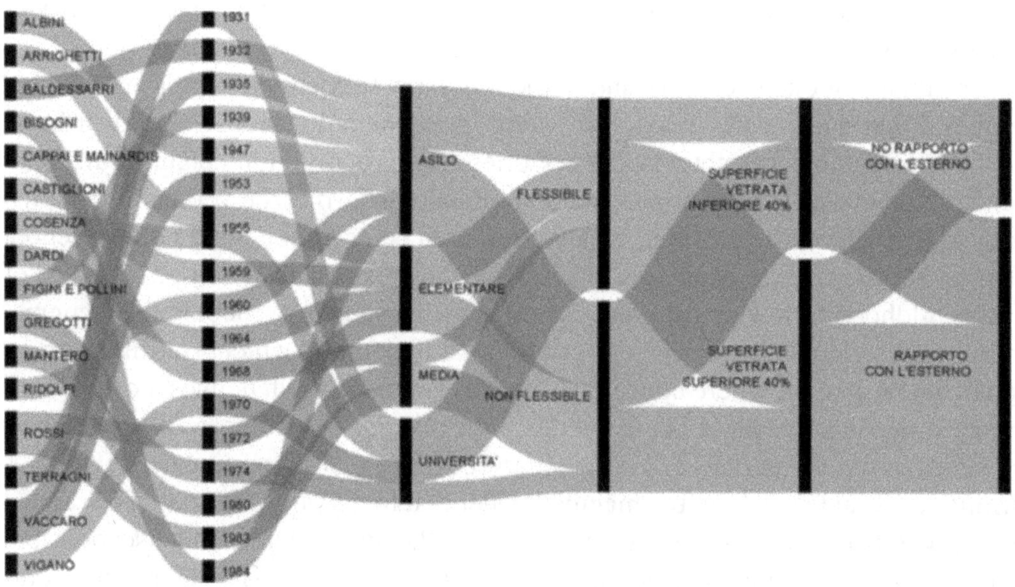

Fig. 6 Research data processing 100 years of schools in Italy (in progress)

the architectural discipline and to the search for the quality of living, which up until now have not been considered. Only a comparative study of the different typologies (in their historical-pedagogical evolution), of the coherence between project and realisation of the form-construction-technology relationship can give back the panorama of the Italian school building research; which today should operate towards integration, recovery and its extension. On the one hand, therefore, the research in progress is aimed at the study and analysis according to a necessary categorization of those remarkable examples that in the twentieth century marked the history of the architecture of the school regarding "every type and level" as is apparent from the infographic above (Fig. 6). On the other hand, the primary objective is unquestionably the design verification of new compositional possibilities which, through the virtuous examples that preceded us, are capable of accommodating the present needs that are dictated by both pedagogical research and by the character of the contemporary city and social demands.

Starting from the experience in 1949 regarding the Study Centre of the Ministry of Education with Ciro Cicconcelli, and, in particular, from the competition announced by the Ministry in which architects were asked to propose their idea of a school without taking into account the then current legislation in order to try to translate from the projects presented the new rules that were actually more effective, there is an evident need for architecture itself to concretely verify the possibilities of giving content as well as shape to a new idea of school, while looking for the most adequate answers to a typological theme, so rich in feasible interpretations.

The research is based on the fundamental interdisciplinary feature, implicit in the theme in question. It is, in fact, impossible to contain the research on school buildings within the architectural discipline alone. The various worlds that gravitate around this topic, which is very central to today's society, require a close dialogue aimed at the concrete verification of a balance between the planning of educational sites and research in the pedagogical field.

At the same time, this interdisciplinary feature does not only concern the relationship between design and pedagogy, it also opens up, from time to time, to new aspects that cannot be considered corollary, rather a base from which to start again in facing the complexity of the problem.

New spaces for teaching, starting from recent experiences that have tried to compare different architectural sensitivities on the subject of educating space, have started a reflection that still needs a concrete verification through the realisation of one or more prototypes of the minimum teaching unit to be carried out on a full-scale (1:1) basis, developing the concrete synergy between quality of space and new learning methods. This experience could start an actual synergy between industry, companies and universities in identifying materials, technologies and spaces suitable for teaching. At the same time, the theme of consolidating and adapting existing buildings can launch virtuous relations between the construction industry, specialized and unspecialized in the specific sector of restoration work, and the world of research that has already been strongly carried out on other typologies.

References

Cicconcelli C (1952) Lo Spazio Scolastico. In: Rassegna Critica di Architettura, n 25, Rome
Ferrari M (2015) Di ogni ordine e grado. L'architettura della Scuola. Rubettino Editore, Catanzaro
Read H (1954) Educare con l'arte, Edizioni di Comunità, Milano

Modernist Schools in the New Rural Landscape of the Pontine Plain

Francesca Bonfante, Nora Lombardini, Emanuela Margione and Luca Monica

Abstract Based on a research and fieldwork carried out in the framework of the EU-funded research project MODSCAPES (Modernist Reinventions of the Rural Landscape, funded under HERA JRP III call "Uses of the Past", Oct. 2016–2019), this contribution focuses on the case study of the Pontine Plain. In the 1930s, as part of Mussolini's ruralization policy, the vast swampy area was converted into a neatly designed countryside hinged on a hierarchy of villages and medium-sized towns such as Littoria (1932), Sabaudia (1934), Pontinia (1935), Aprilia (1936) and Pomezia (1938). How did architecture contribute in shaping a new "place identity"? This chapter questions the role of schools as fundamental collective buildings, helping the settlers put down roots. School buildings offered architects scope to experiment with new spatial layouts and architectural expressions aimed at the widest possible understanding.

Keywords New towns · Civic centre · Rural and urban school · Architectural composition · Typological experimentation

1 The Schools in Agro Pontino: An Introduction

The history of the Agro Pontino in the Twenties and Thirties, extensively investigated by a multidisciplinary literature, is interesting for studying the modern relationship between city and countryside. Moreover, it is worth to be investigated to understand the typological and figurative characters of the public buildings which are organized in a highly hierarchical territorial system and to build the new civic centers where the inhabitants can start new social behaviors.

The school building plays a fundamental role within these new civic centers, conceived as spaces for the collectivity. Also the school is able to emphasize, through its own figuration, that particular relationship between cities, farms and countryside is created in Agro Pontino by the fascist urban policy. In the end, the school is also

F. Bonfante (✉) · N. Lombardini · E. Margione · L. Monica
Architecture, Built Environment and Construction Engineering—ABC Department, Politecnico di Milano, Milan, Italy
e-mail: francesca.bonfante@polimi.it

S. Della Torre et al. (eds.), *Buildings for Education*, Research for Development,
https://doi.org/10.1007/978-3-030-33687-5_5

particularly interesting to study the territorial transformation of this area since it anticipates the strategy of Integral Reclamation. The school, in fact, since the early twentieth century, became the fundamental tool to fight malaria and to improve the living conditions of the population, thanks to the initiatives promoted by A. Celli, A. Fraentzel, G. Cena, S. Aleramo, A. Marcucci and D. Cambellotti.[1]

Following this line of thinking, the school building is here investigated by comparing two main architectural typologies built in the area. The first is the rural school, a forerunner of the concept of Integral Reclamation, and the other is the urban school, one of the main protagonist in the New Towns' civic center.

1.1 The School in the National Debate During the Thirties

To investigate the role of the school in the transformation of the Pontine territory, it is useful to contextualize the issue within the Italian debate started in those years concerning both for the improvement of education and the development of the architectural features of a modern school building. This debate is reflected first in the manuals dedicated to the construction of school buildings. One of the first manuals is the one written by L. Secchi published in 1923. The text is particularly interesting because the author, since the very introduction, well summarizes the condition of the Italian school, highlighting the state of degradation of school buildings in the middle and small urban centers. He wrote:

> If the soul of people is formed and shaped through education, the importance of the building that welcomes youth is immense [...]. The new [modern] concepts were able to increase interest around the school. But although this impulse was strong and powerful, [it] was limited to large cities. So, while the big city had its own typical school building, the middle and small town continued to build without applying the new hygienic and pedagogical concepts (Secchi 1923).

At the beginning of the Twenties, the fundamental role played by education in the creation of new behaviors was recognized. It was also clear how architecture and figuration of space could positively contribute to the education of masses.

The role of school, understood as a tool to create a new society, is again highlighted by M. Piacentini in the preface of the catalogue entitled *Schools*, written by G. Minucci in 1936. The text, defined by Piacentini himself as a *School buildings "Code"* (Piacentini 1936), collects and analyzes "good examples" of modern school buildings in Italy and abroad.

Between Secchi's 1923 publication and Minucci's publication in 1936 (Minucci 1936), the architectural research of a new typology for school buildings achieved an

[1]Angelo Celli (1857–1914), hygienist; Anna Fraentzel (1878–1958), German nurse and philanthropist; Sibilla Aleramo (1876–1960), writer and poet; Giovanni Cena (1870–1917), poet and journalist; Alessandro Marcucci (1876–1968), educationalist; Duilio Cambellotti (1876–1960), artist. They were all members of the "Comitato per le scuole dell'Agro Romano e Pontino" (Committee for the farmers' schools of the Roman and Pontine Agro) founded in 1904.

important stage. Indeed, in 1933, at the *V Trienniale* in Milan,[2] the pavilion by A. Annoni and U. Comolli called *"La Scuola 1933"* was presented. In this experimental building all the characteristics of the modern school were exhibited. The critic A. Pica wrote in 1933 about this topic:

> The contact with nature, favoured by the width of the windows [...]; the variety of forms, realized in the furnishings; the order, reached and supported without the help of symmetries; the abstraction of forms, volumes and wall decorations; they are all elements ordered to awake and excite the imaginations of children's intelligences (Pica 1933).

Among the afore mentioned, two other manuals should be pointed out: one by Del Debbio (1928) which investigates the new typology of the ONB[3]; the other one is the text by P. Carbonara published in 1946 where, once again, the typology of the school building evolves in order to adapt to new *methods of active learning* (Carbonara 1946).

Under the impulse of pedagogical research—Montessori method started to be applied in those years—paralleled by architectural research, numerous laws were issued to define the new educational system and the architectural configuration of the school (Mugnai 1984). The two main laws related to school building were issued between 1925 and 1939.[4] During this time the schools in Agro Pontino represent an important experimental laboratory around the theme of dualism between city and countryside, although being "minor" architectural examples if compared to the national architectural research on the subject.

1.2 The Rural School from the "Hut-School" to the Masonry Building

The long story of the reclamation in Agro Pontino is concluded in the Twenties and Thirties of the twentieth century by the fascist regime under the guideline of A. Serpieri's "Integral Reclamation" strategy that finds its roots into the "Internal Colonization" process.

The whole territory was hierarchically organized following a triad system based on the relationship city–village–farm.[5] In the cities and villages, the public buildings constituted the *core* of the settlements, highlighting the relation both with city and with the countryside.

The school building, among other public buildings in the new settlement *core,* is certainly the typology that actually played a fundamental collective role, even before the start of the fascist project (an example is the Concordia village built around the

[2]Esposizione internazionale delle arti decorative e industriali moderne e dell'architettura moderna (International exhibition of modern, decorative and industrial arts and modern architecture).

[3]Opera Nazionale Balilla (National Young fascist Organization).

[4]D. M. May 4, 1925 "*Rules for the compilation of school building projects*" and R. D. L. May 27, 1939 n. 875 "*Bottai's Law*".

[5]See the Article by Piccinato "Sabaudia Urban Meaning".

"Hut-school"). As a matter of fact, the history of school buildings in Agro Pontino begins in the first years of the twentieth century with the "Committee for the schools for the farmers in Agro Romano and Pontine marshes" activities. The Committee started a work of alphabetization, leading the "Cabin-school" and the "Hut-schools" to be housed in the first masonry buildings, thus adapting teaching spaces to the different needs of the farmers (Campagna 2001).

This evolution yelled a great national success leading the Committee to exhibit their activity in 1911 at the "Esposizione internazionale delle Industrie e del Lavoro" (International Exhibition of Industries and Labor) organized to celebrate the fiftieth anniversary of the Unification of Italy. Here the Hut-School and the Cabin-School both designed by D. Cambellotti were exhibited. The Hut-School (Fig. 1) built mainly in wood, following the tradition of the *lestra*[6] in Agro Romano, was a traditional hut with a circular central plan to be adapted to the needs of education. The Cabin-School, an itinerant structure suited to the continuous movement of the marshes inhabitants, was made in wood but covered with impermeable cloth. Both the structures served as facilities for the teaching of new agricultural techniques in order to create a renewed culture of the land.

The first two rural schools in masonry, that "*constituted the models for the school buildings that arose in the surrounding villages*" (Secchi 1923), were built in the villages of Colle di Fuori and Casale delle Palme. The school in Colle di Fuori (Fig. 2), built between 1912 and 1914, according to a project by A. Marcucci, D. Cambellotti and F. Pierpaoli, *consists of two classrooms: one for elementary school, the other for kindergarten*; *a kitchen and a small teachers' apartment* (Secchi 1923). The same typology was developed in 1933 for the school in Casale delle Palme (Fig. 3). The building was made of two parts: the first was a single floor volume with two classrooms, and a porch in the front end used for outdoor activities; the second

Fig. 1 D. Cambellotti, Project for the Hut-School, International Exhibition of Industries and Labor (Agro Romano Exhibition), 1911 (Cardano 1980)

[6]Typical temporary hut in Agro Romano and Pontino.

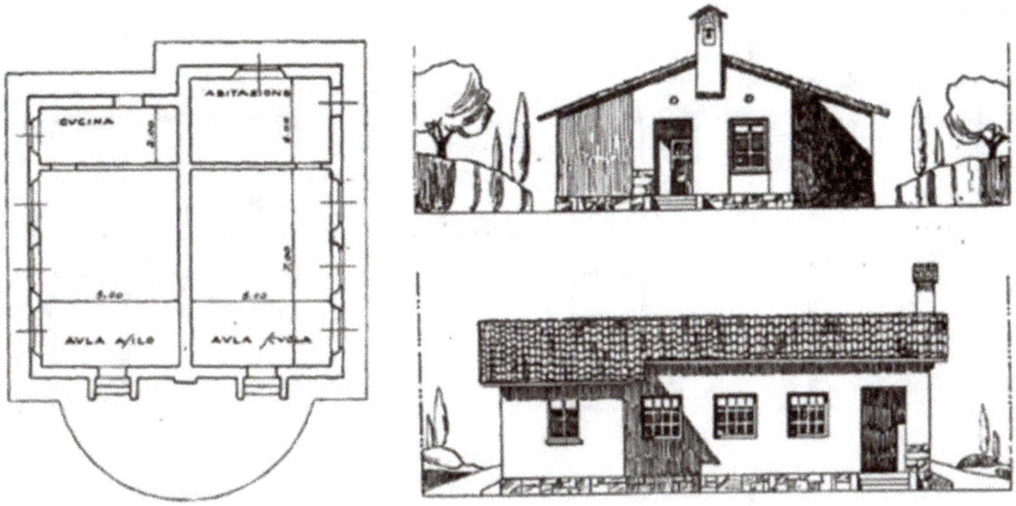

Fig. 2 A. Marcucci, D. Cambellotti, F. Pierpaoli, One room school typology: plan of the prototype; elevation of the school in Colle di Fuori, 1912–1914 (Secchi 1923)

Fig. 3 Rural school, Casale delle Palme, 1922. Redrawing based on the image published in Secchi (1923)

part was a two-storey volume with, on the ground floor, the kitchen, the pantry and the principal's office and, on the first floor, the teachers' apartment.

As anticipated, these schools were used as a model by the ONC,[7] not only for the construction of the rural schools in the villages but also for the construction of the so-called *Farm-Schools*, strategically located within the territory outside the urban centre.[8]

The elementary unit of the first masonry schools is recognizable in the rural schools in villages (Fig. 4). Here the typological experimentation of rural schools

[7] Opera Nazionale Combattenti (Veterans' National Organization established in 1917).

[8] It is worth mentioning as Farm-Schools: Murillo (Littoria-Scalo); Santa Feticiola (Piscinara); Uccellara (Tor Tre Ponti).

Fig. 4 A. Pappalardo: above, Rural school in Borgo Hermada, 1934. (Redrawing based on the materials stored at ACS, Opera Nazionale Combattenti, fondo progetti, 19/24 *Componenti dei progetti: 3. Scuola rurale*, Roma.); below, Rural school in Borgo Montenero, 1935 (Redrawing based on the materials stored at ACS, Opera Nazionale Combattenti, 32/2 *Borgo Montenero. Disegni*, Roma) (1. Classroom; 2. Changing room; 3. Teachers' room; 4. Library; 5. Teachers' apartment)

led to the organization of the space for outdoor activities mainly dedicated for agronomy practice. In fact, since 1914 most rural schools started to be equipped with "experimental fields" and "model school gardens".

1.3 The Urban School and the Opera Nazionale Balilla

In addition to the rural villages, five new towns were built in Agro Pontino: Littoria—in the present-day Latina—(1932), Sabaudia (1934), Pontinia (1935), Aprilia (1936) and Pomezia (1937–1938). Here the public buildings, playing a fundamental representative role, were "*commensurate with the entire population of the open city (urban nucleus and dependent rural area)*" (Frezzotti and Pasqualinucci 1937). These buildings were specifically arranged in order to give life to the main urban *cores*: the political and administrative centre was surrounded by the city hall, the party house—Casa del Fascio—the workers' club—OND—and the cinema-theatre; the handouts centre was surrounded by the church, the maternity and childhood institution and the ONB; and the agricultural core was surrounded by the market and the headquarters of the ONC (Piccinato 1934). The articulated system of the squares was abandoned for the last two new towns—Aprilia and Pomezia—where the 2PTS group—C. Petrucci, M. L. Tufaroli, E. F. Paolini and R. Silenzi—developed the rural villages "urban typology" with all the public services of representation organized around a single centre. The urban projects for the Agro Pontino are therefore an important laboratory on one hand for the development of a "technical-scientific" method related to urban planning and, on the other hand, for the typological and figurative experimentation. We can easily say that the territory in Agro Pontino shows the contradictions between "old" and "new" and between different possible architectural languages, through all its complexity and articulation.

The school localization changed according to the settlement type. In the villages the school overlooked the central square, directly contributing with the other public buildings in the creation of the unitary civic centre, whereas in the cities, the school building is adapted to the different experiments in urban planning. Moreover, the typological variations, adhering to the laws' activity program,[9] depended on the proximity between the school and ONB, an institution which from 1926 to 1937 was in charge of the youth physical education (Teja 2005). Indeed, a characteristic that combines and distinguishes urban schools is the space of the physical education. Therefore, the gym can be used as one of the interpretation keys in understanding the typological evolution of the school building. In Littoria, built few years after the birth of the ONB, the building of the "Casa del Balilla" is a self-standing building, set aside from the school (Todaro 1932). On the contrary in Sabaudia, Pontinia and Aprilia the school cluster is built in physical proximity and architectural design affinity with the building of the ONB. Two examples by O. Frezzotti, are useful to

[9] According to the D.M. May 4, 1925 the schools in the urban centers had to be equipped with: 10 classrooms, medical room, director's house and library.

understand this relationship. Built following the first master plan—designed by the same Frezzotti—the school in Littoria is built around the edges of a block according to a vision still linked to the urban tradition of the nineteenth century. In Sabaudia instead, the ONB–school cluster is located in the site designated by the master plan designed by G. Cancellotti, E. Muntuori, L. Piccinato and A. Scalpelli, setting an urban system independent from the two main squares of the town hall and the church. The school complex is clearly one of the largest blocks within the urban centre, equipped with all the outdoor spaces necessary for open-air activities.

The school building of Aprilia (Fig. 5), designed by architect C. Petrucci, is another emblematic case. Here the school complex consists of pre-school, elementary school and ONB, and with the church it builds one of the main blocks of the urban centre.

The pre-school finds its roots in the rural school typology: the two classrooms on the ground floor are sided by the body of services that houses on the lower level, the kitchen, the refectory and the principal's office, and on the upper level, teacher's apartment. The elementary school, organized with an L-shaped plan, has two levels. On the ground floor there are five classrooms—arranged along the main axes and overlooking two gardens edged by the refectory and the church—the principal's office and the canteen; on the upper floor there are five other classrooms and the library. The ONB volume, accessible from the school only on the ground level, is distributed on two floors: on the lower floor there are the changing room, the clinics and the gym, from where it is possible to reach the garden organized according to the guidelines described by E. Del Debbio; the upper floor, accessible by an independent stair, housed the administrative offices, the conference room and the terraces overlooking the gym.

In Pomezia, after the end of the ONB as an institution in 1937, the school complex will again find its own independence, albeit associated with the building of the GIL (Italian Youth of Littorio), an organization that would have replaced the ONB.

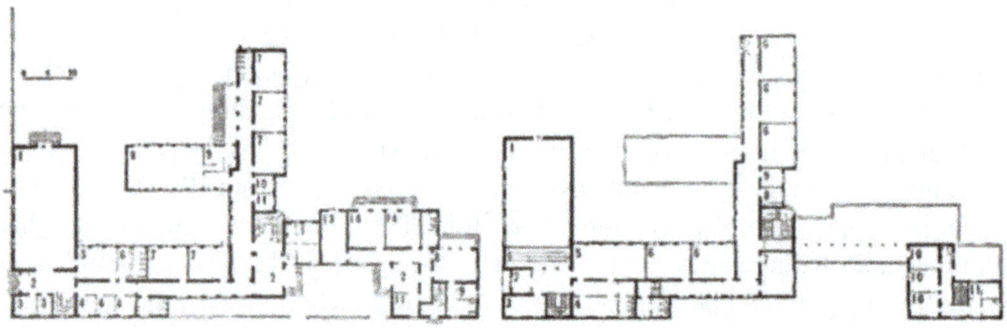

Fig. 5 C. Petrucci, Project for the Aprilia school building, 1936. On the left: Ground floor plan. (1 gym, 2 lobby, 3 clinics, 4 offices, 5 changing rooms, 6 showers, 7 classrooms, 8 refectory, 9 pantry, 10–11 managing offices, 12 guardian apartment, 13 dormitories, 14 kindergarten classroom). On the right: First floor plan: (1 gym, 2 secretariat, 3 presidency, 4 administration, 5 conference hall, 6 classrooms, 7 library, 8 management offices, 9 teachers, 10 bedroom, 11 living room) (Carbonara 1946)

1.4 The Schools in Agro Pontino: For a Possible Future

Studies over the last decade start to pay attention to the renovation of the existing school buildings considering that more than 50% of schools were built before the Seventies. The Italian draft budgetary law of 2019 for its part provides, for the Provinces in Regions with ordinary statute, specific financial funds (from 2019 to 2033) in order to improve safety and quality of space in school buildings. According to the good maintenance practices of public and private buildings, before applying the law it is necessary to collect data and start a survey with a deep analysis of the artifacts. Particularly, for the schools in Agro Pontino, taken into account in this paper, it is important to remember that the buildings were built both before and after the reclamation process according to the social needs related to specific moment of the Italian history (for instance, after the Italian Unity, the schools were built to spread the Italian language and during the Thirties the autarchy gives priority to national materials). Furthermore, the schools in Agro Pontino, from the smallest—spread on the territory and nowadays almost abandoned—to the medium-size buildings—built in villages and new towns—are important documents to be preserved. The schools, in fact, belong to a cultural heritage able to highlight the historical value of the events that have followed over the time in this area, from the ancient times till nowadays.

The different dimensions and localizations of the schools buildings in Agro Pontino require a single-case re-use approach, knowing that the conservation of the rural school is more complicated. Indeed, the original urban schools building, despite some expansions and transformations remained practically intact, unlike to what happened to rural schools which are today in ruins and abandoned, scattered all over the countryside. Two exceptions are represented by the rural school in Casale delle Palme and the one in Latina Scalo. These two examples have been either restored or transformed—as in the case of the school in Latina Scalo—into municipal library.

Today, these buildings spread around the Pontine territory could recover new educational and training purposes for very localized residential settlements. In a territory, where public mobility is marginal, these buildings can be possible places to re-create a network of primary schools or spaces where training the new rural classes for (immigrants or not). This network would make it possible to recover these buildings that are now archaeological traces of an original alternative urban system in contrast with the cities development.

References

Campagna A (2001) Dalle scuole per i contadini alle Scuole Rurali. In: Pennacchi A, Vittori M (eds) I borghi dell'Agro Pontino. Novecento, Latina, pp 181–183

Carbonara P (1946) Edifici per l'istruzione. Scuole materne, elementari, medie e universitarie. Hoepli, Milan

Cardano N (1980) La Mostra dell'Agro Romano. In: Piantoni G (ed) Roma 1911. Catalogo della Mostra alla Galleria Nazionale d'Arte Moderna. De Luca Editore, Rome, pp 179–198

Del Debbio E (1928) Progetti di costruzioni. case Balilla, palestre, campi sportivi, piscine, ecc. Opera Nazionale Balilla, Rome

Frezzotti O, Pasqualinucci O (1937) Relazione al P.R. di Littoria. In: Urbanistica rurale dell'E.F. nell'Agro Pontino. Primo congresso Nazionale di urbanistica, Littoria

Minucci G (1936) Scuole. Hoepli, Milan

Mugnai M (1984) Il progetto della scuola italiana: testi e documenti dalle origini al fascismo raccolti e commentati da M. Mugnai, vol IV. Il periodo fascista. Cesis, Florence

Piacentini M (1936) Prefazione. In Minucci G (ed) Scuole. Hoepli, Milan, pp V–VI

Pica A (1933) La scuola 1933 alla Triennale. In: L'edilizia Moderna, n XI–XII, pp 38–43

Piccinato L (1934) Il significato urbanistico di Sabaudia. In Urbanistica, n 1, pp 10–24

Secchi L (1923) Edifici scolastici italiani primari e secondari. Norme tecnico-igieniche per lo studio dei progetti. Hoepli, Milan

Teja A (2005) L'ONB tra educazione fisica e sport. In: Santuccio S (ed) Le case e il foro. L'archiettura dell'ONB, Alinea Editrice, Florence, pp 13–35

Todaro U (1932) Relazione al progetto esecutivo del centro comunale di Littoria, Archivio Centrale dello Stato - O.N.C. serie Agro Pontino, A.P. 13

Rural and Urban Schools: Northern Greece in the Interwar Period

Cristina Pallini, Aleksa Korolija and Silvia Boca

Abstract Modernism—as cultural and artistic expression of modern core values—is often associated with urban and industrial contexts, in stark contrast to a "backward countryside". Focusing on modernist reinventions of the rural landscape, MOD-SCAPES (funded under HERA JRP III call "Uses of the Past", Oct. 2016–2019) specifically questions these preconceived ideas. In different political and ideological contexts agricultural development schemes carried out in Europe during the twentieth century were pivotal experiments in nation-building policies. In addition, they provided a common testing ground for the ideas, and tools, of environmental and social scientists, architects and engineers, planners and landscape architects, as well as artists. This contribution presents the case study of Northern Greece, focusing on rural and urban schools as a key architectural theme, called upon to express the founding values of a collective identity. The dialectic between tradition and innovation, eclecticism and modernism, uncovers its meaning case by case.

Keywords School architecture · Modernism · Northern Greece · Refugee settlement · CIAM IV

1 Nation-Building and School Architecture

The Kingdom of Greece established in 1829 consisted of Peloponnesus, Mainland Greece, Euboea, the Cyclades and the Sporades, eventually including the Ionian Islands (1863) and Thessaly (1880). After the Balkan Wars (1912–1913), the country almost doubled in size, annexing Epirus, Macedonia, Crete and the Aegean islands. The ambition of a Greater Greece encompassing the coastal regions of Asia Minor collapsed with the Greco-Turkish War (1919–1922), in the aftermath of which 1.3 million Ottoman Christians were forced to cross the Aegean in exchange for half a million Muslims. Asia Minor refugees amounted to almost one-fourth of the population of Greece at the time, a figure favouring cultural homogenization within

C. Pallini (✉) · A. Korolija · S. Boca
Architecture, Built Environment and Construction Engineering—ABC Department, Politecnico di Milano, Milan, Italy
e-mail: cristina.pallini@polimi.it

S. Della Torre et al. (eds.), *Buildings for Education*, Research for Development,
https://doi.org/10.1007/978-3-030-33687-5_6

the national territory. The role of schools has been crucial in the process, endorsing Anderson's (1983) idea of the modern school system as a fundamental component for the rising nation states.[1] The impact of a centralised and standardised school system was crucial in Northern Greece, where foreign and minority schools had long backed cultural propaganda and territorial claims. Here, in the 1920s, Asia Minor refugees replaced earlier Turkish, Bulgarian, Serbian or Jewish settlements. In 1930, when the Refugee Settlement Commission handed over the work to the Ministry of Agriculture, Prime Minister Elefterios Venizelos launched an ambitious programme for upgrading Greece educational asset. The school programme, so crucial to cope with the country's high rate of illiteracy and lack of school buildings, served as a catalyst for a generation of Greek architects, who embraced the revolution of modernism.

2 Branding Hellenism: Late-Neoclassical Proto-Rationalism

In Greece, the first proper schools appeared as late as 1895, following a Royal Decree of 1894, which established their locations and characteristics. These inflected in four standard layouts differing in the number of classes (Fig. 1), defined by Engineer Kallias on the base of French precedents.[2] The classroom constituted a basic spatial unit, aggregated following the principles of symmetry, regularity and hygiene. Depending on location, the school size ranged from one, two, four to six classrooms. Kallias suggested an elevation for each prototype, laying emphasis on the main

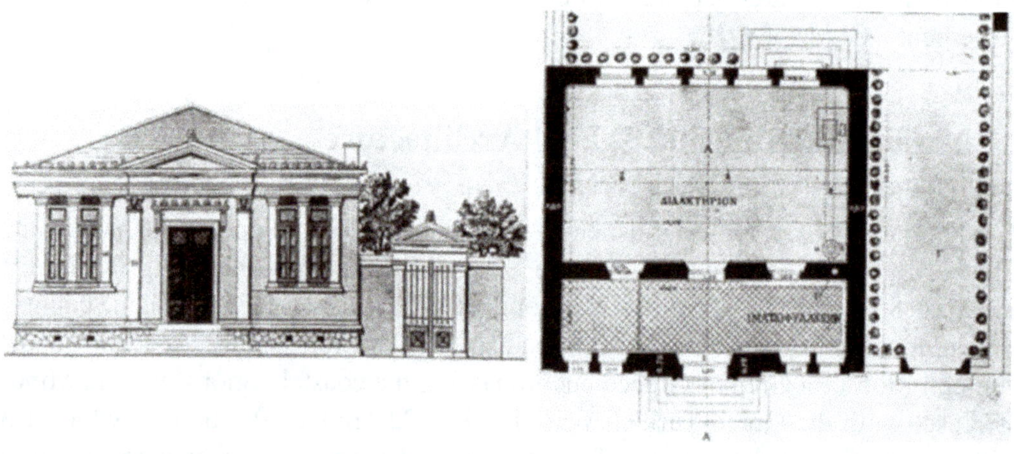

Fig. 1 Single-classroom school by D. Kallias. *Source* Kallias (1906)

[1] Cf. Anderson (1983).

[2] Dimitris Kallias (1858–1939) refers to the French regulation about measurements of classrooms, *Locaux scolaires V* and how during the International Conference on Hygiene and Economicity of School Buildings held in December 1905 in Paris, Greek Schools were highly praised for their features. Cf. Kallias (1906) and Tzonis et al. (2013).

entrance and adopting elements of the neo-classical style introduced in Athens by Bavarian architects. According to the Munich-trained architect Emannouil Kriezis, these schools prioritised discipline over pedagogical criteria, mocking monumentality into urban and rural areas alike, rather than enhancing a sense of place: "pupils should bring back home—into their daily life—something from their school: an idea of beauty rather than mere discipline" (Giacumacatos and Godoli 1985, p. 5).

Combining a functional layout with the idea of Greekness conveyed by neo-classical stylistic elements, Kallias' prototypes acquired a strategic significance for the Greek communities outside Greece, in Alexandria, Smyrna, Istanbul or Thessaloniki. In Macedonia, the epicentre of a conflict between territorial ambitions and inextricable ethnic and linguistic identities, the impact of the standardised Greek school system was particularly effective. The role of Macedonia as a frontier of Hellenization becomes self-evident when considering the proliferation of maps showing the distribution of Greek, Bulgarian, Romanian and Serbian schools in the *vilayet* (province) of Thessaloniki.[3] In this respect, the work of Xenophon Païonidis (1863–1933) for the unredeemed Greeks of the North shows the strategic importance of school buildings. After obtaining a degree in architecture from Munich University (1892), Païonidis moved back to Thessaloniki and succeeded in expressing the architectural ideology of the Greek community, conveying its power, prestige and yearning for independence.[4] A special chapter of Païonidis' career concerns his activity for the progressive metropolitan of Kassandra who promoted works of public utility in the small villages of Chalkidiki damaged by the Turks after the Greek insurrection of 1821. Most of his works were actually schools, where younger generations were to learn Greek and nurture attachment to their distant mother country. Somehow juxtaposed to the church, the school was a symbolic presence of new kind. While church architecture drew inspiration from Byzantine precedents, the school was both a functional and evocative presence marking the village core, often harmonising the neo-classical canon with local building traditions. Following Kallias' prototypes, Païonidis often integrated local building techniques, experimenting with the expressive and values of various materials: different types of stone, exposed brick or marble (Fig. 2).

3 CIAM IV, the Modern School as a Master Problem

In the summer of 1933, CIAM members started their sea voyage from Marseille to Athens on the *Patris II*, where Le Corbusier enthusiastically declared that the Acropolis had turned him into a rebel (Le Corbusier 1933). Later on, they sailed to the Cyclades where, almost unexpectedly, they found traditional villages made up

[3] According to the "Map of the Christian Schools of Macedonia" (*Carte des Écoles Chrétiennes de la Macédoine*) published in the volume *La Macédoine, son passé et son présent* by Phocas-Cosmetatos (1919).

[4] Cfr. Mandopoulou-Panayotopoulou (1997).

Fig. 2 Greek Primary School at Ormilia (Chalkidiki) by Xenophon Païonidis, 1907–1909. *Source* Mandopoulou-Panayotopoulou (1997)

of simple volumes, flat roofs, dynamic sections: the *raumplan* they had been chasing for a decade or so. Back in Athens, CIAM members visited some newly built schools: asymmetrical compositions of pure volumes in perfect harmony with the Attic landscape. On 4 August 1933, the Greek newspaper Neos Kosmos reported on "Foreigners' admiration for the new school buildings, the sign of an advanced civilization". Pierre Chareau congratulated the local architects for finding their own way to modernity (Giacumacatos and Godoli 1985, pp. 9–10). Siegfrid Giedion took pictures of some of the students running on the rooftop of a school at the feet of Acropolis, portraying the scene as true modern public space (Kousidi 2016). Reporting on his attendance to the CIAM, the Italian rationalist architect Pietro Maria Bardi (1933) praised Greece's effort of building 3167 schools in just four years.[5] In Greece, the school became a "master problem" (Sedlmayr 2006) providing a common challenge for architects from different generations and backgrounds, while embodying the collective meaning of architecture. To implement Venizelos' programme, the Minister of Education Giorgios Papandreu established an ad hoc architectural department (1930–1932) including prominent figures like Aristotelis Zachos (1871–1939) and Dimitris Pikionis (1877–1968), as well as younger architects like Nikos Mitsakis (1899–1941), Kyriakos Panayotakos (1902–1982), Patroklos Karantinos (1903–1976) and Thucydides Valentis (1908–1982). Mainly graduated from Athens Polytechnic, these latter played a key part in the design team, adapting modernist principles to the Greek landscape.

Pietro Maria Bardi observed that young Greek architects identified themselves with the Ministry of Education, noticing with much appreciation the works by Karantinos and Mitsakis. According to Bardi, such a massive engagement in school projects had swept away any remains of the Bavarian style. The "Greek spirit" had

[5]The number of schools actually built changes according to the source.

penetrated rationalism and, in a couple of decades, Greece would express its own landscape-oriented architecture, full of vigour and local colour.[6]

Despite limited technical and financial means, the scale of intervention and speed of execution of the new schools marked an undeniable success, which achieved considerable press coverage attracting much contemporary and later scholarly work. According to François Loyer (1966), the school programme served as a catalyst for a movement in the making, allowing its emergence and intellectual definition. The revolution of modernism, in Greece, had a purely formal character: "An intellectual movement of young artists who found the terms of a manifesto in a political circumstance" (Loyer 1966, p. 416).

Limited funding required rationalisation of construction. The standard layout consisted of six classrooms on two separate floors, with the possibility of merging the upper units to form a lecture hall when needed. In the cold and windy northern regions, classrooms were facing south and the corridor, exposed to the north, embedded in the built-up mass. In the warmer regions of the south, instead, classrooms were facing north and the southern façade was shaded by a cantilevered corridor. Karantinos adopted these guidelines in the primary school on Kalisperi Street in Athens (1931), where classrooms turned their back to the Acropolis (Fig. 3).

Fig. 3 Photo of school on Kalisperi Street, with the Acropolis in the background. Photo by P.M.Bardi (1933) from *Quadrante* n.5

[6]Pietro Maria Bardi (1933, p. 13) wrote: "*[...] today, in Greece, we are witnessing the changes that the local spirit brings to the rationalist idea: in twenty years Greece will have its own architecture of environment, of lena, of an entirely local color*".

Experimenting with the thermal zoning between the corridor and the classrooms, the school programme provided excellent opportunities to define energy-efficient design criteria. In addition to orientation, experiments also included building materials capable of storing heat and releasing it gradually, as well as arranging volumes for a better thermal comfort (Mavrogianni and Tsoukatou 2006).[7] Different contexts also meant using brick or local stones as infill walls for the load bearing concrete structures.

In 1938, Patroklos Karantinos published a book on new schools,[8] including standard layouts for small schools of two, three or four classrooms.

The modernist schools of Athens and Piraeus were presented in the opening section, followed by the four schools of Thessaloniki. Unexpectedly, two of these buildings did not comply with the modernist canon, borrowing elements from Byzantine architecture and reinterpreting the traditional Macedonian house. Some schools in Peloponnesus, in the islands, in the frontier regions of Epirus and Eastern Macedonia, featured simplified eclectic forms, bearing a tangible reference to the various architectural traditions still vital in Greece.

4 In the Heart of Rural Refugee Settlements[9]

Significantly, reporting on his journey into Greece, Pietro Maria Bardi did not overlook the critical demographic juncture: "2.600.000 inhabitants in 1907, 5.600.000 in 1921; the arrival of refugees in the aftermath of a gruelling war. Even the efforts by international organisations were not sufficient to organise such a huge avalanche" (Bardi 1933, p.15).

The great majority of Asia Minor refugees repopulated the so-called New Lands of Macedonia and Thrace in Northern Greece, in view of stabilising the borders and unlocking the region's agricultural potential. Following social, ethnic and demographic reshuffling, large-scale reclamation works produced a radical change of the physical features and settlement network. From 1922 to 1930, a special body of the League of the Nations, the Refugee Settlement Commission, undertook this critical process. To foster mutual help and social cohesion, refugees settled by groups, often

[7]The paper gives an overview from the Rationalist period up to present-day. The paper is based on the research project 'The Bioclimatic Dimension of Educational Buildings in Greece' by K. Koukouzi, A. Mavrogianni, M. Tsoukatou under supervision by prof. Evangelos Evangelinos.

[8]The original Greek title is *Τα Νεα Σχολικα Κτιρια*, meaning *The new school buildings*. The book covers the period of the Venizelos government (1928–1932). The volume has been re-printed in 2019 by TEE (Karantinos 2019).

[9]This section is partly based on Vilma Hastaoglou-Martinidis, Cristina Pallini, "Colonizing the 'New Lands': rural settlement of refugees in Northern Greece (1922–1940)", in Clara Architecture/Recherche (forthcoming 2019), and on extensive fieldwork by the three authors, in the framework of the EU-funded research MODSCAPES Modernist Reinventions of the Rural Landscapes. During fieldworks, the authors collected maps and surveyed public buildings, private houses and farms, with a particular focus on the spatial arrangement of the settlement's cores.

by village of origin, according to three alternative solutions: on sites of abandoned Turkish or Bulgarian villages, in new quarters adjacent to existing villages, or in newly built settlements. In early 1930s, 509 new rural communities were founded in Central Macedonia, mostly in the plain, 75 of which in the immediate vicinity of Thessaloniki. The newly established refugee settlement followed a standard layout, characterised by a uniform grid of streets surrounded by field allotments, providing for central public space and rudimentary communal amenities, such as the square with the village hall, the church and the school. The decree law on rural resettlement of refugees specified that each settlement had to "be laid out according to a simple plan and divided into lots" (Kontogiorgi 2006, p. 291) and all public buildings and sites were to be simple and uniform.[10] While affiliation to the Orthodox Church was the reason why refugees had left Asia Minor, fostering a shared Greek identity among the diverse peoples of the New Lands became a priority. Rural refugee settlements clearly show the civic role of educational buildings, juxtaposed to the church at the centre of the village. Most often, the church and the school marked the intersection of the main roads, occupying two adjoining parcels. The village Axos, near Giannitsa, well exemplifies this pattern. Accessed from secondary streets, the school and the church are set on opposite sides of the main road, attracting a combination of collective spaces and playing fields. In the village of Nea Pella, the school and the church are located halfway the main road rising from ancient Via Egnatia to the higher ground occupied by the football field (Fig. 4). At Palafyito, instead, the

Fig. 4 Map of Nea Pella showing the system of public spaces (A, B, C) located along the main axis moving uphill from Egnatia Road (A) The main core (C) includes: the Church (1), the School (2), the Acqueduct (3) and the Sport Field (4). Authors' elaboration

---

[10]According to the Government Gazette, 6/7/1923–11/7/1923, Article 6: "Regarding rural settlement of refugees" (Kontogiorgi 2006).

school—a very simple building dating back to the 1950s—marks the edge between the village and the fields.

The village of Neos Skopos in the Strymon Valley well depicts the eagerness of the community to take an active part in the construction of the main public buildings. It was established in 1923 by refugee families from Skopos in Eastern Thrace, who lived in tents and makeshift huts until the Refugee Settlement Commission drew the plan of the village and built permanent homes. "It was like repotting a plant where the roots begin to grow again, and it continues growing, developing and progressing in its new container" (Naniopoulos 2014, p. 116).

The first Church of Saint Demetrius was a simple wooden structure which served as a school during the week. As early as 1927, a proper school was built on the main square, following the conventional four-classroom layout on opposite sides of the main entrance. Sir John Hope Simpson (1868–1961), a member of the British Parliament, remarked the crucial importance of the church and the school. As the village was taking shape, the inhabitants proceeded to build a temporary wooden church, in view of rebuilding it in stone. Even before being comfortably settled, villagers commenced to agitate for a school. Their demands were so insistent that the Refugee Settlement Commission reserved a plot for the school in every village. In many villages, an extra house was to serve temporarily the purpose of a school. The Commission assisted the population either by making a grant in cash or by providing materials, with the help of which people constructed a school building by themselves.[11]

5 Thessaloniki, Modernism's Fault[12]

In an interview given on 4 September 1931, Minister of Education Giorgios Papandreu announced that 26 elementary schools, 6 gymnasiums and a teacher's college were to be built at Thessaloniki (Giacumacatos and Godoli 1985, p. 6).

If *Patris II* had continued his journey further north, CIAM participants were to contemplate the ruins of the once-thriving Ottoman port city, annexed to Greece in 1912 and destroyed by fire in 1917. Cut off from its Balkan hinterland, Thessaloniki had become the capital of the New Lands whose Greek population had more than doubled by 1926. For this reason, Thessaloniki provides special observatory into modern Greek architecture. In fact, Venizelos' programme for the new schools was part of a wider process of city reconstruction, rendered even more difficult by massive refugee settlement. In a context where foreign and minority schools had long backed cultural propaganda and territorial claims, the new school buildings became

[11]Cfr. Simpson (1929).

[12]This section is partly based on "Colonizing the 'New Lands': rural settlement of refugees in Northern Greece (1922–1940)" by Vilma Hastaoglou-Martinidis and Cristina Pallini (forthcoming 2019), and on extensive fieldwork by the three authors, in the framework of the EU-funded research MODSCAPES Modernist Reinventions of the Rural Landscapes.

strongholds of a future urban topography and cultural makeup. The Neo-Byzantine style codified by French planner Ernest Hébrard,[13] the main author of the reconstruction plan, was to qualify the future city centre, marking a clear break with the Ottoman past to recapture the city's Hellenic identity. As documented by Karantinos' book, two of the new schools in Thessaloniki moved away from the modernist canon. One is the Aghia Sofia school complex designed by Nikos Mitsakis (1928–1932) to host a Jewish school, a Greek elementary school and a high school. Hovering between a rational volumetric articulation and an eclectic approach, Mitsakis experimented with elements of Byzantine architecture—arch, column, pilaster strip, capital—simplified and adapted to modernist syntax.[14] A few blocks away, Dimitris Pikionis built the famous Experimental School (1935–1936) which marked his shift from the architecture of the islands to traditional Macedonian architecture (Fig. 5). This "Macedonian diorama" exemplified Pikionis's notion of "re-invention":

> *Form is the result of many efforts by many souls. Architects should not invent short-lived forms, they should instead "re-invent" existing forms to meet our current needs. Form can join our souls in an ideal symbol. [...] Architects and artists should not invent ephemeral forms, rather should they reinterpret the perfect forms of tradition in line with current needs and constraints. This is not just a mental exercise, it also involves emotions. A text from ancient Greece describes three kind of creations: (a) the "backward-looking creation" indicating our link to the past; (b) the "prevident creation" indicating our way of dealing with the present and (c) the "lovable creation" indicating our feelings as opposite and complementary to logic.* (Pikionis 1991, p. 6).

Fig. 5 Maquette of the Experimental School in Thessaloniki by D.Pikionis *Source* Karantinos (1938, p. 132), *Ta Nea Scholika Ktiria*

[13]Ernest Hébrard (1875–1933), architect, archaeologist and town planner. Hébrard received the Prix de Rome in 1904, for which he produced, as head of the French Army Archaeological Service, a conjectural reconstruction of Diocletian's palace at Split. Hébrard was in Thessaloniki in August 1917 when the fire occurred.

[14]Cfr. Paiousaki (1999).

References

Anderson B (1983) Imagined Communities. Reflections on the origin and spread of nationalism, Verso, London

Bardi PM (1933) 'Cronaca di Viaggio', *Quadrante- Mensile di arte, lettere e vita* no. 5, pp 1–35

Giacumacatos A, Godoli E (1985) L'architettura delle scuole e il razionalismo in Grecia. Modulo, Firenze

Kallias D (1906) On the Greek measures for the hygiene of school buildings. Archimidis 2, p. 1 and pp 14–17 [in Greek]

Karantinos P (2019) Τα Νεα Σχολικα Κτιρια (The New School Buildings). TEE, Athens

Kontogiorgi E (2006) Population exchange in Greek macedonia. The rural settlement of refugees 1922–1930, Oxford Historical Monographs, Oxford

Kousidi M (2016) Through the Lens of Siegfried Giedion. Exploring Modernism and the Greek Vernacular in situ. RIHA Journal Special Issue: Southern Modernism (0136) [Online] https://www.riha-journal.org/articles/2016/0131-0140-special-issue-southern-modernisms/0136-kousidi

Le Corbusier (1933) Air, son, lumière. Annales Techniques IV, pp 44–46

Loyer F (1966) Architecture de la Grèce contemporaine, PhD thesis, Univ.de Paris-Faculté des Lettres et Sciences humaines

Mandopoulou-Panayotopoulou T (1997) Xenophon Paionidis, an outline of his work. 6000 years of writings and sources on Thessaloniki. Thessaloniki Cultural Capital of Europe, pp 114–126 [in Greek]

Mavrogianni A, Tsoukatou M (2006) The bioclimatic dimension of educational buildings in Greece. In: Proccedings of the international workshop on energy performance and environmental quality of buildings, Milos Island, Greece, July 2006 [Online]. http://www.inive.org/members_area/medias/pdf/Inive%5CMilos2006%5C37_Mavrogianni_6P.pdf. Accessed 10 June 2019

Naniopoulos A (2014) Skopos diaspora. A Volume Honoring and Celebrating the Skopinon Community in the United States of America, Orpheus Society, Neos Skopos

Paiousaki H (1999) Nikolaos Mitsakis 1899–1941. Museum Benaki, Athens

Phocas-Cosmetatos SP (1919) La Macédoine, son passé et son présent. Payot & Cie, Lausanne

Pikionis A (1991) Vita, opere e pensiero di Dimitris Pikionis. Controspazio 5:3–7

Sedlmayr H (2006) Art in Crisis, the Lost Centre. Transaction Publishers, Piscataway

Simpson JH (1929) The work of the Greek refugee settlement commission. J Royal Inst Int Aff 6 (8):583–604

The Schools as Heritage and a Tool for Political and Cultural Integration. The Buildings of the *Plan de Edificación Escolar* in Buenos Aires

Maria Pompeiana Iarossi and Cecilia Santacroce

Abstract The school buildings represent an important testimony to the social and cultural policies adopted in geographically and historically determined contexts. Today, these complexes constitute a broad and diversified patrimony, largely worthy of protection and enhancement. In the framework of a wider research program with Universidad de Belgrano-Buenos Aires, this chapter presents the results of a research about the 43 schools built in Buenos Aires between 1885 and 1904 as an implementation of the *Plan de Edificación Escolar*, showing how the school architecture can constitute a tool for the social and cultural integration, and simultaneously a decisive element to outline the face of the great *Capital federal*.

Keywords Schoolhouses · Buenos aires · Architectural heritage · Immigration in Argentina

1 Introduction

Concerning the development of modern states and in the majority of geographical' backgrounds, the architectural definition's choices for primary school buildings reveal features that a specific nation decides to give to itself. Often, not only the pedagogical principles are used like a model for distribution and functional layouts but also constructed buildings through their architectural language reveal a clear declaration about the cultural origin chosen to represent and define the national identity. Examples of these are the adoption of Lombard neo-Romanesque style by Camillo Boito for post-unitary Italian schoolhouses and, for the Germany between the two wars, the strict and, at the same time, domestic proto-rationalism type of school buildings by Heinrich Tessenow.

This iconic role of "future citizens' knowledge factory" assigned to the schoolhouses gives them a huge documental and testimonial value, suggesting the necessity of safeguarding this extended heritage that are often threatened by more than from

M. P. Iarossi (✉) · C. Santacroce
Architecture, Built Environment and Construction Engineering—ABC Department, Politecnico di Milano, Milan, Italy
e-mail: mariapompeiana.iarossi@polimi.it

S. Della Torre et al. (eds.), *Buildings for Education*, Research for Development,
https://doi.org/10.1007/978-3-030-33687-5_7

the passage of time, to the obligation to mitigate between the necessary sanitary updates and the lack of available public resources.

An area of characteristic interest is represented by the extended compendium of scholastic complexes built in Buenos Aires in a historical crucial moment for the young South America republic, corresponding to the beginning of the massive migrant wave in 1880 (AA.VV. 1906; Zuccarini 1909; Capocaccia et al. 2016; Carcacha, and others 2016), which will continue until the middle of the last century, and to the transformation of the city into *Capital federal*.

2 The Schoolhouses Design as a Tool for Political and Cultural Action

The Argentinean historiography designates a group of liberal, conservative and positivist cultural formation politicians and intellectuals like *Generación del Ochenta*, who, between the 1880 and the 1916, were in-charge of the young South American republic, making the destiny of it, choosing to incarnate the *imprinting* of the Nation on three axioms: national secularity, Europe-like cultural model and immigration, especially the European one, like a resource.[1]

A natural result of the last one was the *Ley de Educación Comun n. 1420* of 1884—which sanctioned the obligatoriness, universality and the gratuitousness by the State of primary education up to 14-years-old[2]—an essential tool to govern, first of all through linguistic unification, the process of integration of huge and multiethnic migrant flows, which poured out on the nation (and above all on Buenos Aires) until the middle of the twentieth century, except the interruptions during both the world conflicts (Fig. 1).

1420 Ley's inspiration were Domingo Faustino Sarmiento's theories, who was the President of the Nation from 1868 to 1874, and whose theories, explained in *De la Educación popular*,[3] addressed the strategic role of primary education and the necessity to propagate it through an organic implementation framework, the training of teachers and the definition of high architectonical and sanitary facilities *standards* for school buildings.

This organic vision was reflected in establishing, both the *Escuela Primarias*, destined to be a district's basic service, and the *Escuelas Normales*, imagined like laboratories where the direct observation of students' daily life was an essential part of the teachers' training course.

The immediate application of the law determined in the first two-year period 1884–1886 the construction of 56 new schools in the Nation, which design was

[1] Gerchunoff and Llach (1998) Italianos en la Arquitectura Argentina, 2004.

[2] Art. 2 of the *Ley* says "*La instrucción primaria, debe ser obligatoria, gratuita, gradual y dada conforme a los preceptos de higiene*" [primary school must be obligatory, free, gradual and in compliance with sanitary precepts] in: Collección de Leyes y decretos, Tomo 1, p. 282.

[3] Sarmiento (1849).

Fig. 1 *Escuela Elemental de Niñas* during their construction, photo S. Boote, 1889

defined by Italian architects and engineers, delineating the *Escuela-Palacio* [school–palace] model, in which the layout was organized around courtyards and the façades were in *Neo-renacentista* style, like the Escuela Normal Superior en Lenguas Vivas "Mariano Acosta", designed by Francesco Tamburini between 1883 and 1885.

3 Buenos Aires as an Experimental and Coding Lab of Schoolhouse Architecture

In 1899, in view of a more organic realization of the law, the *Plan de Educación Escolar* was promulgated, with the goal of supervising the construction of school buildings on the whole national territory and drawing the guidelines for their design, testing them at first in Buenos Aires.

Meanwhile in the city—chosen in 1880 as Capital of the Federal State—processes of transformation were in place, changing its original structure of *Gran aldea* [big village]—based on the extensive application of the checkerboard foundation principle, fixed in the sixteenth century by the *Leyes de Indias*[4]—in the current metropolis (Figs. 2 and 3).

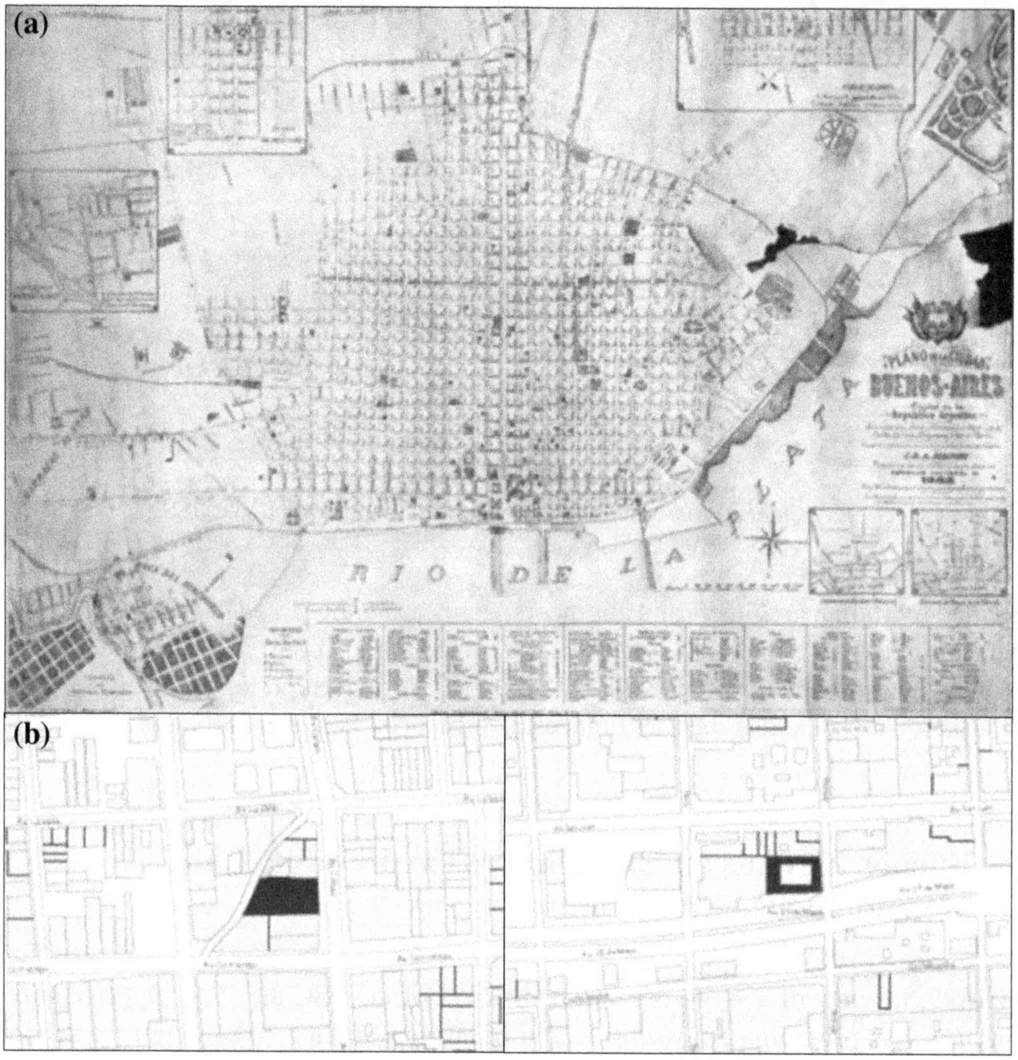

Fig. 2 Relationship between urban layout and schoolhouses: **a** City's plan by Bianchi, 1882; **b** Buildings' position in a block: along a street (left: *Escuela Normal Superior n°9* "Domingo Faustino Sarmiento", 1886) and in the corner (right: *Escuela Normal Superior n°7* "Bernardino Rivadavia", 1902)

[4]The *Leyes de Indias* is an imposing juridical *corpus* composed of the sum of all laws released by the Spanish Crown, between 1512 and 1680, to regulate the different aspects of political, economic and social life in the New World. In 1680–1681 they were revised and collected in the *Recopilación de Le Leyes de las Indias*, consisting of nine books, of which the fourth, title 111–130, includes procedures about settlement, fields' division and public works' fulfilment in new conquered areas. Through proportional rules within *cuadras, calles y solares* [blocks, streets and parcels], it was meant to guarantee the urban form's normalization, as a sign of Spanish supremacy all over the New World. See: España (2017).

Fig. 3 Example of facades for schoolhouses along a street (left: *Escuela n°3* "Juan Maria Gutiérrez", 1901) or in the corner (right: Escuela "Presidente Mitre", 1902)

In the *porteño* urban landscape, the implementation of the *Plan* set two questions: a typological one—concerning the necessity to define some layout solutions based on sanitary facilities requisites—and a semantic one, linked to the need to identify the school buildings through their architecture, viewed as the tangible and unequivocal presence of the State, also in the Buenos Aires monotony, generated by the infinitive repetition of the grid block.

To answer these requests, *CNE-Consejo Nacional de Educación* nominated the Italian engineer Carlos Morra as general supervisor, who between 1898 and 1904 designed, only in the *Capital*, 23 schoolhouses (D'Amia and Iarossi 2018). Morra defined three different solutions, based on school educational level, on students' class and on building's position in the block (Grementieri and Shmidt 2010):

Type A: for elementary school: with classrooms on ground floor and the principal's accommodation on first floor.
Type B: for *Escuelas Normales*: with classrooms on two floors and residence on the other side of the parcel.
Type C: for school in the corner of the block.

However, the most important of Morra's contributions was the definition of a specific language for the building's façade, determining an elements' abacus—like entrance portal or thermal window—that identified the building's functional parts and a syntactic rule system, and which application had determined a façades' composition that instantly allows to recognize schoolhouses among the other buildings (Fig. 4).

4 Survey and Analysis of Study Cases

Researches were carried out in CeDiap's archive in Buenos Aires showing that between 1880 and 1910 in the Capital, 46 schoolhouses were built; these have been collected in a database that gathers and compares, for each schoolhouse, historical archive data with the observed ones on site and recorded in individual census'

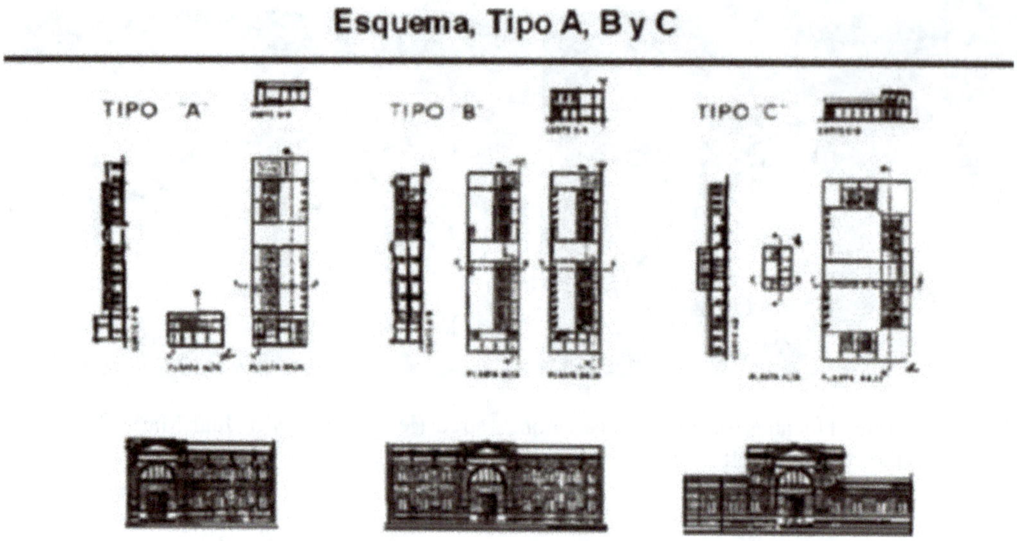

Fig. 4 Scheme of building types of school buildings developed by the *CNE-Consejo Nacional de la Educación* under the direction of C. Morra (by Brandariz 1998, p. 88)

cards, supplied with photographic documents concerning the current state and, when available, stock photos (Fig. 5).

The main sections in which the database was divided were building's name, architect, builder, historical address, actual address, *Barrio* [neighbourhood], construction year, references. In this first step of the research, it was verified that the main schoolhouses had been designed by Italian architects and engineers like Carlos Morra, Francesco Tamburini and Gino Aloisi.

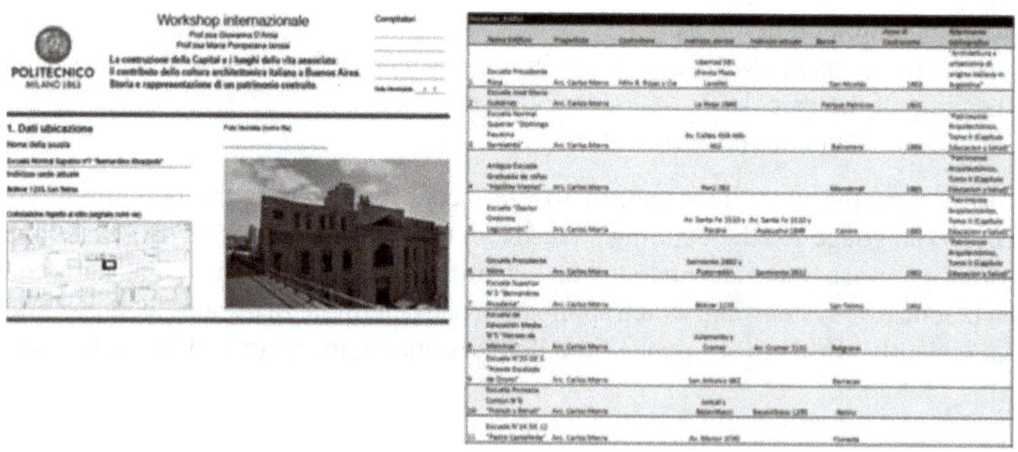

Fig. 5 An extract of the buildings' census card and the database used for the classification the schoolhouses

During a survey and census campaign carried out in Buenos Aires,[5] 23 of the 46 schoolhouses were designed by Carlos Morra. They represent today an important architectural bequest, deserving of knowledge, safeguard and enhancement.

Finally, in the collection of schoolhouses by Carlos Morra, four case studies were selected, chosen to guarantee a variety in terms of location, size, dating and pedagogical destination.

In fact, two buildings were built before and two after 1899, when *PPE* started. In addition, the second and fourth example were designed to host primary schools, whereas the first and third were *Escuelas Normales*, institutions like cornerstones for the implementation of the Sarmiento's reform and of laical education, where primary school was included as an internship laboratory for future teachers.

The selected buildings are:

(A) *Escuela Normal Superior n°9* "Domingo Faustino Sarmiento" (1886);
(B) *Antigua Escuela Graduada de niñas* "Hipólito Vieytes" (1880–1885);
(C) *Escuela Normal Superior n°7* "Bernardino Rivadavia" (1902);
(D) *Escuela n°3* "Juan Maria Gutiérrez" (1901) (Fig. 6).

The analysis—based on surveys carried out in Buenos Aires, complemented with original archival materials, kept at the *CeDiap-Centro di Documentación and Información*—has been developed in order to identify four aspects, characterizing the architecture of each building:

- Urban and architectural characters: position in the block, size and layout articulation on the ground floor;
- Ordering principles of the façade: structural organization, symmetry, modularity and overhang/retraction of its parts;
- Proportioning of the façade and geometric layouts;
- Constructive system (trilithic system and/or masonry wall) and components of the façade, summarized and compared in a specific abacus.

This analysis highlighted significant differences between schoolhouses built before and after the *Plan*.

The two examples referred to the first period—although they are very different in size and destination, being one a primary school and the other an *Escuela Normal*—show typological and linguistic adhesion to the model of the *Escuela-Palacio* [palace–schoolhouse]. Connections are assured by monumental staircases, corridors and open galleries, marked by often coupled columns. The facades are

[5] The results of this campaign—carried out in March 2017 during the International History and Representation's Workshop, in partnership with School AUIC of Politecnico di Milano and FAU of Universidad de Belgrano, under the scientific direction of M. P. Iarossi and G. D'Amia—were improved through researches and thesis expound on schoolhouses' topic. They are still object of detailed studies, since they have converged in an innovative educational project and in a research about Italian-Argentinian heritage in Buenos Aires (Iarossi et al. 2017), in partnership with Dept ABC of Politecnico di Milano (coordinator, M.P. Iarossi) and FAU-Universidad de Belgrano (coordinator L. Bonvecchi) supported with Erasmus+ Ka 107 funds.

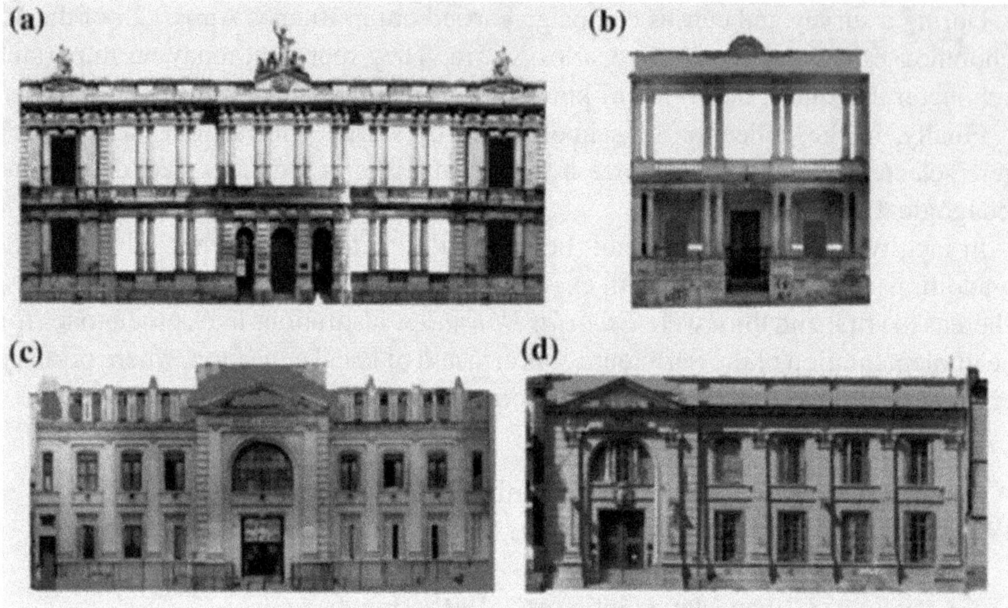

Fig. 6 Case studies, façade orthomosaics (elab. by Agisoft Photoscan): **a** *Escuela Normal n°9* "Domingo Faustino Sarmiento"; **b** *Escuela de niñas* "Hipólito Vieytes"; **c** *Escuela Normal n°7* "Bernardino Rivadavia"; **d** *Escuela n°3* "Juan Maria Guitiérrez"

Neo-renacentista style, elevated on a stylobate and completed by an attic wall as a sculptural crowning.

The composition of the front, especially in the case of the grandiose *Escuela n°9*, in complex and hierarchically organized, with altimetric overlap of the elements and planimetric articulation, by protrusions and indentions of the façade (Figs. 7 and 8).

Fig. 7 *Escuela Normal Superior n°9* "Domingo Faustino Sarmiento". The facades are *neo-renacentista* style, elevated on a stylobate and completed by an attic wall as a sculptural crowning

Fig. 8 *Escuela Normal Superior n°7* "Bernardino Rivadavia". The façade has tectonic wall. Columns replaced by pilaster strips in giant order, to mark the sequence of wall modules of the front

Instead, the two examples following the *PPE*—in addition to the foreseeable surrender to this model in favour of adherence to the typological schemes by Morra itself codified in implementation of the *Plan*, with a close division of administrative spaces, always facing the road, from those for teaching, articulated around one or more courtyards—show the development of a specific language for the scholastic architecture, paying specific attention to define the elements of the façade, which is entrusted with the task of representing, in relation to the city, the value of the educational institution and the presence of the State. In addition, they demonstrate a foreseeable surrender to this model in favor of an adherence to the typological schemes by Morra, codified in the implementation of the Plan, with a close separation of the administrative spaces, always facing the road, from those for teaching, articulated around one or more courtyards.

This objective is pursued accentuating more the aspects of the tectonic wall compared to the trilithic order. Thus columns, as an autonomous element of articulation of the front, disappear and they are replaced by pilaster strips or, at most, by semicolumns, in giant order, to mark the sequence of wall modules of the front. These are of only two kinds: the window-module and the portal-module.

The latter, whether placed at the centre of the façade, to identify a symmetry axis, whether at an extremity, to conclude it asymmetrically, always consists of pilasters in giant order, resting on a base and surmounted by a tympanum, with the name of the school inscribed, to frame, at the ground floor, the entrance door and, above, a thermal window, an element that, reproposed by Morra in his project of *Biblioteca Nacional*, will become a distinctive element of his architecture (Figs. 9 and 10).

Fig. 9 Model of the *Escuela-Palacio*. Connections are assured by monumental staircases, corridors and open galleries

Fig. 10 Architectural elements composing the Morra's façade: tympanum, thermal window, pilasters in giant order, cornice

5 Conclusions

The study carried out highlighted the documental and the architectural intrinsic value one of the schoolhouses' heritage built in Buenos Aires between the nineteenth and the twentieth century.

But, above all, the pragmatic value of Argentinean experience in building schoolhouses is shown, and also how they can be an efficient political and cultural action tool, especially helpful and necessary in contests where the occurrences of massive migrant phenomenon asks for the use of social and cultural integration's strategies, able to set up real development perspectives for the Nation.

References

AA.VV. (1906) Gli Italiani nella Repubblica Argentina all'Esposizione di Milano 1906. Stabilimento grafico della Compañía general de Fósforos, Buenos Aires. Censo general de población, edificación, comercio e industrias de la ciutad de Buenos Aires levantado en los dìas 11 y 18 de septiembre de 1904, 1906. Compañia sud-americana de billetes de banco, Buenos Aires

Brandariz G (1998) La arquitectura escolar de inspiración sarmentina. FADU-UBA, Buenos Aires

Capocaccia F, Pittarello L, Rosso Del Brenna G (eds) (2016) Storie di emigrazione: architetti e costruttori italiani in America Latina. Stefano Termanini Editore, Genova

Carcacha G, and others (2016) Conformación del sistema educativo argentino: capital social y formación de atributos productivos de la fuerza de trabajo (1880–1930). In: Jornadas de economía crítica. IX Coloquio de SEPLA. 25–27 agosto del 2016, UNC-Córdoba Argentina. https://www.academia.edu/28206275/Conformación_del_sistema_educativo_argentino_capital_social_y_formación_de_atributos_productivos_de_la_fuerza_de_trabajo_1880-1930

D'Amia G, Iarossi MP (2018) Carlos Morra e il disegno degli edifici scolastici a Buenos Aires come strumento di azione politica e culturale. In: Salerno R (ed) Rappresentazione Materiale/Immateriale. Gangemi, Roma, pp 615–622

España (2017 [1681]) Recopilación de leyes de los reinos de Indias: mandadas imprimir y publicar por la Magestad Católica Don Carlos II. Tomo 4. Biblioteca Virtual Miguel de Cervantes, Alicante. http://www.cervantesvirtual.com/obra/recopilacion-de-leyes-de-los-reinos-de-indias–mandadas-imprimir-y-publicar-por-la-magestad-catolica-don-carlos-ii-tomo-4

Gerchunoff P, Llach L (1998) La generación del progreso (1880–1914). El ciclo de la ilusión y el desencanto. Un siglo de políticas económicas argentinas. Ariel, Buenos Aires, pp 13–59

Grementieri F, Shmidt C (2010) Arquitectura, educación y patrimonio. Argentina 1600–1975. Pamplatina, Buenos Aires

Iarossi MP, Mele G, Rossini M (2017). Architetture italiane a Buenos Aires. Censimento, descrizione, analisi e valorizzazione di un patrimonio a rischio, In AA.VV. (ed) Territori e frontiere della rappresentazione. Gangemi, Roma, pp. 1325–1334

Italianos en la Arquitectura Argentina (2004). Buenos Aires: Boletín CEDODAL

Sarmiento DF (1849) De la educación popular. Imprenta de Julio Belín, Santiago de Chile

Zuccarini E (1909) Il lavoro degli Italiani nella Repubblica Argentina. Leggende studi e ricerche. La Patria degli Italiani, Buenos Aires

Origins and Development of the American Campus: The "Academical Village" of Thomas Jefferson

Mariacristina Loi

Abstract The study intends to analyze the many influences that led Thomas Jefferson, the third President of the United States, to conceive and realize a very innovative project for universities in America. The research, started many years ago and still ongoing, is based on the very large amount of original documents and on the ever-growing bibliography. It was carried out partly in the USA, thanks to funding from the Robert H. Smith International Center for Jefferson Studies, in Charlottesville, Virginia. Through this, relationships were established with other institutions, such as the New York University Casa Italiana Zerilli-Marimò and the School of Architecture of the University of Virginia.

Keywords North America · Campus · Models · Antiquity · Academical village

The University of Virginia is the most celebrated project among the numerous ones that Thomas Jefferson developed during his long and intense life (Fig. 1).[1]

[1]The following are the main stages of the intense life of Thomas Jefferson (1743–1826). After completing his studies in Law, he entered politics in 1768, with the election to the Virginia Assembly. In 1774, he wrote "A Summary View of the Rights of British America", a document which was a prelude to the Declaration of Independence. In 1775, he was elected to the Continental Congress of Philadelphia, in 1779 he was governor of Virginia, and between 1784 and 1789 he was in Paris, acting as Minister of the United States in France. In 1790, he was appointed Secretary of State by George Washington. Defeated by Adams in the 1796 presidential election, he was named vice president. In 1801, he was elected as the third President of the United States, a post he held in a second term, from 1808. In 1803, he redeemed Louisiana from France. Among the proposed laws are the "Bill for religious freedom", approved by the Virginia assembly in 1786; the ban on importing slaves from Africa, promoted in 1808.

Main source for each study on Thomas Jefferson are the collections of his writings and documents: *The Writings of Thomas Jefferson*, edited by A. A. LIPSCOMB, A. E. BERGH, Washington 1905; *The Works of Thomas Jefferson*, edited by P. L. FORD, New York 1905, *The Papers of Thomas Jefferson*, edited by J. P. BOYD, 35 vols., Princeton since 1950. Many of these collections,

M. Loi (✉)
Architecture, Built Environment and Construction Engineering—ABC Department, Politecnico di Milano, Milan, Italy
e-mail: mariacristina.loi@polimi.it

S. Della Torre et al. (eds.), *Buildings for Education*, Research for Development,
https://doi.org/10.1007/978-3-030-33687-5_8

(a) (b)

Fig. 1 Charlottesville, University of Virginia, bird-eye view

It is a complex project, which had a long gestation and saw the collaboration of several figures. It can be considered the quintessence of his ideals, and one of his most important and significant works, as the words inscripted on his tomb recall:

> Thomas Jefferson, author of the Declaration of American Independence, of the law for religious freedom in Virginia, and father of the University of Virginia.[2]

The project for the Charlottesville campus was developed in approximately two decades and can be regarded as Jefferson's last great work. He devoted much time to architecture, both in the private and public sphere, as shown by the extensive and documented catalog of his projects (some of which were executed, while others remained on paper). From his residence in Monticello to the competition for the President's House, from the Capitol of Richmond to the Capital on the Potomac and the University of Charlottesville,[3] Jefferson tends to all aspects, from the furnishings to the entire plan of the city and the territory.

The University of Virginia is not just an architectural work. It is the concretization of a larger project on education, and its design is emblematic of his multiple interests and of Jefferson's deep commitment to educating the young nation.

---

together with extensive and up-to-date bibliographies, are now available online. See *Thomas Jefferson Papers*, Library of Congress; *Thomas Jefferson Papers*, an Electronic Archive, a selection of the most important documents kept at the Massachusetts Historical Society of Boston. See https://www.monticello.org, website of the Robert J. Smith International Center for Jefferson Studies; https://founders.archives.gov, website of the National Archives. The original documents are preserved mainly at the Coolidge Collection of the Massachusetts Historical Society of Boston; the Special Collection Department section, University of Virginia Library; the Rare Book and Special Collection Division and the Manuscript Division of the Library of Congress, with about 27,000 documents.

[2]See full transcription in https://founders.archives.gov.

[3]On Jefferson as an architect after the fundamental works by Fiske Kimball many studies have followed. For the main bibliographical references, see https://www.monticello.org/site/research-and-collections/tje/architecture; https://www.encyclopediavirginia.org/Jefferson_Thomas_and_Architecture.

In 1778, he presented “A Bill for the More General Diffusion of Knowledge” to the House of Delegates:

> … And whereas it is generally true that people will be happiest whose laws are best, and are best administered, and that laws will be wisely formed, and honestly administered, in proportion as those who form and administer them are wise and honest; whence it becomes expedient for promoting the publick happiness that those person, whom nature hath endowed with genius and virtue, should be rendered by liberal education worthy to receive, and able to guard the sacred deposit of the rights and liberties of their fellow citizens, and that they should be called to that charge without regard to wealth, birth or other accidental condition or circumstance ….[4]

The proposal was not approved. It was submitted another time in 1780 and again by James Madison when Jefferson was in France, and it was finally approved in an amended version as “Act to Establish Public Schools” in 1796.[5]

The intent to provide culture and education for all is based on the same Enlightenment ideals that inspired the Declaration of Independence, which recognizes the inalienable right to happiness, achievable only by a population free from tyranny. Education is the essential foundation of this principle, and the University of Virginia represents the culmination of such great a project.[6]

This confluence of thoughts and ideals emphasizes how it is impossible to think of Jefferson’s individual architectural project as unrelated to his work as a politician and a man of the law.

Along with the ideal of the spreading culture came a concrete plan, too. In the projects dedicated to the organization of the West territories—the Land Ordinance of 1784 and its revision of 1785—Jefferson designed buildings dedicated to public instruction for each “township”.[7]

The University of Virginia satisfies both symbolic representative and practical needs. It celebrates and makes use of universal ancient models, while also keeping the local tradition alive, especially in terms of the materials used and the scale of the buildings.

---

[4]See https://www.monticello.org. On this particular aspect of Jefferson’s activity, see also “A Bill for Establishing a Public Library”, 18 June 1779, in https://founders.archives.gov.

[5]The proposal was again re-elaborated in the second decade of the nineteenth century. See “Thomas Jefferson’s Draft Bill to Create Central College and Amend the 1796 Public Schools Act”, ca. 18 November 1814, in https://founders.archives.gov.

[6]Jefferson also provided, with an important acquisition campaign, collection of books for the Library. On the subject see Loi, M. C., *La biblioteca di Thomas Jefferson*, in *I Libri e l’Ingegno*, ed. by Curcio, G., Nobile, M. R., Scotti Tosini, A., Caracol, Palermo 2010, pp. 203–210, with a detailed bibliography.

[7]The projects for the organization of the territory established the creation of township, settlement of 6 miles per side. Cfr. Maumi, C., *La griglia del National Survey e la democrazia Americana*, in *Jefferson e Palladio, Come costruire un mondo nuovo*, ed. by Beltramini, G., Lenzo, F., Officina Libraria, Milano 2015, pp. 95–105, and the bibliography there indicated.

1 State-of-the-Art

The project is known and very well documented. Countless studies have been dedicated to it by American and international scholars and more research are in progress.[8] The large amount of documentation made it possible to trace the history of the university. Today, these documents have been in large part transferred in electronic format: original documents, writings, letters, projects, and, of course, a large number of drawings have been allowed to trace and deepen the understanding of the history of the University and the creative process that brought it to completion (Figs. 2 and 3). It should be emphasized, however, that the general guiding principles of the project—the formal organization of the campus, the language adopted, the distribution of the functions—have always been well recognizable since the very beginning of these studies. They emerged clearly already in the pioneering studies of the late nineteenth and early twentieth centuries, and have remained substantially unchanged.

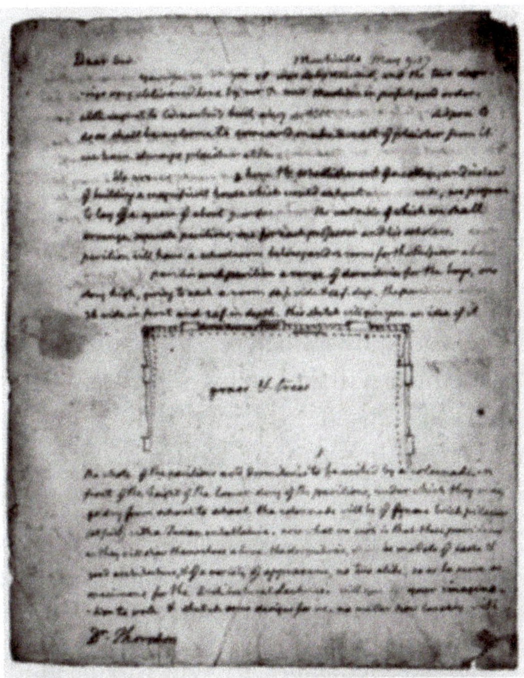

Fig. 2 Letter from Thomas Jefferson to William Thornton, May 9, 1817. N300. MSS 171, Albert and Shirley Small Special Collections Library, University of Virginia (Courtesy of the University of Virginia Library)

[8] One of the most recent publications, Wilson, R. G., *Thomas Jefferson's Academical Village: The Creation of an Architectural Masterpiece,* University Press of Virginia, Charlottesville 2009 (revised edition), offers essential materials for the study of the project. Furthermore, the Robert H. Smith International Center for Jefferson Studies, provides a continuously updated and detailed bibliography and reports on all aspects of Thomas Jefferson's studies, in which the history of the University of Virginia plays an important role. For the drawings related to the project see http://www2.iath.virginia.edu/wilson/drawings.

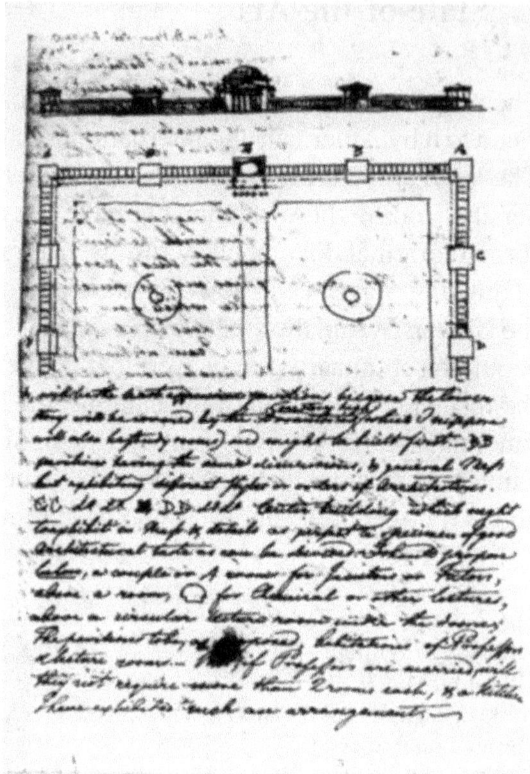

Fig. 3 Letter from Benjamin Latrobe to Thomas Jefferson, July 24, 1817. N304, K213. Coolidge Collection of Thomas Jefferson Manuscripts. Massachusetts Historical Society (Courtesy of the Massachusetts Historical Society)

2 The "Academical Village"

As is well known, the core principle guiding the realization of the university is the "academical village". This idea constitutes the springboard for the entire project. Jefferson himself explained it in a famous letter to Littleton Waller Tazewell, a member of the Virginia House of Delegates who had manifested his approval for Jefferson's ideas regarding the new university:

> Large houses are always ugly, inconvenient, exposed to the accident of fire, and bad cases of infection. A plain small house for the school and lodging of each professor is best… In fact, a University should not be a house but a village.[9]

The same letter already states all the guiding principles of Jefferson's project, from the financial and administrative aspects to the study programs and the architectural project. The idea of the academic village had already been formulated in the "Rockfish Gap Report" of 1818:

> … They [commissioners] are of opinion that it should consist of distinct houses or pavilions, arranged at proper distances on each side of a lawn of a proper breadth, and of indefinite extent, in one direction, at least; in each of which should be a lecturing room, with two to four apartments, for the accommodation of a professor and his family; that these pavilions should be united by a range of dormitories, sufficient each for the accommodation of two students

[9]Thomas Jefferson to Littleton Waller Tazewell, January 5, 1805, in Peterson, M.D., *Jefferson Writings*, 1152.

only, this provision being deemed advantageous to morals, to order, and to uninterrupted study; and that a passage of some kind, under cover from the weather, should give a communication along the whole range … It is supposed probable, that a building of somewhat more size in the middle of the grounds may be called for in time, in which may be rooms for religious worship, under such impartial regulations as the Visitors shall prescribe, for public examinations, for a library, for the schools of music, drawing, and other associated purposes.[10]

Pavilions for classes and professors' residencies, dormitories for the students, all surrounding the lawn, and a larger building to host the library and other public purposes: this text de facto describes all the core elements which would then appear in the final project.

What are the implications of this innovative concept, unprecedented in schools both in the colonies and in England, and destined to become the model of a new architectural typology for universities?

The campus designed by Thomas Jefferson symbolizes a series of strong, innovative and "revolutionary" ideas in both education and architecture.

It symbolizes an important and innovative pedagogical principle: to establish a new relationship between student and teacher, based on mutual respect, a relationship *inter pares* to be experienced in a space both solemn and human-sized. The focus of this ideal space is the lawn. The role Jefferson gives to this space, at the center of the entire complex, is unequivocally connected to the ideal of a rural, uncontaminated America. This principle was present in all Jefferson's projects. In the *Notes on the State of Virginia* he wrote:

Those who labour in the earth are the chosen people of God, if ever he had a chosen people, whose breasts he has made his peculiar deposit for substantial and genuine virtue.[11]

Jefferson developed a utopian, anti-urban idea for the young rural nation based on physiocratic principles. His vision was pro-agrarian and anti-urban. He intended to contain the inevitable acceleration of American financial and industrial capitalism, and exemplified these ideas in his project for the new capital.[12] Jefferson's naturalistic

---

[10]Rockfish Gap, August 4, 1818, Report of the Commissioners for the University of Virginia. Reprinted in Cabell, N. F., ed., *Early History of the University of Virginia,* Richmond; text in https://founders.archives.gov. For the various influences in designing the plan, besides the already mentioned bibliographical references, see Loi, M. C., *Thomas Jefferson, Roma e l'antico*, in *American Latium*, ed. by Johns, C., Manfredi, T., Wolfe, K. (to be printed), and its bibliography; Benoit, M., Wilson, R. G., *Jefferson and Marly: Complex Influences*, in *Bulletin du Centre de recherche du château de Versailles* 2012, http://journals.openedition.org/crcv/11936.

[11]Thomas Jefferson, *Notes on the State of Virginia*, Query XIX, "Manufactures". The relationship between city and countryside is central to the theoretical debates of eighteenth century Europe. Jefferson took part in this debate in its crucial years in Paris, from 1784 to 1789. However, the European model could not yet be introduced into the New World, which was slowly coming into existence on the new continent.

[12]On Jefferson's project for the new capital see Reps, J. W., *The Making of Urban America*, Princeton 1965; Id. Monumental Washington, Princeton 1967; Tafuri, M. Progetto e Utopia, Bari 1973, p. 35 and *passim*. See also Loi, M. C., *Gennaio 1902: i progetti per il centro di Washington della Mc Millan Commission,* in *Il Disegno e le Architetture della città eclettica*, a cura di Mozzoni, L., Santini, S., Liguori Editore, Napoli 2004, pp. 127–162.

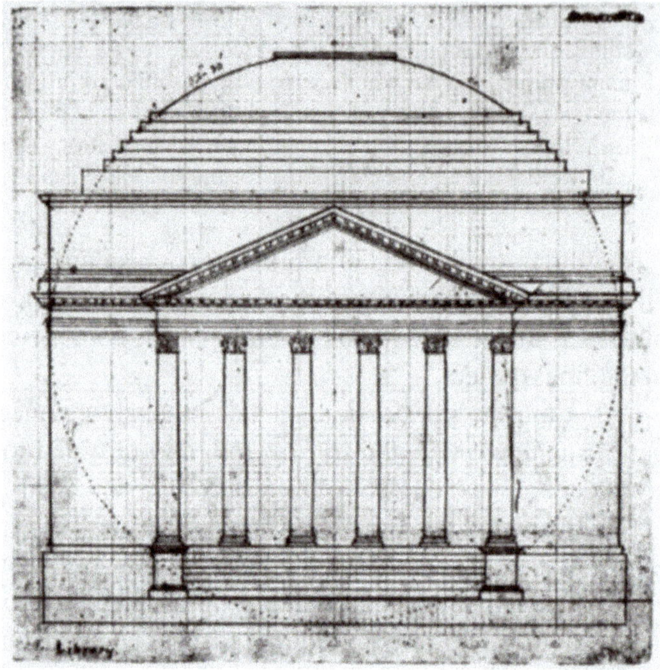

Fig. 4 Thomas Jefferson, The Rotunda, Façade. N328, K8. MSS 171, Albert and Shirley Small Special Collections Library, University of Virginia (Courtesy of the University of Virginia Library)

ideology had strong political connotations. In this context architecture, "the most useful of all arts"[13] became a tool to communicate the values of independence and freedom of the young nation. All these ideas found expression in his first projects for the University.

Once he laid the groundwork to create better conditions for learning, Jefferson incorporated *exempla* of Antiquity into the project. Not only did he use different elements of the classical orders in the pavilions' facades (almost like an architectural treatise in bricks and painted wood[14]) but he placed a temple-shaped building in a privileged position. His "americanized"—in size, material, organization of the interior spaces—Pantheon became the Library (Figs. 4, 5, 6, 7 and 8). This "temple of knowledge" stood at the center of shorter side of the lawn[15] (Fig. 9).

Despite the abundance of studies on the history of Thomas Jefferson's project for the University of Virginia, there are still uncertainties regarding the specific role he actually played in defining the final project. As it is well known, several different

[13]"Jefferson's Hints to Americans Travelling in Europe", 19 June 1788, https://founders.archives.gov.

[14]He used the Doric order in Pavilion I (Diocletian Baths), IV (Temple in Albano), VII (Palladio's *Quattro Libri dell'Architettura*), X (Theater of Marcello); the Ionic order in Pavilion II (Temple of Fortuna Virile, Rome), V (Palladio's *Quattro Libri dell'Architettura*), VI (Teatro di Marcello), IX (Temple of Fortuna Virile, Rome); the Corinthian order in Pavilion III (Palladio's *Quattro Libri dell'Architettu*ra), VIII (Diocletian Baths) and the Tuscan order for the colonnade which linked all the pavilions.

[15]Thomas Jefferson wrote: "*Rotunda, reduced to the proportions of the Pantheon and accomodated to the purposes of a Library for the University with rooms for drawing, music, examinations and other accessory purposes*." See http://www2.iath.virginia.edu/wilson/drawings, N. 331.

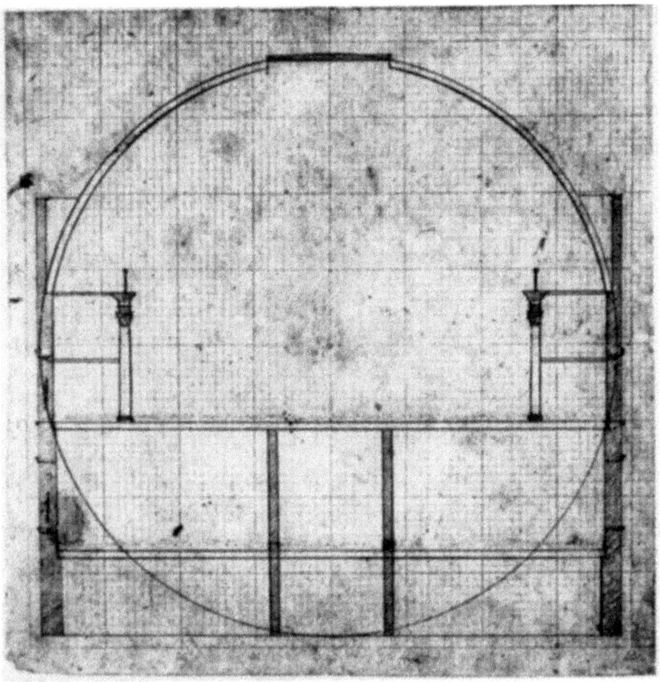

Fig. 5 Thomas Jefferson, The Rotunda, Section N329, K9. MSS 171, Albert and Shirley Small Special Collections Library, University of Virginia (Courtesy of the University of Virginia Library)

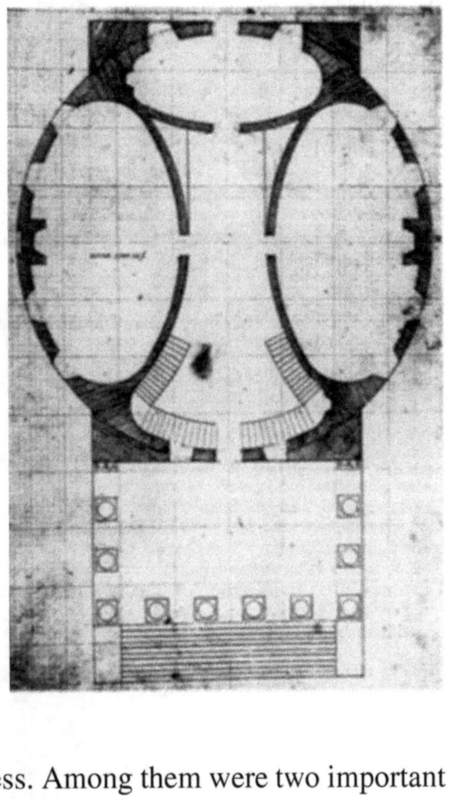

Fig. 6 Thomas Jefferson, The Rotunda, Plan, ground floor N330, K10. MSS 171, Albert and Shirley Small Special Collections Library, University of Virginia (Courtesy of the University of Virginia Library)

people were involved in the entire design process. Among them were two important architects: William Thornton and Benjamin Latrobe.

The exact role played by Jefferson and his main collaborators in the final design is still object of study and debate. The research aims to re-analyze the network of

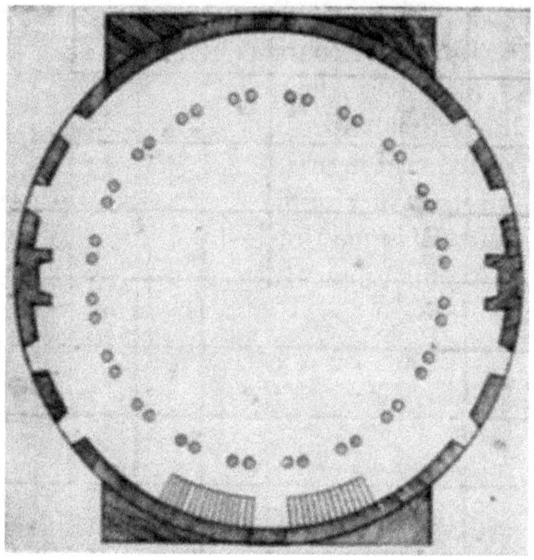

Fig. 7 Thomas Jefferson, The Rotunda, Plan, Library Room N331, K11. MSS 171, Albert and Shirley Small Special Collections Library, University of Virginia (Courtesy of the University of Virginia Library)

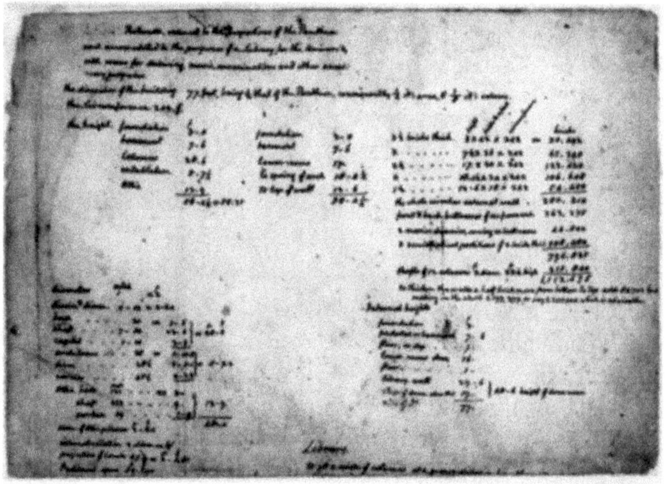

Fig. 8 Thomas Jefferson, The Rotunda, notes N331 verso, K11 verso. MSS 171, Albert and Shirley Small Special Collections Library, University of Virginia (Courtesy of the University of Virginia Library)

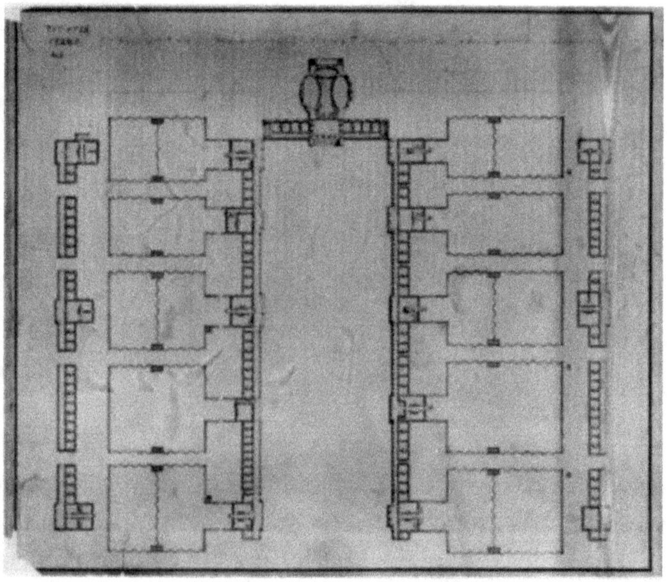

Fig. 9 Charlottesville, University of Virginia. Study for 1822 Maverick Engraving Library of Virginia (Courtesy of the Library of Virginia)

relationships and possible influences that converged in the project, focusing especially on the years Jefferson spent in France as Ambassador, when he came in contact with artists, politicians, pedagogists and philosophers. It also analyzes his role in the foundation of the West Point Military Academy in 1802. In that instance, he was not involved in the architectural project, but his participation speaks to the importance that military education had for Jefferson, which is then reflected in the hierarchical layout of the University project. In addition, it is worth mentioning that in the same year while the project for the campus in Charlottesville was taking shape, other schools and military complexes were under construction, both in Europe and in North America. Those too may have influenced him.

Jefferson conceived such a project, thanks to the convergence of multiple ideas, people and influences. Together, not only did they shape the design for a new space devoted to learning, but they effectively gave birth to a new architectural typology of the American campus, an innovative model for the decades to come.

Bovisa: A Park for Work and Research

Domenico Chizzoniti, Luca Monica, Tomaso Monestiroli and Raffaella Neri

Abstract Some projects for the "Goccia di Bovisa" (Drop) area and the neighbouring district are shown here: they fall within the urban transformation areas of the Milanese Territorial Government Plan (PGT). The Goccia area is involved in the plan to transfer part of the university campus to this site. This hypothesis was promoted and supported at the end of the 1980s by a group of researchers of the Faculty of Architecture. The idea was related to a sort of polycentric territorial city. Specifically, what they promoted was a new function for the area to set a new city centre in order to redeem its peripheral condition. The main reason for this option was the availability of dismissed industrial areas and the great rail accessibility provided for by the city plans that would have guaranteed urban and territorial relations. In agreement with the Municipality in 1990 the Faculty of Architecture developed a masterplan for the new settlement, which was blocked for decades due to pollution issues in the areas. The following projects derive from a new agreement promoted by the Municipality of Milan and Politecnico in 2016.

Keywords Urban design · Architectural composition · Bovisa · University Campus

1 The Settlement Places

The settlement project of the new Politecnico in Bovisa is primarily the project of a part of the city; a part that has long been destined to the outskirts, linked to the factories and their residences, emblematic of the last century's disorderly and separate growth from the city.

D. Chizzoniti (✉) · L. Monica · T. Monestiroli · R. Neri
Architecture, Built Environment and Construction Engineering—ABC Department, Politecnico di Milano, Milan, Italy
e-mail: domenico.chizzoniti@polimi.it

S. Della Torre et al. (eds.), *Buildings for Education*, Research for Development,
https://doi.org/10.1007/978-3-030-33687-5_9

The far-sighted choice of the late 1980s to place the new Politecnico, an important institution in the city, there and link it to the public transport infrastructure system by rail should have redeemed this state, bringing regional-scale public activities, services, residences, public places and green spaces to it: a new centrality, therefore, was able to reorganise all those activities that are typical of the urban settlements around it. This is a great opportunity to redesign this part of town, to give it a new identity and a corresponding shape.

This change in purposes and activities located in Bovisa should have corresponded to a transformation in the character and the quality of the places able to represent the new role and the new identity of this part of the city: places for study, research, living, community life, leisure, entertainment, sports and so on, urban places, to represent the life of an important part of the city.

This transformation has not yet taken place for several reasons, and instead the process of decline has already begun due to the dismantling of the School of Civil Architecture, located in the Bovisa district for nearly two decades.

More importantly, Bovisa does not have a recognised architectural quality nowadays in terms of its places and buildings which correspond to the potential that the new Politecnico settlement could offer, together with other institutions (the Mario Negri Institute, for example). The study and research activities were positioned by renovating existing buildings, or by building them one by one again without a specific design, without modifying or modernising the overall urban structure, without adding the quality that a settlement of such selected activities could involve, and without defining new areas for the city.

Railway connections have been enhanced, making this area highly accessible from different parts of the entire Lombardy region and more, making the potential of this area even more obvious; but the Drop area has remained as an enclave, still isolated and closed within the railway fence.

The redesign of the large area included in the UTA of Bovisa and the possibility to design the vast area of the Drop, the original settlement site for the Politecnico, offers the opportunity to give a new definition to the places of this part of the city, which has the scientific and technological faculties of the Politecnico di Milano campus as a centre of value.

The following projects are just some of those deriving from the consultation promoted by the Municipality of Milan and Politecnico in 2016. Compared to the theme of the lower-level school, the settlement and the project for a university campus have to deal with other problems. Assuming a role of a major urban body the campus deals with the issue of building a central core in the city, a collective institution able, through its architecture, to define the identity and condition the destiny of a large urban district. The project is supposed to address the idea of the settlement and clarify its compositional principles to organise buildings and places that define it.

2 The Waste Land. the Politecnico di Milano at the Bovisa District in the Gasometri Industrial Park Area, 2016

Design goup: D. Chizzoniti, L. Monica, R. Gabaglio, G. Guarisco, L. Jurina, M. Bocciarelli, V. Donato, S. Recalcati, S. Riva, H. Pessoa

With: S. Cusatelli, P. Galbiati, O. Meregalli (Laboratorio informatico di architettura), D. Orlandi Arrigoni, I. Sgaria, R. Zucco

The theme of the project "A park for research and work in Bovisa" is, from the earliest hypothesis of 1974, the theme of the projects and studies that have taken place over the years and that have made the area of gasometers an unresolved urban area. The area is still waiting for the definition of balances between the need for redemption in terms of architecture and landscape, compared to the potential opportunities for qualified training, research and economic development. The creation of a new Politecnico headquarters in the Bovisa area is an important opportunity to help overcome a crisis in the city, a crisis that threatens to alter its physiology. The hypothesis adopted is that of an industrial settlement, articulated over time as a place of production, of transformation and of exchange: a place that maintains these propulsive characteristics, transferring them to the sectors of production and culture. Starting from the awareness of the Politecnico history, which since 1974 sees the area of Bovisa as a place for research and education, the project tackles with the complexity of the intervention by reorganising it according to some strongly integrated functions. There are some basic aspects to be taken into consideration:

(a) the need for the consolidation and expansion of the Politecnico in the area of gasometers, with the integration of the Science Park;
(b) the opportunity to build a large urban park, equipped and connected to the campus rediscovered in the "waste land" of the gasometer area.
(c) Closely connected to these, the other functional nodes are achieved:
(d) the recovery, restoration and reuse of the abandoned heritage of the industrial complex of gas production, which continues today as a well-structured fragment of a city-factory;
(e) the integration between the future enlargement and that already implemented (Politecnico Lambruschini-La Masa). This structure requires consolidation, with interventions on urban public spaces for greater pedestrian accessibility and facilities for the campus life;
(f) the organisation of a linear residential park to complete the campus, with integrated services, facilities for accessibility with a new architectural structure of the FNM Bovisa railway station and the new shopping area.
(g) the renewal of accessibility and public transport for full compatibility between urban scale and new functions, with consequent decongestion of local traffic. This hypothesis is based on checking the sustainability on the urban scale of the amount of traffic generated by the new functions, by means of model simulations able to verify the local aspects and those of the areas surrounding (Figs. 1, 2, 3, 4, 5).

Figs. 1, 2 Axonometric view of the 'Goccia' and Perspective view

Figs. 3, 4, 5 Perspective views

3 Bovisa 2016: A New Urban Settlement

Design group: R. Neri, T. Monestiroli, I. Boniello, F. Menegatti

With: O. Meregalli (Laboratorio LidAr), F. Guzzetti, M. Ronconi (topography), Isabella Balestreri (architectural history), S. Croce, T. Poli (environmental sustainability), M. Morena, T. Truppi (economic sustainability), E. Garavaglia (structures), C. Campanella (restoration). Technical advisors: S. Recalacati, V. Donato, S. Riva (city planning), and: G. Uboldi, E. Cuogo

We believe that the Drop area should not remain closed off, to be allocated exclusively for the university campus, but it should become an integral part of the city, open to life and to the inhabitants of Bovisa; it should promote a mix of activities and include collective places, open spaces and green places, measured empty spaces, corresponding to the squares and gardens of the ancient city. These open spaces, the commercial places of the city par excellence, are the elements that structure the settlement according to a hierarchical principle that distinguishes its role, character, shape and size.

The central point, organising the system, is the large rectangular lawn developed along the south-east/north-west axis, which is a recurrent orientation in Milan, where the road follows through to the previous location of the gasometers. It is concluded to the north by the library building and to the south by two gasometers, the remains and symbol of Bovisa's recent past. Inside are some of the site's most interesting industrial archaeology buildings.

All the parts that make up the new settlement converge on this site: on this side and beyond the lawn, the buildings for study and research belonging to the new Politecnico and its residences face one another.

A square with the services, commercial activities and other community spaces for the university and the neighbourhood intersects across the park and marks the central place. From the Villapizzone station, the square goes along a porticoed services building, including other buildings recovered from the former settlement and ends in another place, a triangular square with the large auditorium for the campus and the city, which the science park and business start-up accelerator buildings look out onto.

Beyond the Bovisa station, in relation to the existing district, lies a new mixed settlement that includes residences, workspaces, small shops and services. It is built along an inner central pedestrian spine, and almost parallel to Via Bovisasca, a green communal area which houses and workplaces look out onto.

This sinuous place ends in a square which the buildings of large retailers in the vicinity of Piazzale Lugano look out onto, whose road system must be properly reorganised on the basis of the definition of the shopping centre (Figs. 6, 7, 8, 9).

Fig. 6 General plan

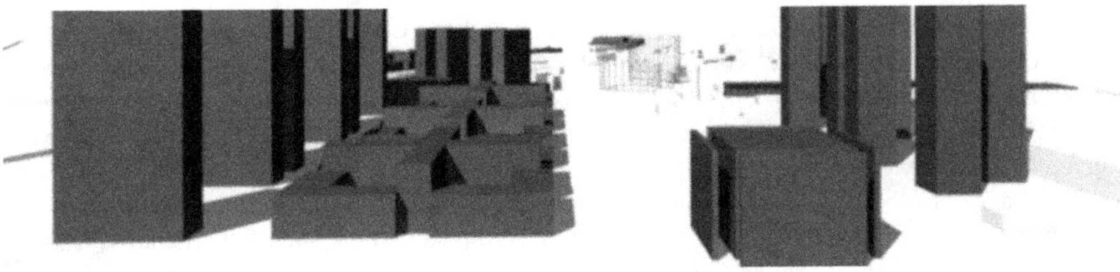

Fig. 7 Perspective view from the central lawn towards the Gasometri

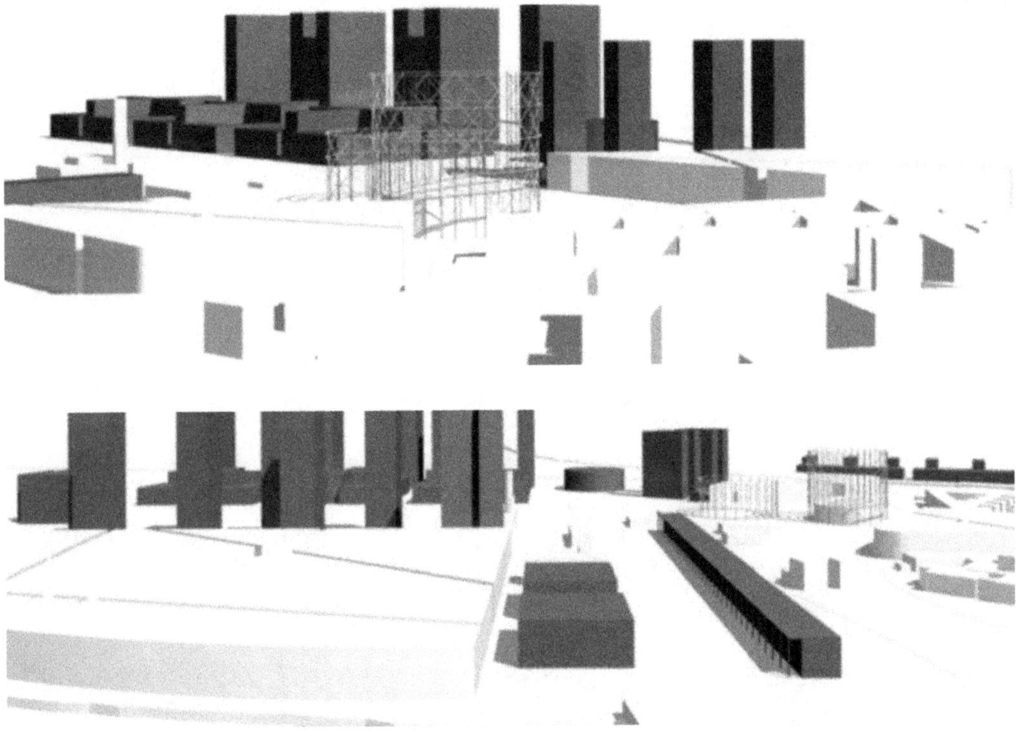

Figs. 8, 9 Perspective views from the Gasometri and from the collective plaza

The City's New Road. The Fundamental Role of Nature in Urban Transformation Processes

Adalberto Del Bo

Abstract As with numerous contemporary situations regarding industrial suburbs, nature and artifice, alternating over time, have provided Bovisa—a strategic area in Milan—with a legacy full of contradictory signs such as the negative effects left by the productive apparatus, the vigorous return of nature and the marked evocative value of the remaining buildings, living deposits of a not-too-distant past. The development of the project, presented in the context of a Call for Ideas (Moro in Bovisa, un parco per la ricerca e il lavoro. Maggioli, Milan, 2017), involves three main parts that correspond, in order, to the phases indicated for the transformation:

- the first regards buildings outside the Goccia area, intended for residence and commerce;
- the second is the link between the stations and the construction in the area of the gasometers of the Library and the Congress Centre;
- the third, along with setting up the park, entails, on its sides, the construction of the New Politecnico/Science Park and the residential system of the Strada Nuova, two diverse settlements, albeit conceived according to similar formal and constructive logic.

The ideas expressed here, according to a single project aimed at enhancing the area's features, propose a transformation process that will enable the construction of a formally defined and complete urban part, characterized by high accessibility and

The proposal is the result of a Call for Ideas launched by Politecnico di Milano for a vision on the future of Bovisa. Through this experience many professors and students of the ABC Dept. PhD Course have positively tested their polytechnic approach, here featured by the author of the contribution, who coordinated the team. Professors Roberto Camagni, Roberta Capello, Alberto Franchi, Paola Ronca, Pietro Crespi, Massimo Ferrari, Maria Cristina Loi, Cristina Pallini, Sara Protasoni, adjunct professors Daniele Bignami Samuele Camolese, Mario Maistrello, Stefano Perego, Claudia Tinazzi, Alessandro Zichi and PhD students Daniele Beacco, Annalucia D'Erchia, Derya Erdim, Manar El Gammal, Marta Ferretti, Alessio Passera, Manuela Scamardo, Marco Zucca all participated together in the development of the project.

A. Del Bo (✉)
Architecture, Built Environment and Construction Engineering—ABC Department, Politecnico di Milano, Milan, Italy
e-mail: adalberto.delbo@polimi.it

S. Della Torre et al. (eds.), *Buildings for Education*, Research for Development,
https://doi.org/10.1007/978-3-030-33687-5_10

reduced vehicle traffic, built in a green environment according to principles of sustainability and equipped with activities, residences and services of a metropolitan nature. As a result, the general idea of the project tends to express, in organized and orderly terms, the traces of the phases which over time have created the Goccia di Bovisa area: the original phase of nature, the industrial urban service phase, the phase where nature returns to occupy the abandoned work-related spaces and the phase being proposed here in which the elements present are valued in relation to the characters and the size of the set of activities envisaged. At the basis of all this is the search for a balanced relationship between artifice and nature, where the latter, an original and fundamental element, makes up the primary place in which to structure the proposals of the new city.

Keywords Urban project · Bovisa · Technology park · New Politecnico · Residentials

1 Bovisa Area. The Relationship with the City and Its Parts

It is an ancient geometry, that which initially marks the ground disposition of the urban layout imagined as a new research settlement, an underlying plot that places in order the individual parts of the project, the recognition and compensation of an orientation that, despite the contemporary confusion, for centuries has been drawing directions to the north-west of Milan where the reasons, which are not only political and social but even more so geographical and orographic, behind the rooting in the territory that coincides with the Lombard culture, are still evident (Figs. 1–2).

Inside and outside the fence of the train tracks, a nineteenth-century limit imposed by the culture of mechanical work, the suspension of the city's consistently slow transformations still highlights the isolation of the area balanced between memory

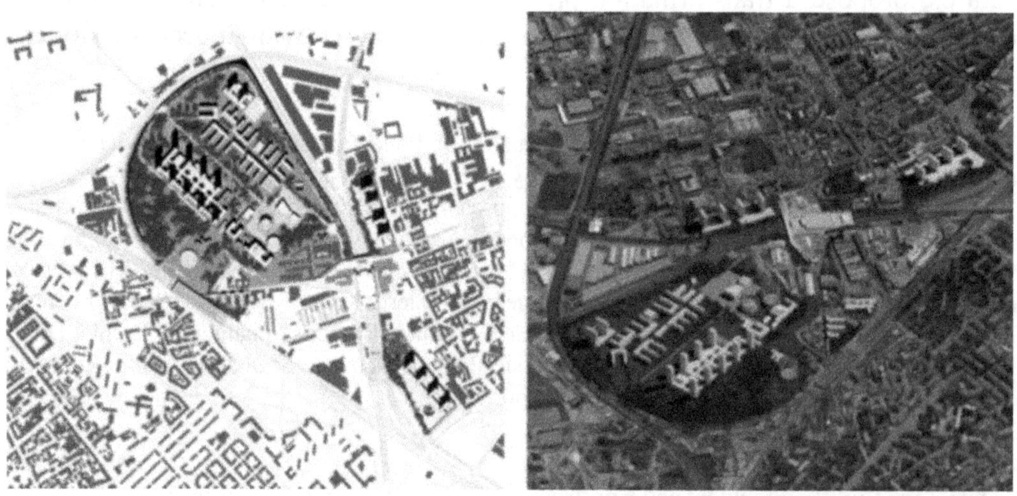

Figs. 1–2 Project plan and general view

of its own forced identity and the possible homologation to the city built around it; a theme that has been the main feature of the project. Moreover, the same ideal axis along which the project area is located, confirms, starting from the Garibaldi-Repubblica area, passing through the former Farini railway station up to, past the Gasometres area, the most recent area occupied by Expo 2015, a contamination between green passages and public and research places.

The proposed solution, developed between isolation and connection, chooses the enhancement of green areas found within the area, as a value to be preserved, to be integrated, to be implemented starting from the relationship with the pre-existing structures which initially traced, in their orthogonal layout, the more natural and original story of settlement in the area.

The imagined greenery is therefore connected to an ideal system which, from the centre of Milan, has the possibility of enjoying public routes in nature; a tamed and equipped nature in which the same research sites offer contact points between the not-too-distant neighbourhood life and the daily technical-scientific training and cultural learning.

A continuous change in horizon, scale, which gives the neighbourhood the most intimate and somehow introverted dimension, yet at the same time creates and searches for ideal, distant relationships, new points of view that are compared to the territorial scale with a rich environment of memories and values capable of recalling the ideal effort for the construction of a single cohesive landscape, whose identity was already imagined in the words of Cattaneo (1844).

2 Accessibility to the Area and Internal Mobility

The demand for mobility generated by the new university Campus and complementary functions is the cornerstone for defining the infrastructure layout serving the new functions. The site is expected to host a daily population of around 13,500 people, including teaching staff/employees and students. The current 13,300 users who gravitate around the existing campus must be added to the above figure. The demand for mobility, generated by the new settlement, is estimated at roughly 22,000 journeys (commuting) (Fig. 3).

The largest share of journeys will be managed by public transport (tram + rail); in this regard, the extension of the tram line becomes the strategic connection with the local mobility system of the north-west quadrant of the city, while the train service, with stops at Bovisa and Villapizzone, allows a connection at both a suburban level (S lines) and regional level. In order to guarantee a direct relationship with the city centre, the Garibaldi station and the transformation of the Farini airport will be connected to the Goccia's interior via a new overpass to the existing tracks. Once the planned work has been implemented, there will be three primary networks with public roads, north, south and west, and these will be connected to each other by a partially underground or trench road that develops between the park and the new university buildings. A second underground network crosses transversely the gasometers area.

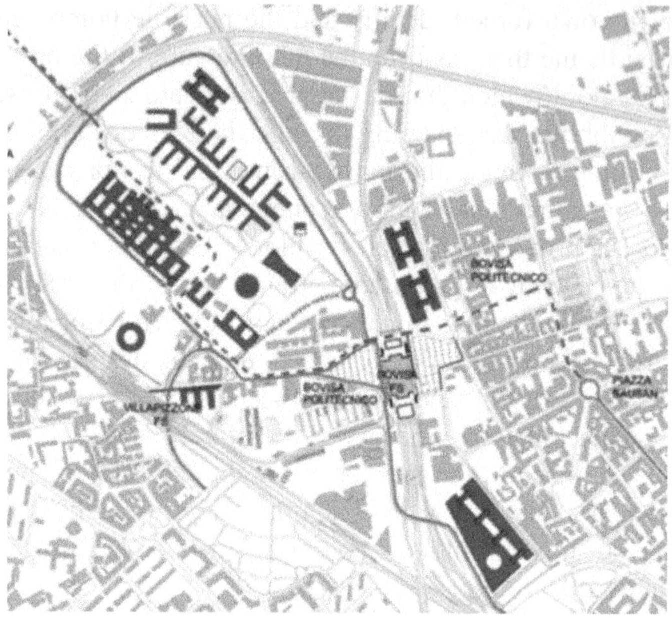

Fig. 3 Public transport (dashed) and project mobility. The new system of covered parking spaces of the Bovisa Station

The decision to lower the street level will ensure continuity of pedestrian spaces, providing for a continuous path without obstacles/dangers between the new Campus, the gasometres area and the Bovisa and Villapizzone stations. A secondary perimeter road in the works area will enable to serve the residences, thus guaranteeing access to the appurtenant parking areas and relevant public areas.

3 The Project: Identification of the Parts

3.1 The New Politecnico Campus and the Technology Park

As the main aspect of the overall project, the university settlement directs and indicates all the further alignments within the urban design, while emphasizing the original pre-existing footprint. In terms of size and position, the research centre qualifies and identifies the new destination of the enclosed space, aligning itself to the existing urban areas, maintained and reused, yet increasingly confirming the natural division into quadrants within the protected enclosure. Separation and connection are the two souls that coexist within the interpretation of the proposed theme which alternates pauses and proximity in its views that relate to the general composition, choosing to allocate the perpendicular sides to the east/west axis in direct comparison with the natural park while emphasizing, furthermore, to the south, the possibility of a continuity capable of interpreting the necessary connection with the Politecnico settlement, already present in the area. Separation and connection are, simultaneously, the keys to interpreting a settlement principle that also places in order and

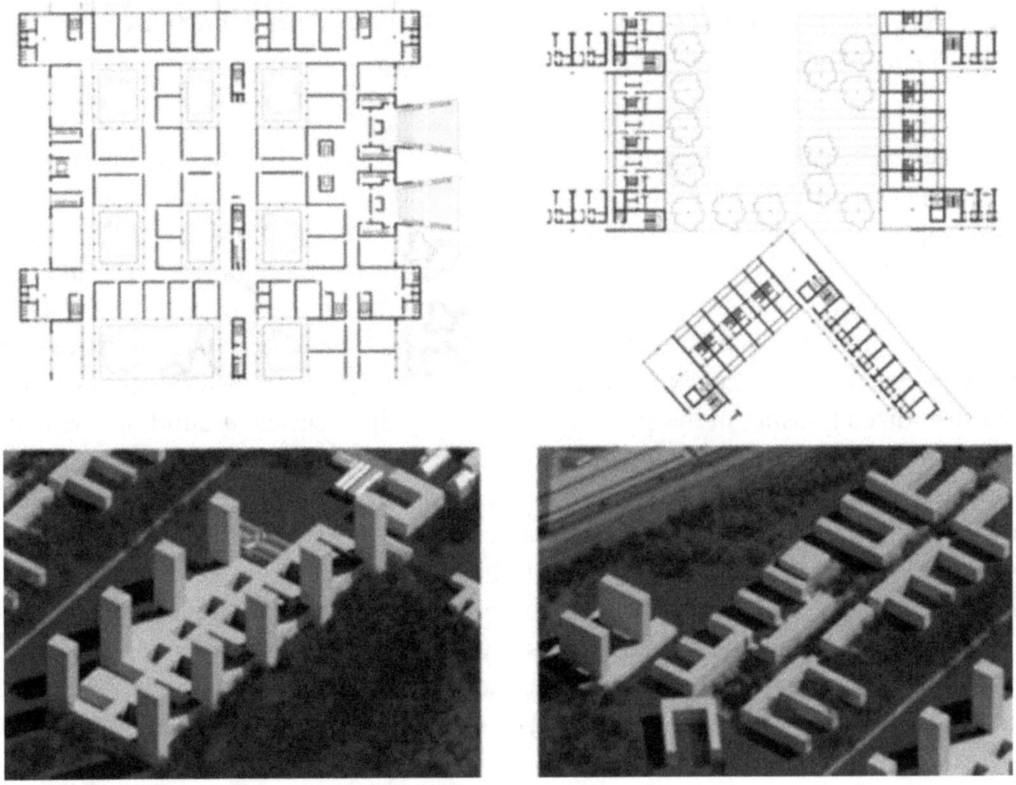

Fig. 4 Plan and axonometry of the New Politecnico Campus and of the residences inside the Goccia Area

chooses, through hierarchy, the distances and the adhesions needed for studying and for university research (Fig. 4).

At the northern border of the entire project area, the spaces structure for higher education and specialization has been thought of as a complex machine, an elongated distribution designed to be built in consecutive phases capable of accommodating the study and research facilities of the various Departments inside the Milan Politecnico. Transferring the School of Design from the nearby Campus Durando, placing its Department next to the Departments of Electronics, Information and Bioengineering, Physics and Computer Science currently located in the Leonardo Campus, will give life to a coherent macro-laboratory regarding the skills related to the industrial sector in an enthusiastic structure inhabited by researchers. A superimposed composition made up of courts, cross vaults and towers that allocate the ground floors to studying and teaching of the various disciplines with a considerable degree of interrelation between the parts, placing, in height, the slender towers of singular and specialized research. Through a realistic hypothesis, this follows numerous European and global experiences, the expansion of the research centre within a purely university context, envisages the possibility of accommodating specific business realities which, with regard to the close relationship to the technical-scientific disciplines found on the Campus, give life to a Technological Park.

3.2 Residential Settlements and the Internal and External Trade System Within the Goccia Area

There are two types of residence within the project; two different housing models whose qualities stem from their expected role and from the provision established with respect to the context in its future prefiguration. These collective dwellings are comparable to a mid-density neighbourhood, within the large urban park, and the thin tall buildings that are placed at the edges of the area, in relation to the city, mark the boundary towards the train tracks. The first residential model, distributed in the greenery that is already present in the area, mainly assembles student housing and subsidized housing in the succession of a diversified series of buildings with an open courtyard which, starting from a central spine with an urban character, looks outwards towards the houses. Court houses, a student residence and two residential towers fulfil the requirement for a total area of 100,000 m^2 of residence.

Outside the area, fenced off by the railway, a single settlement system places the tall buildings on the opposite side, following a broader scale logic of reference: slender slabs leaning on capacious bases capable of containing the services needed by the neighbourhoods in question as well as proposing new ones for the imagined settlement (30,000 m^2). Sporting, leisure, entertainment, and cultural activities define a road within the protected, yet open, block in which to find the closest neighbourhood dimension. The bodies in line, overlapped and supported by these massive bastions that raise the horizon line above the disordered proximity of the train tracks and by casual urban choices, oriented north-south, create a discontinuous system that punctually confronts the built city through its most evident positions, while drawing, with their principle, a recognizable broken line which is a memory, once again, of the design of the infrastructures engraved in the ground (Fig. 5).

3.3 The Park: The Green System

In its current layout, the "goccia" landscape appears as a still unstable result of a profound change that has affected spaces, natural elements and structures. The dismantling of the production plants and the total segregation of the area have established the return of a colonizing nature with an unexpected and evocative result (the formation of actual biodiversity protection) of an original condition lost in almost all the other areas throughout the city. The green project and the design of open spaces were intended to work around this specific identity of the location through two different types of action.

1. An overall re-naturalization plan which, starting from the existing vegetation covering, directs the spontaneous processes towards the formation of a complex system of landscape units starting from the heritage of tree and shrub species which, after the disposal of the installations, have created woods composed of

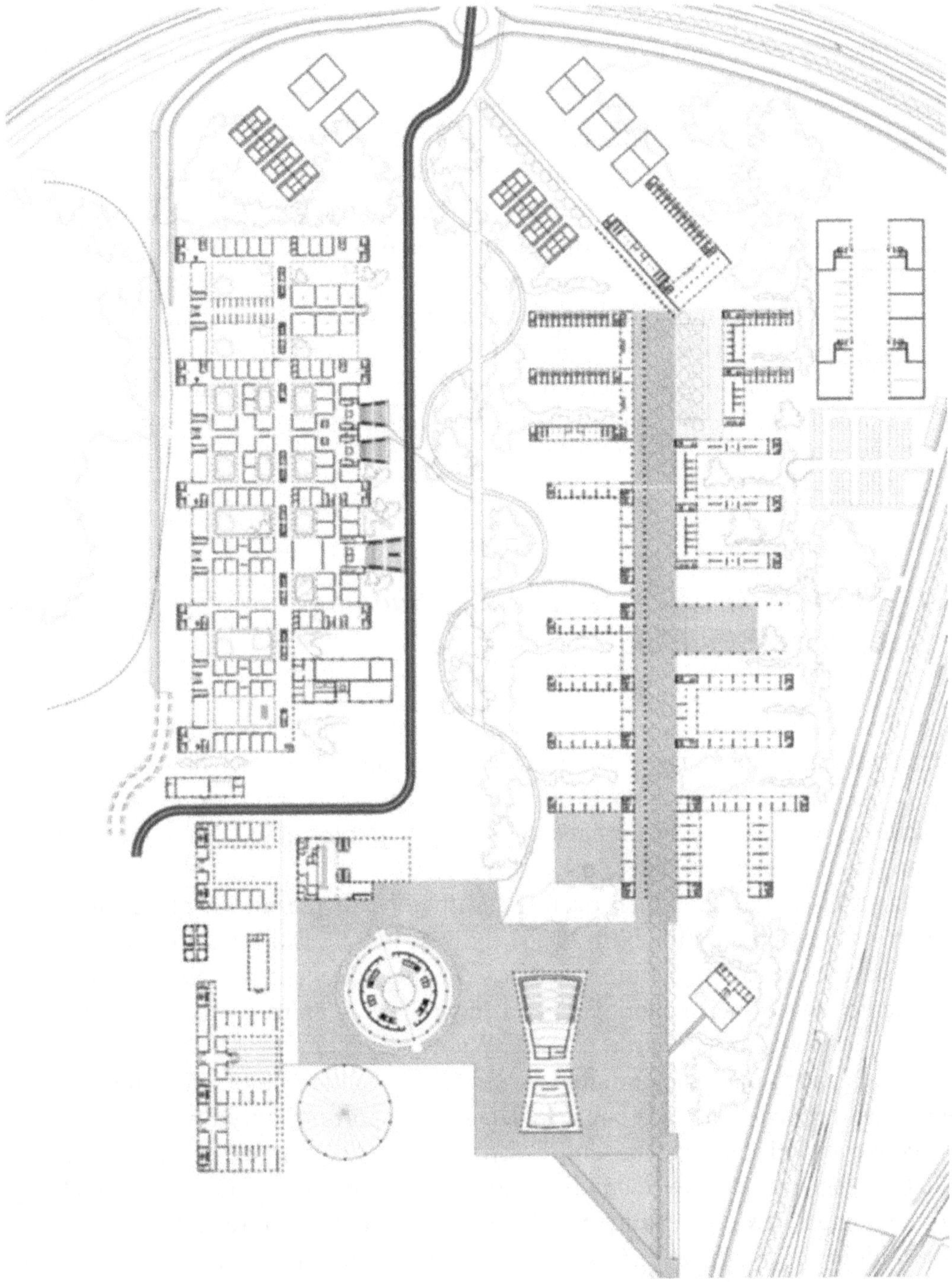

Fig. 5 General plan of the intervention

Platanus orientalis, Tilia cordata, Celtis Australis, Paulonia tomentosa and Populus Nigra. Three forms of aggregation are envisaged in this plan: belt types, brushes and neat rows.[1]

[1]Reference is made to the specific Action COST dell'Unione Europea (COST 837 Plant biotechnology for the removal of organic pollutants and toxic metals from wastewaters and contaminated

2. The design of a system of green open spaces closely integrated with the structure of the city, each with a different layout and character, in relation to the various parts of the project:

   - The New Road as a pedestrian axis with a double row of Judas Trees (*Cercis siliquastrum*) planted according to a quincunx arrangement.
   - The Central Park, a large meadow (*formed by Dichondra repens*) dotted with large size trees so as to form shaded areas for summer comfort, with relative undergrowth of herbaceous and shrubby species.
   - The gardens of the residence created by a sequence of tree-lined courtyards in continuity with the wooded area of the Central Park
   - The Politecnico gardens with a more distinct character, are delimited and shaded by rows of Tilia plathyphyllos and isolated specimens of Magnolia grandiflora.

3.4 The Gasometers Area: Existing Buildings and New Collective Buildings

The urban landscape of this part of the city reveals, despite the progressive state of abandonment, some features of a history which, unfolding over the centuries, has been characterized through some fundamental passages: from a rural area, whose structure is witnessed by historical cartography beginning at the end of the sixteenth century, to an important industrial-worker district.

This transformation, which occurred at the end of the nineteenth century and which was favoured by the birth of the railway, is followed by a process of decay and abandonment starting from the 1970s. Revitalized via the establishment of the Politecnico two decades later, the area is now once again ready to welcome major intervention towards its rebirth.

In the general plan, the Gasometers Park takes on the role of a fulcrum towards which its different functions come together to support not only the district's more collective activities but also the entire city of Milan. The planned permanence of the two Gasometers and historic buildings, along with their reuse, opens the possibility of imagining a new composition in which the two large structures dialogue with the new buildings, in particular a large Auditorium juxtaposed in dialogue with the new redesigned Campus library inside one of the two cylindrical volumes.

---

sites) who was in charge of the state of experiences and research, at a pan-European level, in this sector.

Figs. 6–7 Dry-connection system of the prefabricated elements: reinforcement of the pre-fabricated element, connected to the node made with steel plate and bolts and pre-fabricated element ready for installation

3.5 Standardization and Economy of the Construction System

In accordance with the idea behind the architectural project, a single flexible structural system was designed and consequently easily adaptable to all the different functions found within the project itself. This system involves the use of frame structures, consisting of prefabricated elements made of high-strength reinforced concrete.

Starting from the module at the base of the geometric-design development (7.2 m), two standard types of recurrent structural meshes have been identified which, given the repetitiveness of the elements, have allowed to opt for a prefabrication technique, minimizing the costs and time necessary for carrying out the work. Furthermore, the high-strength concrete allows for the use of structural elements with reduced sections compared to those made with ordinary concrete, thus obtaining a considerable saving in terms of raw material used. With regard to the exposed concrete elements, a self-cleaning "photocatalytic" white cement can be used that "purifies the air", spontaneously eliminating organic and inorganic contaminants. The coverings, given the reduced weight loads, will be made of laminated wood type "X-lam" (Figs. 6–7).

4 The Project: Evaluations

4.1 Risk Assessments

Within the growing importance of Disaster Risk Reduction—DRR strategies, Urban Risk Assessment action is characterized by specificity. If the DRR is defined as "the concept and practice of reducing disaster risks through systematic efforts to analyze and cause causal disasters, including reduced exposure to hazards, less vulnerability of people and property, wise management of land and the environment […], in parallel the concept of Corrective Disaster Risk Management is considered as "management

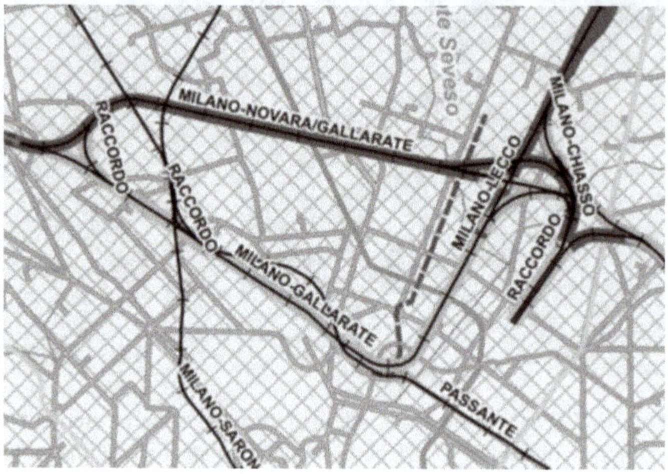

Fig. 8 Dangerous transport of dangerous substances on the railway network

activities that address and correct or reduce disaster risks which are already present". It is the task of urban planning and architecture to seize DRR opportunities, in a corrective way, particularly in the case of territorial transformations where risks are known.

With regard to Bovisa, it is worth highlighting that the area could be involved in railway accidents, since it is considered as being a "high" hazard and a "high" risk location due to the trains passing by on the Milan-Novara-Gallarate line. What emerges therefore is the indication of a project that offers spatial arrangements that pursue invariance in the risk levels. It is suggested, as far as the project is concerned, to increase the quota starting from the 30 m zone, in order to contribute to or limit any possible release of harmful substances or to absorb possible pressure waves or energy flows. This solution camouflages the presence of the railway, providing landscape benefits, while also protecting it from noise, dust, vibrations, etc. and offers a synergy with the needs of cycle-pedestrian transport and with the creation of the greenery needed to guarantee sustainability and quality of life in the new district (Figs. 8 and 9).

4.2 Evaluation of Energy Sustainability

The strategies chosen for energy and environmental sustainability were imagined so as to intervene in a coordinated manner, both at the neighbourhood and building scale. In order to compare the quality level of the building work to be carried out, it was decided to evaluate the project with the aid of two rating systems: "GBC Quartieri" and "LEED Italia Nuove Costruzioni e Ristrutturazioni". Thus, by displaying the qualitative plaque, the quality of the technological centre, the residences and the shopping centre will become easily visible to users coming from all around the world. At the same time, the innovative task becomes a potentially replicable experiment.

The quality level expected in the preliminary project would be:

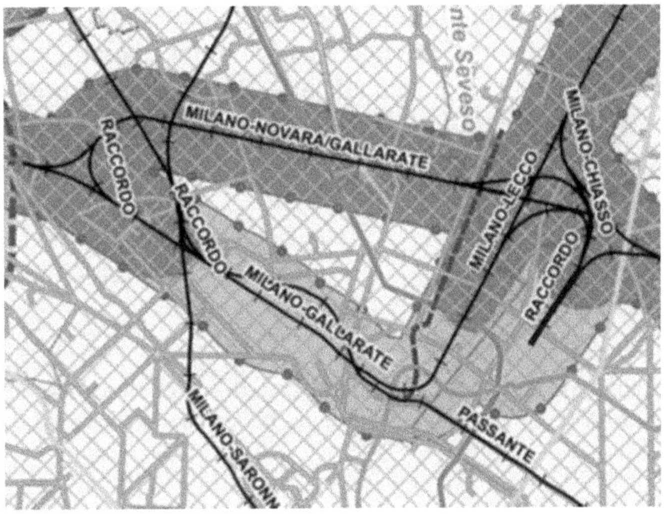

Fig. 9 Risk for the population of the transport of dangerous substances on the railway network

- Gold, achieving 69 points with the 'GBC Quartieri' system;
- Gold, achieving 72 points with the 'LEED Italia Nuove Costruzioni e Ristrutturazioni' system, for the building.

From the neighbourhood point of view, a balance has been established regarding the urban planning aspects, density of services and creation of infrastructures that enable the production of thermal and electric energy from renewable sources, the recovery of water sources and action on the microclimate. With regard to water management, it will be possible to reduce the use of water, both for irrigation and within buildings, for over 40% of typical consumption thanks to the infrastructures.

On a building scale, the typological mix, the use of easily renewable and low-emissivity materials, the content of recycled material, and the reduction of water consumption, will allow for the creation of buildings with low environmental impact without increasing costs and, above all, they will have reduced management costs.

Integrated Energy Management

Action was taken both on an urban scale and on a building-services scale, with the aim of providing users with the highest environmental quality within the buildings during the various seasons of the year, while simultaneously creating an innovative pole relevant to the management of energy resources.

The plant will be built as follows:

A rotating furnace and air flow control gasifier that generates SYNGAS, mainly consisting of hydrogen and carbon monoxide, used by an alternative internal combustion engine connected to an electric generator for combined production of both electricity and heat. During the summer the heat is sent to an absorption machine, which will enable to produce chilled water for the settlement's needs. The wood material will be transported via rail to Villapizzone and storage will take place out in the open in the former quarry area. At the building-services level, in addition to optimizing the zenithal orientation, the winds and insulation, work was carried out on the installation of a water-to-water heat pump system; the plant-building is made to work at low enthalpy. The heat exchange is entrusted to foundations that function as an accumulator of thermal energy produced by the tri-generation plant. The biomass tri-generation plant located between the area occupied by the old gasometer adjacent to the railway station and the old quarry, allows for the use of wood-type virgin plant biomass to generate the overall requirements of the settlement.

References

Cattaneo C (1844) Notizie naturali e civili su la Lombardia. Tip G Bernardoni, Milano
Moro A (2017) Bovisa, un parco per la ricerca e il lavoro. Maggioli, Milano

The Quality of the Project and the MIUR Standards for the Control and Funding of Buildings for Education and Training

Giovanni Castaldo, Matteo Gambaro, Elena Mussinelli and Andrea Tartaglia

Abstract This paper reports the results of an activity of scientific advice for the winning project of the national competition for the construction of a new building for the *Fondazione Collegio delle Università Milanesi*. Starting from the project submitted to the IV MIUR funding tender in 2016, the attention is focused on the issue of the quality of the project and its development up to the detailed design phase within the current regulatory and procedural models defined by the MIUR and the Public works laws for the design and construction of temporary residences for university students.

Keywords Environmental design · Temporary residences · University colleges · Public works · Dimensional standards

1 Temporary Residences for University Students

The university student residency has been progressively attracting the interests of real estate investors over the last few years, appearing as a growing market, particularly in a city like Milan, which is increasingly showing a European propensity with the growing presence of foreign students.

The offer of accommodation in facilities specifically designed to host university students is considerably insufficient compared to the real demand, both in quantitative and in qualitative terms, obligating the users to face the free housing market.

To cope with these issues, the research is developing innovative design solutions and management tools, in order to interpret changing needs and lifestyles.

In light of these considerations, with the aim of encouraging the construction of new student residences, in November 2000 the Italian Parliament passed an ad hoc law, aimed at universities, public bodies and foundations, which provides for the co-financing of specific interventions concerning existing buildings, extensions,

---

G. Castaldo (✉) · M. Gambaro · E. Mussinelli · A. Tartaglia
Architecture, Built Environment and Construction Engineering—ABC Department, Politecnico di Milano, Milan, Italy
e-mail: giovanni.castaldo@polimi.it

S. Della Torre et al. (eds.), *Buildings for Education*, Research for Development,
https://doi.org/10.1007/978-3-030-33687-5_11

new buildings and purchases of buildings to be used as residences for university students.[1] The Law 338/2000 is the first national example of an organic program aimed at encouraging interventions towards different building scales: from the removal of architectural barriers, to building improvements for hygiene and safety, to maintenance, restoration, expansion and new construction, with also the possibility of purchasing areas and buildings for residences (Del Nord 2014).

This is an ambitious and innovative program aimed at increasing the number of dormitories and at improving the design and construction quality through proper and detailed regulations focused on the "Minimum dimensional and qualitative standards and guidelines relating to technical and economic parameters concerning the construction of housing and residences for university students".[2]

Within this framework, the policy for university residence of the Milanese universities has profoundly changed and the panorama of their offer has been articulated also in relation to these building programs and their development.

The case of the *Fondazione Collegio delle Università Milanesi* is one worthy of not in this regard, a Merit College legally recognized by the Ministry of Education, University and Research—MIUR, which in the recent years has started a significant construction program aimed at increasing the number of accommodation facilities.

Precisely within these activities, an international design contest published by the Fondazione was the chance for a group of researchers from the ABC Department—who focus much of their activity on the topic of social housing and special residences[3]—to transfer theoretical knowledge and research results on techno-typological innovation, environmental quality and technical construction solutions within a design process in which these elements represent a significant added value. Analysis of trends, typologies and innovative international approaches, together with advanced tools and methods typical of the environmental technological project, as well as studies on the sustainability of building materials in a circular economy logic, have become important added values to support the project proposal which was then the winner of the competition.

---

[1]Law 14th November 2000, n. 338 "Indications regarding the accommodation and the residences for university students". The implementation of the law consists of the publication of National tenders: for each tender specific decrees are published for the definition of the modalities of presentation of the co-financing requests, the required documents, the spatial and functional standards, the procedures and the constraints for the co-financed initiatives.

[2]D.M. 28th November 2016, n. 936 "Minimum dimensional and qualitative standards and guidelines relating to technical and economic parameters concerning the construction of housing and residences for university students as prescribed by the Law 14 November 2000, n. 338".

[3]Cf. research projects: "Policies, projects and techniques of rehabilitation and transformation of urban suburbs" MURST 1998, operative coordination Elena Mussinelli; "Innovation and project for residential buildings", AUPREMA soc,. coop., Elena Mussinelli (2005–2010); "To live tomorrow. Technological innovation and sustainability in the residential building project", Fondazione Politecnico di Milano, coordinator Elena Mussinelli (2007–2008); "Hybrid modular architecture for emerging housing behaviours" PhD research, supervisor Elena Mussinelli, tutor Andrea Tartaglia (2015–2019); "Vivere e abitare l'università. Bilancio nazionale sulla residenzialità universitaria—Living in the university. National analysis on university residency" conference, scientific coordination Oscar Bellini and Matteo Gambaro (2019).

2 The Scenario for the Experimentation

The *Fondazione Collegio delle Università Milanesi* is a non-profit institution supported by seven universities and important public and private bodies of the city of Milan. The program of the *Fondazione* includes the provision and the management of student residence for temporary housing in a highly multicultural context, starting from the enhancement of an interdisciplinary and international method: a concept of "social intelligence" to promote life skills as well as cognitive supports such as extra-curricular course credits, which are the basis of this new educational approach. The current headquarters of the *Collegio* is in the south-west part of Milan, more precisely in a building designed by Marco Zanuso in the Seventies. This is site of great interest from both an architectural point of view—with a remarkable example of organic architecture promoted by *Cariplo* for hosting a center for financial aid to African countries—and an environmental point of view—with the presence of a high-quality garden that surrounds the area. The *Collegio*, with reference to the undergoing programs of expansion and consolidation of the campus, as well as to new development projects relating to accommodation, such as the Expo area, deals, in terms of scientific research, with the topic of university residence and of collegiality, with reference to the change of needs and to new cultural models and lifestyles. The activities of the *Fondazione* are aimed at the dissemination and at the promotion of college life, at the enhancement of the culture of merit, the internationalization of the university system and the integration of local realities. Through the study of the dynamics relating to temporary residency within multicultural contexts, it is also proposed as an incentive lever for social mobility and active citizenship.

In line with this approach, in 2008 it promoted an invitational competition for a first expansion of 53 new residential units. The winner project was by the Piuarch Studio, a choice that confirms that the *Fondazione* was well aware of the legacy of Marco Zanuso's architecture (Nannerini 1974). The work is financed with the III tender of the 338/2000 law, started in 2016 and completed in 2019.

In April 2016, the *Fondazione*, before the publication of the IV tender of the law 338/2000, promoted a "Competition for the preparation of a preliminary project for the construction of a new building for the *Collegio di Milano*" for the second expansion, which saw a large participation of architects and engineers. The theme of the competition was the construction of a new autonomous building, with access from Via Ovada, to be used for accommodation for university students, with the relative common and service spaces. The competition was organized also with the aim of participating, with the winning project, in the selection procedure called by the Ministry of Education, University and Research (MIUR) of the IV three-year program of co-financing of student residences, in the framework of the law n. 338/2000. It also required the consideration, in addition to the dimensional and qualitative constraints defined by the framework of the 2011 decree, of the volumetric, typological, functional and technological characteristics of the work to be carried out. In order to meet the objectives of the promoter, the type of accommodation to be developed was that of a "hotel", with a corridor distribution system and preferably single rooms with private toilets. The collective residential services were to be concentrated in areas separated from the at least 50 rooms of the residents. In order to respect the qualitative standards and the functional program, the evaluation criteria for the identification of the winning project were indicated by the tender, summarized in: "aesthetic and functional aspects" with particular reference to landscape integration and the relationship with pre-existing buildings of the campus; "economic aspects" with particular reference to durability and the control of the construction and management costs; "general aspects" in compliance with the conditions of the tender and the use of advanced technological solutions.

The jury of the competition awarded the project of *Centro Studi TAT*, coordinated by Fabrizio Schiaffonati[4], stressing in the motivation the original approach related to environmental integration and the effectiveness of the techno-typological solution. The winning project, even if it was in continuity, due to its morphological and typological characteristics, with the two previous interventions, also coherently interpreting the Zanuso's legacy, to which it explicitly refered without any manneristic satisfaction, was able to implement and qualify the open public space both with enviromental and social values. It consisted of a linear building with a north-south orientation, articulated in two sections of different width, a double body and a triple body.

The expansion project, which has obtained already the funding by the MIUR, is based on principles of environmental compatibility, typological and functional optimization of spaces and maximization of maintainability, substitutability and durability of materials and technologies adopted, as suggested and derived also by the preliminary studies carried out by scientific consultants.

3 The Design Experimentation

The campus of the *Collegio di Milano* has an area of 22,400 square meters, of which only 4,000 are occupied by the original intervention dating back to the early Seventies designed by Marco Zanuso and the recent expansion by the Piuarch Studio, with a total capacity of around 170 university students. In 2016, the area was expanded with a new contiguous plot measuring 4,600 square meters, already allocated by PGT to university residence, which therefore led to a total area of 27,000 square meters, with a total capacity of 220 beds (Fig. 1).

[4]The winner group consists of CSTAT, and refers to architects Fabrizio Schiaffonati, Arturo Majocchi, Giovanni Castaldo, with Elena Mussinelli, Andrea Tartaglia and Matteo Gambaro as scientific consultants for the techno-typological aspects, technical innovation and environmental sustainability, and the collaboration of Roberto Castelli, Federico Cecere, Gregorio Chierici and Francesca Scrigna. The winner was also entitled for the development of the definitive and detailed design as well as for the related commitments for the obtainement of the authorizations and permits for the construction (Tartaglia 2018), and for the definitive and detailed design phases the team included also BCMA, Broggini and Carrera Studio for the structural design and Casassa and Cigliutti Studio for the systems design.

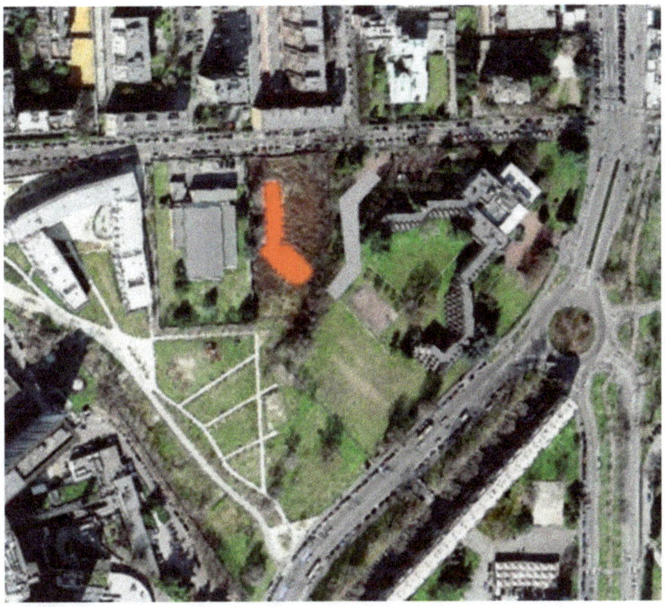

Fig. 1.1 Aerial photograph of the campus with the first expansion highlighted in gray and the second in red

Environmental Compatibility

The intention of the project was to combine the criteria of high architectural quality with the environmental and economic sustainability of the intervention. An approach that takes into account the constraints of the context, distances from the existing buildings, heights of the surrounding buildings and orientations (Schiaffonati et al. 2011). Also enhancing the potential of the campus characterized by a remarkable amount of greenery, including a sports area, with free pedestrian and scenic paths that confirm the attention for the landscape paid by the designers of the previous interventions.

The new volume originates from the observation of the overall texture of the context, the matrix of which is represented by Zanuso's organic plant building, which Piuarch took into consideration in the morphology and alignments of their intervention, and which the new extension confirms completing a coherent articulation of the entire building complex. So the basis of the project concept is the role that the new building will play in completing this urban environment. The Zanuso building, consisting of two arms connected by a central nucleus, defines a "C" shape open towards south-west. The body of the building designed by Piuarch represents a further arm, which, with the building proposed by CSTAT completes the plant by enclosing a new space into a new "C" facing north. This morphological and functional recomposition defines the succession of two "C"s, one open towards the south (building by Zanuso) and the other towards the north (Piuarch building and the new addition) (Fig. 2).

Therefore, the project is configured as a building divided into two parts, the first with a double body in a north-south orientation with one façade facing via Ovada and the second, with a triple body, rotated of about 30° towards south-east. This choice reflects the objective of minimizing distribution spaces, particularly in the triple body, the surface to volume ratio of which is also particularly efficient from

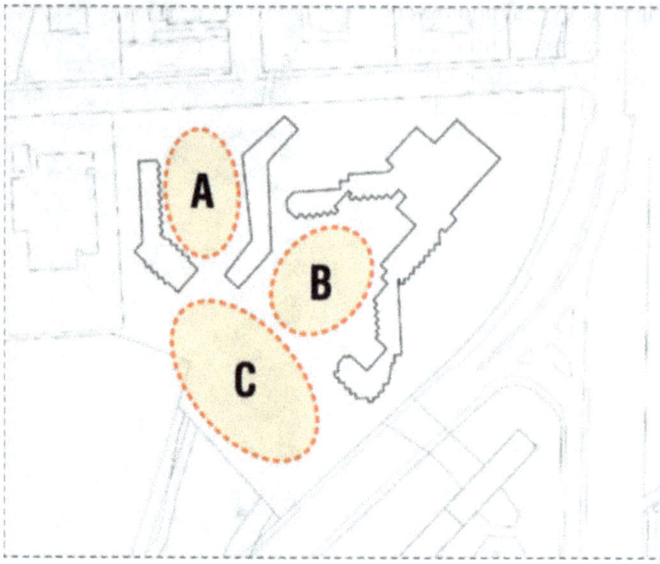

Fig. 1.2 General plan with the identification of the three open spaces defined by the buildings

an energy point of view. The building consists of 4 floors above ground: the ground floor, with shared access and services, and three residential floors with some services. The total surface area is 1,927 m^2, with a gross volume of 5,781 m^3. There are 51 rooms, including 3 for disabled use, and all the required additional services.

The environmental compatibility is the result of planivolumetric, morphological, typological, system and technological choices, as well as of alignments in terms of orientation and optimization of the sunlight (Schiaffonati et al. 2015). The soil consumption is limited, with a small footprint of the building that maximizes the permeable surface. Even the open spaces, with prevalent lawn portions, confirm the objective of limiting the environmental impact of the intervention in terms of hydraulic invariance and permeabilty. The green described above helps to mitigate and compensate for the intervention.

The open space facing the building is configured as a new square, partly paved and partly green, which visually and functionally connects the different buildings; this space is characterized by the presence of plants and trees, with the function of mitigating and increasing the environmental quality, as well as by the presence of chairs. The compact building shows on the facades the regular rhythm of the windows of the rooms and of some wider openings for common services. A number of volumetric additions and subtractions aim to maximize the energy performance of the building, as well as to express spatial relations with the context and to characterize the building in terms of recognizability. In this sense, the arcade on the ground floor operates as a covered connection between the new pedestrian entrance of via Ovada and the atrium of the building (Fig. 3).

Building Typology and Distribution Characters

The "hotel" building typology distributes the (single) rooms partly along the double body and partly (24 rooms) along the triple body with a central corridor. This

Fig. 1.3 Perspective view from the south of the new building

"hybrid" distribution system aims to optimize the corridor space, to enhance the orientation and to harmonize the new building within the pre-existing morphological and environmental context. Thus, the new volume seeks to dialogue with existing and under-construction buildings through its alignment and dimensions. The articulation of the distribution system also offers optimized views to the rooms: the corridor is in fact located on the west side in the double-body portion and centrally in the triple-body portion. The correct orientation of the rooms contributes to increase the quality of the spaces of the residence, as well as the overall energy efficiency of the building.

The distributional rigor and the optimization of the relationship between served/servant spaces are also sought at the accommodation scale. Each room has an area of 17.9 m^2, including a bathroom of 3.9 m^2. All the rooms have a 4 m^2 balcony and an entrance that serves the bathroom and the room. The arrangement of the furnishings, even if with a certain degree of flexibility, is designed to guarantee high levels of rationality and usability of the spaces. The wide windows of each apartment allow for a correct solar gain and the visual fruition of the context. The window is smooth, packable on one side, with a maximum opening of more than 2.40 m. When fully open, the balcony becomes an extension of the interior space. On the privileged fronts for sun exposure (east and south-west) there are 39 rooms, only 12 facing north-east (Fig. 4).

The residential spaces are completed by the services prescribed by the National tender and by the decrees. The environmental units for services envisaged by the project are: cultural and educational services (study rooms, multi-purpose spaces

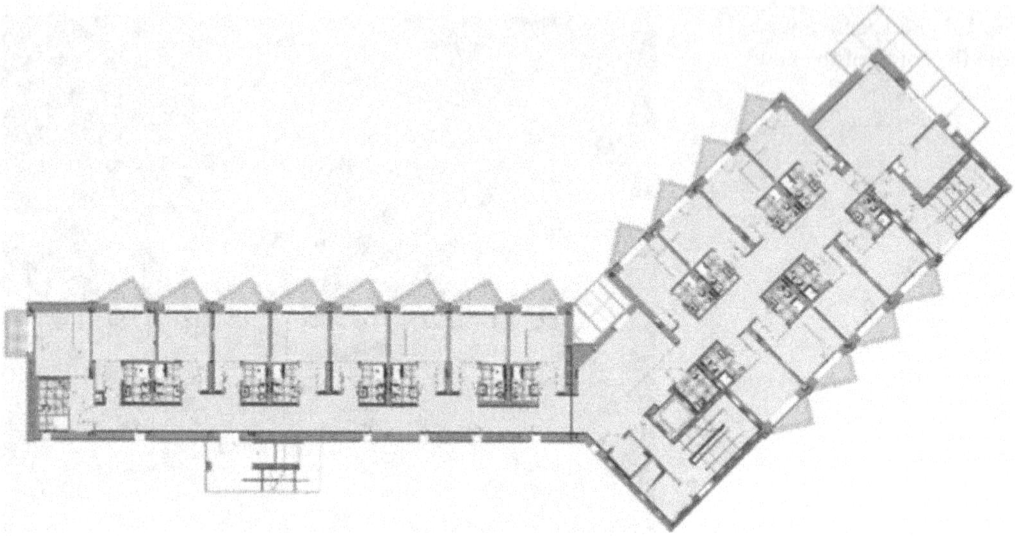

Fig. 1.4 Type floor plan of the new building

for conferences and educational activities: 265.9 m^2); recreational services (multipurpose spaces, lounge spaces: 96.5 m^2); support, management and administrative services (laundry, warehouses, office space: 118 m^2); access and distribution functions.

Material Alternatives, Maintainability, Substitutability, Durability of Materials and Construction Technologies

The material simplicity and the volumetric rigor of the facades are a salient feature of the image of the new building, also to ensure a high degree of maintainability, durability, substitutability of the various components. The external envelope is characterized by the provision of a ventilated façade, of which the last layer is made up of large vertical GRC panels, which, in addition to guaranteeing adequate energy performance, are optimal for conservation and maintenance issues. The large translucent vertical closures of the rooms and of the common spaces provide for the adoption of doors and windows with profiles with a thermal break.

Another key-element of the facades is the triangular-shaped balconies of the rooms, which represent an expansion of the living space and an important view on the surrounding greenery. The "jagged" image of the balconies of the rooms is an explicit reference to the architecture of Marco Zanuso. The parapet of these balconies is partially opaque, realized with a GRC panel and partially transparent grate of metal rods.

On the west elevation there is an external safety staircase designed with a central reinforced concrete core that supports cantilevered ramps, stairs and horizontal connections, with transparent metal parapets.

Overall, material choices have been made in harmony with the main colors identifiable in the surroundings, without any sophisticated contrast. The elevated structure is in reinforced concrete. In the competition, the proposal included also the possibility of using concrete blocks made with aggregates, produced with the use of the waste from recycled glass processing: solutions developed by the research "Ethical concrete" which also saw in 2015 the participation of a number of researchers from the ABC Department (Tartaglia et al. 2016).

Standards and Laws Legge 14 novembre 2000, n. 338 Disposizioni in materia di alloggi e residenze per studenti universitari.

D.M. 28 novembre 2016, n. 936 Standard minimi dimensionali e qualitativi e linee guida relative ai parametri tecnici ed economici concernenti la realizzazione di alloggi e residenze per studenti universitari di cui alla Legge 14 novembre 2000, n. 338.

D.M. 7 febbraio 2011 n. 27 Standard minimi dimensionali e qualitativi e linee guida relative ai parametri tecnici ed economici concernenti la realizzazione di alloggi e residenze per studenti universitari di cui alla Legge 14 novembre 2000, n. 338.

References

Del Nord R (2014) Il processo attuativo del piano nazionale degli interventi per la realizzazione di residenze universitarie. Edifir

Nannerini G (1974) Un centro per l'assistenza finanziaria ai paesi africani a Milano. L'industria delle costruzioni 41:3–8

Tartaglia A (2018) Progetto e nuovo codice dei contratti. Maggioli Editori

Tartaglia A, Terenzi B, Ubertazzi A, Cecconi R, Ronchetti A (2016) Ethic concrete. Environmental impact reduction and enhancement of mechanical and thermal performances of building components in concrete re-using waste. Italian Concrete Days 2016:212–220

Schiaffonati F, Mussinelli E, Gambaro M (2011) La Tecnologia dell'architettura per la progettazione ambientale. Techne, Journal of Technology for Architecture and Environment 1:48–52

Schiaffonati F, Mussinelli E, Majocchi A, Tartaglia A, Riva R, Gambaro M (2015) Tecnologia Architettura Territorio. Studi ricerche progetti. Maggioli Editore

Education as Reconstruction. School Typology in Post-earthquake Reconstruction in Central Italy

Enrico Bordogna and Tommaso Brighenti

Abstract The text investigates the role of education and school typology in reconstruction strategies in the different settlement and socio-economic contexts of Central Italy affected by the 2016 earthquake.

Keywords Earthquake reconstruction · School typology · Norcia · Amatrice · Camerino

In keeping with the long case history of earthquakes in Italy (Messina 1908; Belice 1968; Friuli 1976; L'Aquila 2009; Romagna 2014), the earthquake that struck the territories of Central Italy in the Summer–Autumn of 2016 caused damages on many different fronts: to monuments, to the urban residential fabric, to scattered private buildings, to production facilities, to the school and services systems, to the road and access networks, and so on.

In all these areas, beyond the necessary distinction between the time of the emergency, to be tackled with rapid reversible interventions, and that of the reconstruction, which, on the contrary, requires thoughtful and prospectively stable interventions aimed at restoring and relaunching the form and life of the cities and territories affected, the reconstruction strategy can be, and has been, variable.

If, for example, in the case of the Friuli earthquake, the reconstruction tended to favour precise individual interventions in order to revive the production activity and the reconstitution of a residential fabric certainly not "as it was and where it was", but conforming to the morphology and the characteristics existing prior to the earthquake; and if in the case of the Belice earthquake, the appraisal can only be varied but was, on the whole, positive, balanced between the extremes of the "foundation

E. Bordogna (✉) · T. Brighenti
Architecture, Built Environment and Construction Engineering—ABC Department, Politecnico di Milano, Milan, Italy
e-mail: enrico.bordogna@polimi.it

S. Della Torre et al. (eds.), *Buildings for Education*, Research for Development,
https://doi.org/10.1007/978-3-030-33687-5_12

city" of Gibellina Nuova (with Burri's wonderful invention of "The Great Cretto" in memory of the historical Gibellina) and more circumscribed interventions but of definite quality like those in Salemi; unquestionably less convincing, or explicitly negative, is the experience of the so-called "new towns", decentralized and of poor architectural quality, built after the earthquake in L'Aquila.

In the case of the earthquake in Central Italy in 2016, restricting the field to the municipalities of Norcia, Amatrice and Camerino, the diagnostic analysis and the reconstruction strategies were necessarily diversified.[1] If, in Norcia, apart from the monumental buildings of the centre for which a philological restoration is anticipated, it would appear that the most urgent sector for reactivation and relaunching (with some similarities to the case of Friuli) is the widespread fabric of small commercial and production units linked to the agri-food sector; if in Amatrice, so dramatically affected that there is practically nothing left of the ancient nucleus, the most urgent exigency appears to be the reconstitution of a new urban centre consisting of civic services and residences, to complete an episode of valuable morphological definition and architectural buildings like the one carried out by Arnaldo Foschini between the 1930's and 1960's (a unitary complex with an orphanage, hospice and separate church); if in Camerino it seems that any intervention cannot neglect the important university and cultural structures present; in all three of the contexts examined, on the contrary, beyond these differentiated situations, the system of basic and secondary education has been hard hit, thus becoming a common priority field for reconstruction.

1 Norcia. A Campus for Basic Education and Sports Facilities as a Part of the City

At the end of 2016, around 800 pupils, from nursery to secondary school, found themselves deprived of the opportunity to take advantage of their school buildings which had been destroyed or seriously damaged by the seismic shocks. The existing school system had an arrangement that was markedly bipolar: one school complex located immediately beneath the ancient city walls, just outside the main gate from the territory to the Old Town, given over to an elementary school, a junior secondary, and a comprehensive series of sports facilities; a second complex, further north, in a valley just outside the walls, was entirely dedicated to secondary education (Classical, Scientific, and a Technical Institute for Surveyors).

[1]The essay being published here refers to research work executed, starting from the month of September 2016, within the Architectural Design Workshop of the Master Course of the School of Architecture, Urban Planning, Engineering of constructions (AUIC) of the Politecnico di Milano, in the academic years 2016–2017, 2017–2018, 2018–2019 (Professors Enrico Bordogna, Tommaso Brighenti, Vito Maria Finzi (Technology), Mauro Madeddu (Structures); with the collaboration of the newly graduated architects Marco Pinna, Silvia Faravelli, Marco Frisinghelli, and Nicolas Decima.

Both complexes, while in the current post-earthquake state of compromise, are characterized by a plan that is approximately that of a "campus", with the individual school buildings for different levels and subjects interrelated and connected to the sports facilities and the surrounding green areas, to form structures that are morphologically unitary and integrated.

In particular, the lower schools, close to the main gate in the city walls, are characterized by the elementary school building, a typical C-shaped structure with two storeys from the late 1950's, aligned with the main thoroughfare to access the city from the surrounding territory with classrooms arranged in series along corridors overlooking the inner courtyard. A typology that was frequent at the time, evidently derived from pre-war school building, but featuring a clear layout and dignified architectural forms, with a pitched roof, pale plastered façades, and a regular and uniform pattern of openings.

Continuing upwards from one wing of the C-shaped building is another structure with three storeys above ground, and typical forms of the 1960s in unclad reinforced concrete, infill walls in brick and a gable roof, which houses the junior secondary school. In the open space delimited by these two buildings, in anonymous rectangular structures with only one storey above ground, a gym and other service spaces in a prefab are housed, while all around, in a richly wooded green environment characterized by repeated variations in height, are a large range of sports facilities open to the citizens in addition to school use, comprising a football pitch, tennis courts, and outdoor swimming pools with associated locker rooms.

All in all, a small, well-organized "campus for education and sport", probably built up over time without an explicit original design, but of undoubted quality and with its own identity, which the 2016 earthquake seriously maimed but did not completely destroy, although it did leave the elementary and junior secondary school buildings unusable.

With this scenario, the choice of this project was to confirm the layout of the existing campus with even greater force and a wealth of facilities. To this end, the elementary school building was restored, its plan and measurements being taken as the generating element of the proposed morphology, while the junior secondary school building was replaced since it was considered no longer recoverable, as were the small anonymous pavilions for the gym and related services (Figs. 1–2–3).

The two arms of the C-shaped elementary school building were extended with two in-line blocks interconnected by a path covered by a portico, the latter also acting as a retaining wall while delimiting the internal open space by creating a court in part left green, more reserved but intentionally permeable, so as to also function as a small urban square. Within this court, a small outdoor wooden theatre can accommodate dramatic and musical performances by the students, and other activities and events of the entire school complex and the community. In the wider in-line structure is the gym and a swimming pool, while in the narrower structure two storeys above ground house the junior secondary school. On the side of the C-shaped building, facing south, lying along the internal pedestrian axis which crosses the whole of the campus, is a small school for infants including a nursery school and a crèche, with a

Fig. 1–2–3 Norcia: aerial photogrammetry with inserted project; plan; axonometry of the exhibition center and covered market (Students: V. Boffo, A. Bugatti, A. Sposetti, 2018)

square ground plan arranged around a small courtyard overlooked by the classrooms with communal outdoor spaces for the children to play.

A possible variant of the project indulges in greater freedom than the existing situation, confirming the elementary school building but introducing above it an entrance plaza of an urban value, rhomboidal in shape, surrounded by three new buildings for the nursery school, junior secondary school and gym, in turn the object of more accentuated typological and expressive research. Also the sports facilities, occupying part of the football field area moved not far away, are more consolidated,

With more substantial indoor and outdoor amenities (gym, swimming pool, basketball, volleyball and tennis courts), with a stand, locker rooms, and other service areas for the public.

The principle common to both the solutions proposed is however the confirmation of the campus' layout, using the sloping orography of the site to make the area of the intervention visually attractive and more permeable since, being situated immediately outside the historical walls, it acts as a filter between the ancient compact fabric of the centre and the charming surrounding natural landscape which characterizes the Municipality of Norcia and its countryside. In this sense, in both solutions, the design brief, rather than investigating the internal typological configuration of the various school buildings, attempts to lay out a part of the city which is specifically intended for functions of education, sports and leisure, combined, beyond the historical access route from the territory, with the proposal of an intervention intended as a trade fair area (exhibition halls with a wooden structure, arranged in a grid and with square pyramidal pitched roofs) and an in-line block fronting the street, opposite the elementary school, intended for shops and market structures for the traditional local agri-food products (Figs. 4–5, 6, 7–8–9, 10–11–12).

2 Amatrice. Hotelier Institute and Cooking School for a New Urban Centrality

In Amatrice, the 2016 earthquake completely destroyed the Old Town (Fig. 13): all that currently remains is the central road axis which diagonally structured this typical ridge settlement from gate to gate, from west to east, regrettably now surrounded by flattened areas of rubble clearance that make the original morphology of the ancient centre unreadable. On the contrary, on the same axis to the west just outside the centre, we can still find, entirely recognizable despite the substantial damage suffered, the urban area created by Arnaldo Foschini[2] between the 1930's and 1960's, a unitary complex with an orphanage, hospice and separate church, of a clear morphological

[2]In the early Twenties, Arnaldo Foschini created one of his first works for Amatrice, the Institute for War Orphans, built between 1921 and 1923 for the Opera Nazionale per il Mezzogiorno d'Italia, a building for 150 children and five other minor buildings for schools of art and crafts. At the same time, he was developing his project for the church, revised in 1938, interrupted because of the war and then brought to fulfilment in various phases until it was finally finished in 1961, enriched by numerous works of art, such as the large bas-relief in travertine of the façade. In the same years, also the large building of the hospice was completed, whose U-shaped plan and three storeys concluded the entire ground plan. See Pirazzoli (1979), pp. 86–89.

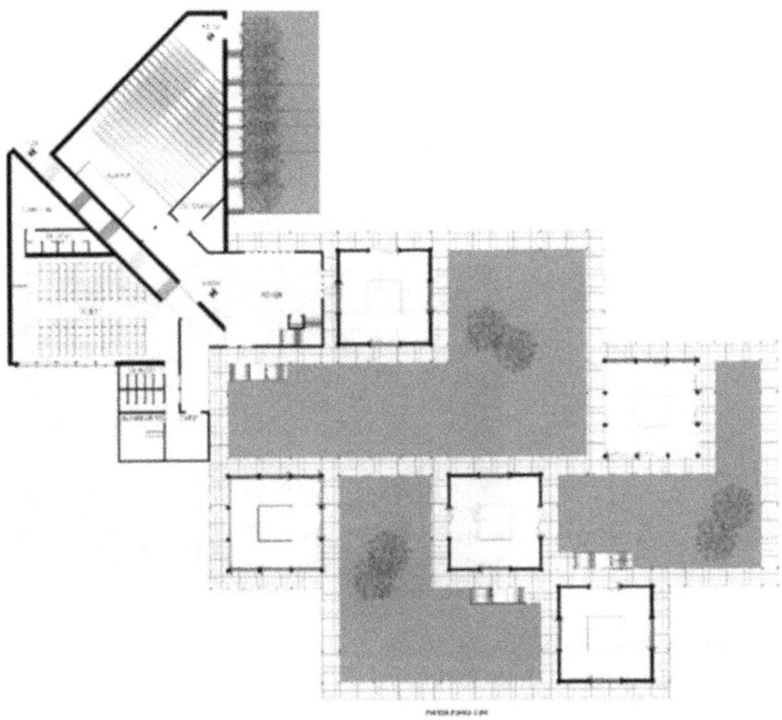

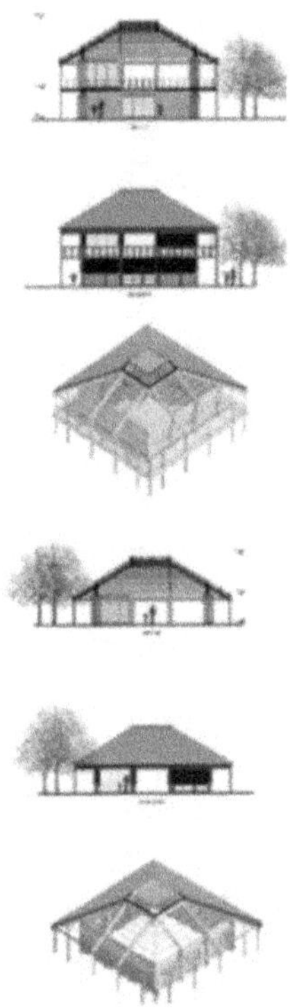

Fig. 4–5 Norcia: Trade fair plan; sections, elevations and axonometry of the pavilions (Students: V. Boffo, A. Bugatti, A. Sposetti 2018)

Fig. 6 Norcia: plans and elevations of the covered market (Students: V. Boffo, A. Bugatti, A. Sposetti 2018)

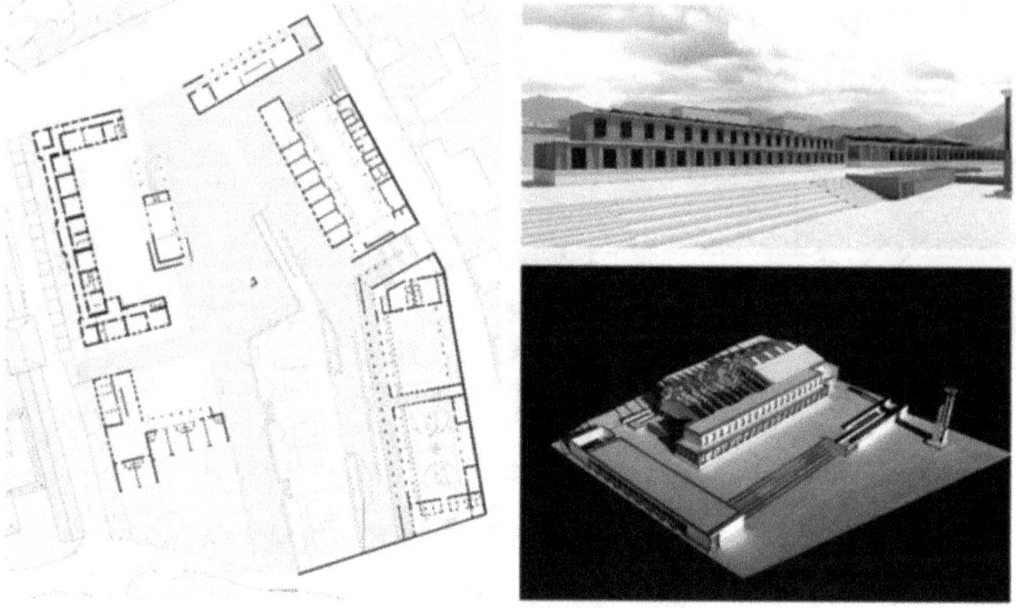

Fig. 7–8–9 Norcia: plan of the school campus; perspective view of the kindergarden; photo of the model (Students: M. Colombo, P. Escoriza, M. Iotti 2019)

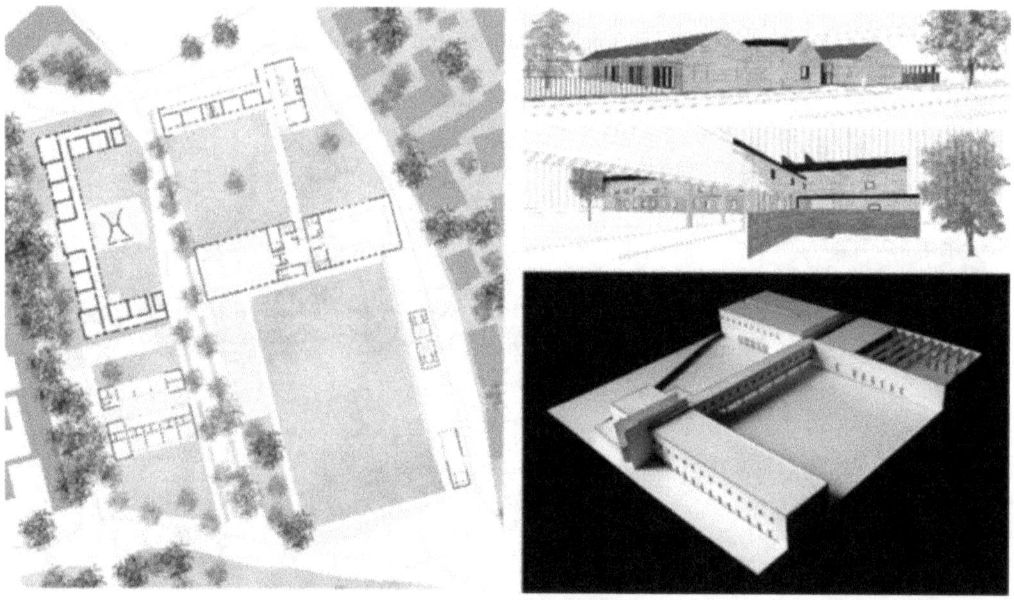

Fig. 10–11–12 Norcia: plan of the school complex; perspective view of the asylum and school building; model photo (Students: S. Angeli, S. Angrilli 2019)

Fig. 13–14–15–16–17 Amatrice: view of the municipality after the 2016 earthquake; aerial photogrammetry with the design of public buildings to complete the urban section created by Arnaldo Foschini; general project axonometry; axonometric cutaway of the covered market; elevations and axonometric cutaway of the "anatomical" theater of the cooking school (Students: L. Bongiolatti, A. Giamboni, C. Landoni 2018)

definition and a sober expressive quality, to which a series of other facilities including a football pitch and an indoor gym was subsequently added lower down, on an orography characterized by significant changes in height. Instead, between the old nucleus and Foschini's complex, the buildings of an elementary school and a hotel school that was important for the economy of Amatrice were entirely destroyed. The hotel school, which was attended by around one hundred and thirty students, mostly from outside the town and the province, has been temporarily transferred to Rieti after the earthquake. In the same way, in a contiguous area, four simple pavilions arranged parallel, with one storey above ground, which once served as barracks, were totally destroyed.

In this scenario, by resorting to certain Muratori-style antecedents, in particular the square of the Cortoghiana workers' centre in the mining district of south-west Sardinia, the project has focused on configuring a new urban centrality, arranging,

in correspondence with Foschini's buildings now restored and given new destinations (a municipal seat in the former orphanage, healthcare services and special residences for the elderly, students, and young couples in the former hospice), two orthogonal squares arranged as an "L", with, on the one side, an in-line block with two and three floors for council housing overlooking the piazza sloping down towards the valley, a portico on the ground floor and continuous eaves, and, on the opposite side, near the area of the former barracks, the hotel and cookery school, as an ideal extension and completion of Foschini's scheme. Meanwhile, on the side towards the valley overlooking the former orphanage, there is a double in-line block for commercial use, with below it a stand and changing rooms for the existing sports field. In Amatrice, as in Norcia, in the project (and also in the successive variations) what has prevailed is the attention to the urban dimension of the intervention rather than a specific typological investigation of school building. Arguably, the only exception is the cookery school. In this case, the plan of the former barracks pavilions was reused, with teaching spaces and external roofed environments for services and a restaurant, but for the specific needs of a cookery school a new building was inserted, in front of Foschini's church but isolated from it, and with a central plan that incorporates the evocative typology of the anatomical theatre, here reconverted for culinary education (Figs. 14–15–16–17).

3 Camerino: Rescue, Recovery and Restoration of Works of Art Affected by the Earthquake

In Camerino, the project sought to address a single theme, but equally important, that of rescuing, recovering and restoring works of art affected by the earthquake, also in consideration of the presence at the local university of a course of studies in Cultural Heritage with which to establish useful synergies of mutual exchange, involving the students in the laboratory work and internships for the restoration of works of art or research into local art.

This has resulted in an original typology, conditioned on the one hand by the particular characteristics of the context and designed on the other hand to match educational and museum/laboratory needs.

The choice was to intervene outside the Old Town, today only partially accessible, consolidating with the new addition, a complex consisting of buildings for a university residence and a departmental library created at the beginning of the 2000s by the architect Raffaele Mennella.[3]

On a gentle slope, north of the Old Town, the project envisages three linear blocks opening into a trident, which climb the contours of the hilly terrain in continuity with the existing university residences. The enclosed green areas that link the various parts of the intervention in a system of public and collective spaces, designed to enhance the characteristics of the site, overlook the landscape of the Esino Valley, towards the Primo and San Vicino mountains.

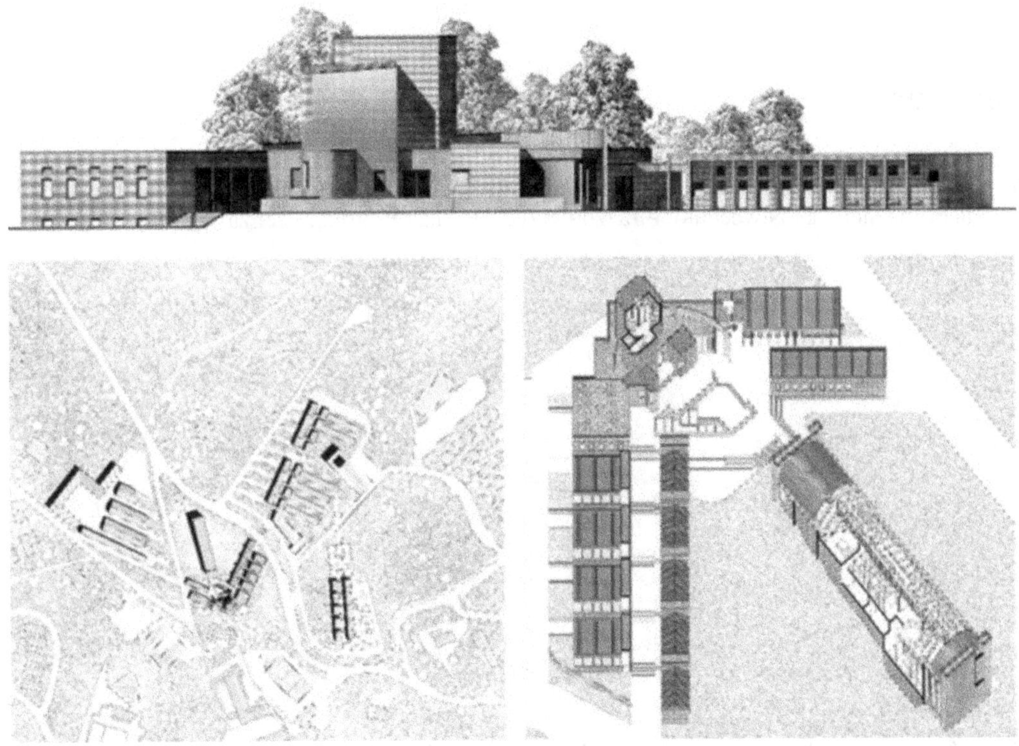

Fig. 18–19–20 Camerino: Center for the recovery and restoration of cultural assets affected by the earthquake, internal elevation, plan, axonometric cutway (Students: S. Faravelli, M. Frisinghelli, Graduation 2019)

[3]The project area is located at the end of Via Madonna delle Carceri, in the north of Camerino. This trajectory arises inside the centre from the main street that runs through the historical settlement and which, near the Museum of San Domenico, splits in two to define the main axes of the town's suburbs. Past the university's science centre, the Church of Madonna delle Carceri and the commercial area of the supermarket, the trapezoidal-shaped zone chosen lies strategically at the end of the sequence of statements positioned along this trajectory.

The three buildings have separate destinations. The first wing, in continuity with the existing university residences and in turn split, houses standard university functions: teaching, research, work spaces for students; the central wing is given over specifically to a museum, with rooms for storage, restoration workshops, exhibition spaces for works salvaged from the territory; the third wing, of smaller dimensions and facing towards the recently constructed shopping mall, houses the service structures, with spaces for local associations, offices, a projection room, and a hostel.

The three in-line buildings converge in a sort of slab that is articulated in plan and elevation, whose underground floor is intended for the deposit of works of art awaiting restoration, while the roof is a public square at the service of the entire complex. At the summit of the slab is a tower with a composite disposition of the masses, containing a specialist library and complementary service spaces, acting as a formal hub of the whole design, configuring a systematically concluded intervention that is centroidal with respect to the existing university structures (Figs. 18–19–20).

Reference

AA.VV (1979) Atti del Convegno. Arnaldo Foschini. Didattica e gestione dell'architettura in Italia nella prima metà del Novecento, Pirazzoli N (ed), Faenza Editrice, Faenza

Design for Schools

Domenico Chizzoniti, Luca Monica, Tomaso Monestiroli, Raffaella Neri and Laura Anna Pezzetti

Abstract Some competition projects concerning the construction of new schools are collected in this article. These projects deal with different themes depending on the level of education: high schools and secondary schools are generally integrated with collective services open to the city etc.—open to the city to seat new important urban centers. On the other way round, primary schools and kindergartens are generally more introverted buildings, to protect the safety of children, so as the organization of space is more linked to educational patterns. In all cases, the schools are buildings that include open spaces, courtyards or gardens, which relate to the features of the different sites.

Keywords New schools · Education · Urban role · Social role

1 Introduction

Schools are civil institutions and, as such, can become urban landmarks facilitating the construction of new public places. This role determines the order of problems facing the school project: first, it is necessary to clarify the relationship with the places where it is located and with the other elements of the city. Architecture must always be considered as a tool for building places, the features of which, in the case of schools, vary considerably depending on their rank, and therefore their possible opening up to the city.

In this sense, the typological question becomes decisive: the school is a building having common spaces and spaces reserved for smaller groups, with classes which we could say are more private, just like a home, a collective home. It is therefore essential to define the central place which identifies it, its common space, one which will most likely establish a privileged relationship with its external urban spaces; and then the relationship of the collective space with the classrooms, its internal organisation, which is to be considered in terms of the educational system.

D. Chizzoniti · L. Monica · T. Monestiroli (✉) · R. Neri · L. A. Pezzetti
Architecture, Built Environment and Construction Engineering—ABC Department, Politecnico di Milano, Milan, Italy
e-mail: tomaso.mnestiroli@polimi.it

S. Della Torre et al. (eds.), *Buildings for Education*, Research for Development,
https://doi.org/10.1007/978-3-030-33687-5_13

The last but crucial issue relates to the construction methods. These must make these relationships tangible, if one considers that the building is not only a technical element but also the means of expression of its architecture, through which to express the building's identity, the very idea of the school and the quality of the places that it defines.

2 Primary School and Lower Secondary School in the Muncipality of Monreale (PA) 2016[1]

@Scuole Innovative—National Competition, 1st Prize

The area allocated to the new school district in Monreale is situated south of the historic centre. It is a rugged area fully overlooking the natural landscape of the Oreto River plain and the Gulf of Palermo. These features make the area an impressive place which can be further enhanced by the settlement of a public building whose value goes beyond its own use but benefits from the environmental quality of the place where it stands.

The specification in the competition notice regarding the need to open the school to the community was the starting point of the whole project for us.

The school building becomes one of the urban centres of reference for the citizens of Monreale: the auditorium, gymnasium, educational gardens, the Stoa, the library, theatre spaces, the square and the cloister are all places open to the city independently from the school's operation.

The school's new square overlooks the edge of the historic centre and the impressive monumental complex of the cathedral.

The typological choice underlying the project is the construction of a "Stoa" facing the landscape, which is the distribution axis of the internal school complex. The individual elements of the school are connected to the Stoa: to the north, in direct relation to the city and creating the new street facade is the Auditorium, the block of workshops open to the city, the entrance hall and the gymnasium; to the south, in correspondence with the entrance hall, lies the large teaching cloister around which the primary and secondary school classrooms and laboratories are organised.

The idea is to build the school as a communal place of integration between cultures and thus, in all respects, like a small town. A structure that is built starting from the relationship between public places and private places, streets and squares and covered spaces and open spaces, all in direct contact with nature and the landscape.

Thus, the Cloister, which all classrooms and laboratories overlook and is located at a lower level to the entrance, becomes the student's square, an Agora of culture where collective activities can be shared and fun and recreational activities held (Fig. 1).

[1]Team work: Tomaso Monestiroli, Luca Cardani, Riccardo Nana, Giovanni Uboldi.

Fig. 1 General view and general plan

3 Innovative School Complex for Monreale, Palermo, 2016[2]

@Scuole Innovative—National Competition

School Building represents a fundamental architectural resource to set up a place around which the civic dimension of the community can gather. An essential element for the design of Monreale's "Innovative School" (2017) is the introduction of public spaces (such as a forecourt and a accessible roof) as founding element of the layout in an area otherwise isolated by the main settlement. The design is also assumed as a component to turn the criticality of the steep slope into an opportunity to enhance the landscape character through the association of the section and spatial configuration to the diagram of activities, reconnecting also with the lower valley settlement. The innovative characters of the learning spaces are not limited to the introduction of spatial flexibility, 3.0 devices and furnishings, but integrate them in a holistic reflection on the capacity of the architectural space to be educator itself. Reacting to the concept of innovation as un unstructured learning environment, the design learned from the Monreale's Cathedral to define a clear prototype for the complex programme, centred on a nave and two circular "cloisters" and based on additional recognisable "space-places" and a plurality of "centres of attention". Those are attractors of multiple activities for informal socialisation and learning and are equipped with various types of patios. Sustainability and bioclimatic principles have generated a logical coherence of technical, constructive and material choices, offering an eco-efficient performance of the building in all phases of the Life Cycle: recyclable materials, integration of renewables (80% of energy needs for heating with photovoltaic), passive systems of natural ventilation and control of visual comfort, spatial flexibility and acoustic comfort. The reduced structural masse of the wooden structural system, the box-like structure and the shapes intrinsically provide the building with an effective anti-seismic behaviour (Figs. 2, 3–4, 5, 6).

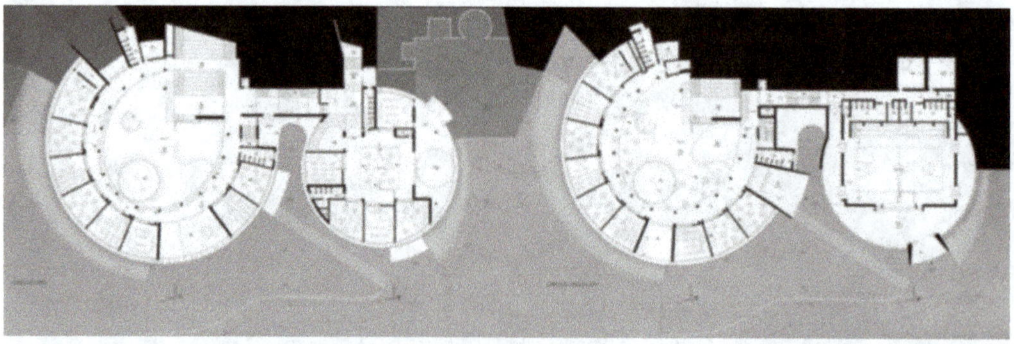

Fig. 2 Plans of the school complex (middle and primary school and gym) at levels +283 and + 259

[2]Team work: Laura Anna Pezzetti (architectural design); Carol Monticelli and Alessandra Zanelli (technology); Claudio Del Pero (energy); G. Piantato (structures).

Figs. 3–4 View of the middle school's roof terrace looking towards Palermo and the sea

Fig. 5 View of the middle school's "core centre" with the equipped ring-shaped space and the stepped aisle of the auditorium

Fig. 6 Site plans of the school complex at +266, 85 level (ground floor); Section, scheme of the bioclimatic operation (summer) and the passive cooling of the middle school building, chosen as strategies to reduce operational energy consumption

4 Middle School at Sorbolo, Parma, 2016[3]

@Scuole Innovative National Competition

The project for the new middle school at Sorbolo is part of the existing sports complex, acting as the pivot of a system of cardinal axes. However, it is also intended as part of a larger project on an urban scale that, with the addition of the Elementary School, will structure the entire school "campus": the future new educational center of a rural river consolidated landscape. The school's architecture is based on two main aspects: the geometry of distribution and spatial system; the timber construction technique and its aesthetics.

Considering the opportunities of opening in the whole day and for extracurricular activities, we wanted to propose a scheme that can be sectioned into two parts that can be used independently. An entrance area contains activities that are most oriented to the external use (atrium, conference room refectory, tower-library, extracurricular activities, teachers, laboratories). A second area contains only the classrooms for the concentrated activity. A third educational area is outdoors, with botanical flowerbeds and an earth embankment as theater (Fig. 7, 8, 9, 10).

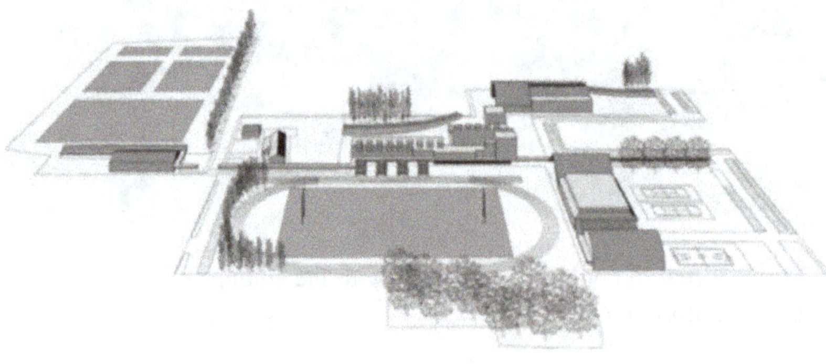

Fig. 7 General view from the east of the school complex

[3]Team work: Luca Monica, Elena Bonelli, Luca Bergamaschi.

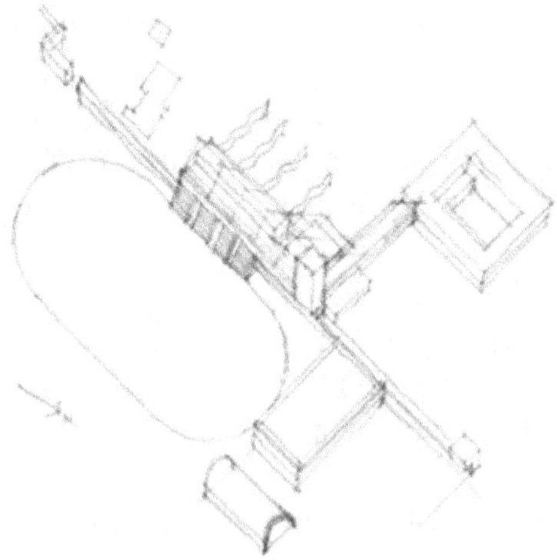

Fig. 8 Early drawing of the school complex

Fig. 9 View from the east, from the athletic field

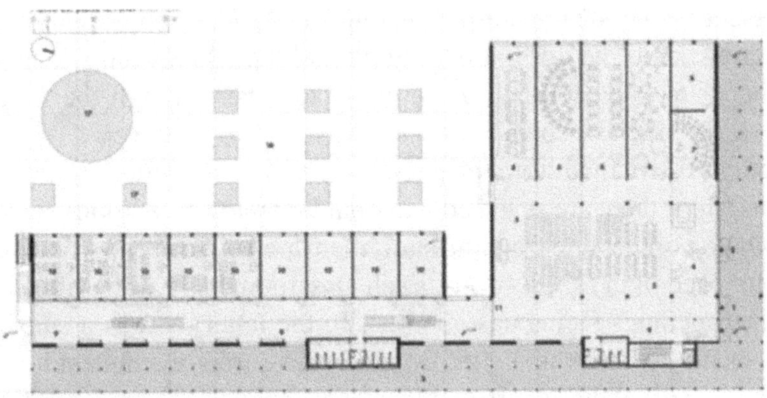

Fig. 10 Plan of the ground level

5 New Lower Secondary School in the Municipality of Casatenovo (LC), 2016[4]

The area allocated to the new secondary school in the Municipality of Casatenovo is located south of Via San Giacomo, a highly important thoroughfare parallel to the

[4]Team work: Tomaso Monestiroli, Luca Cardani, Riccardo Nana.

Roggia Nava canal, which feeds a large area of agricultural parks where historical and architectural villas are located.

It is a slightly elevated area which enjoys a beautiful landscape especially to the north, earmarked as a resource of great environmental value. These features make the area significant, and it can be further enhanced by the placement of a public building whose value goes beyond its own use, as the area also benefits from the environmental quality of the place where it stands.

We now know that the type of a building is defined by the relationship between its internal and external factors. External factors are related mainly to the characteristics of the place, the road network and the access system. Internal factors are related to the building's functionalities.

It is necessary to establish a harmonious relationship between these factors in order to create a suitable place for the activities that take place there, but at the same time to be able to embrace the character of the environment in which it stands, especially when, as in this case, it is a place of great beauty.

The choice at the core of the typological layout is to distribute all the parts of the school around a long gallery facing north-east with a continuous series of large openings facing the school's green area, partly intended for sporting activities and partly for a large garden that strengthens and gives shape to the many existing plants. The gallery is two stories high. Upstairs, a continuous balcony distributes the classrooms and looks out onto the landscape.

This gallery is the backbone of the entire typological layout; it is the distribution point of the building's two floors but also, and no less importantly, the meeting place of the entire school population. Along the gallery south-facing classrooms and laboratories lie, to the north are the gym and the auditorium which, located at a distance opposite to each other, are brought together through the school's large garden. The series of classrooms ends with the refectory to the east.

The functions distributed from the gallery are as shown in the competition notice. On the ground floor there are seven educational laboratories, a library and rooms for the administration offices and teaching staff; there are eighteen classrooms on the first floor, three for each of the six sections. Each classroom has a wall with a continuous series of large openings facing east or west. In the latter there is a sunshade and a mechanical shading system as shelter from the sun during the hottest hours.

The north-east side of the gallery contains: the ample-sized gym for different sports (volleyball, basketball, handball etc.) and other equipment for freestyle exercises. A flight of steps on the long side of the gym overlooks, in addition to the indoor and outdoor playgrounds, the garden, and at the bottom of the latter, the auditorium. The auditorium is one big square-plan classroom that distributes the audience across three sides with a wooden staircase, under which the necessary storage area is to be found.

The orderly succession of one singular element, a pillar with a circular section that is repeated every 2.60 m and runs along the entire perimeter of the building and the two collective buildings, the gym and the auditorium, has this dual role: to give a grand formal unity to the entire building and to allow for large distribution flexibility within (Figs. 11 and 12).

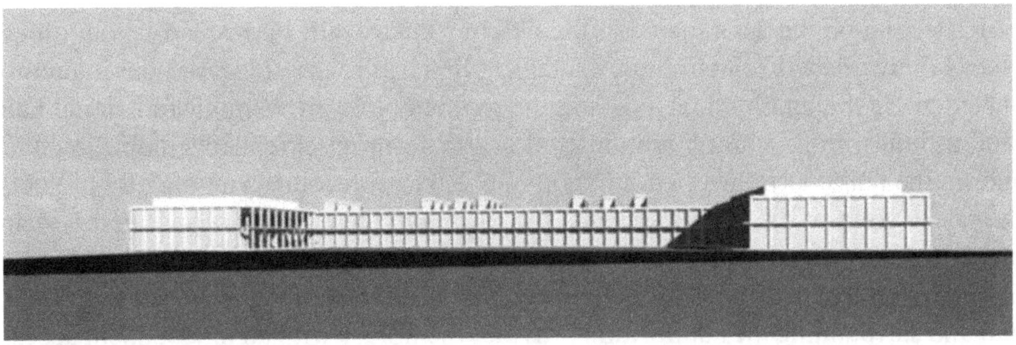

Fig. 11 View of the northern courtyard

Fig. 12 General plan

6 Design for Nearly Zero Energy Building (NZEB) Middle School at Casatenovo (LC), 2016[5]

The design project's layout responds to the purpose of "founding" a suburban-rural site, as well as to the complexity of school buildings acting as civic centres and to the challenge of integrating the issues related to 3.0 teaching innovations with the continuous reflection on learning architecture.

[5]Team work: Laura Anna Pezzetti (architectural design); Claudio Del Pero, Mario Maistrello (energy); G. Piantato (structures).

Reflecting on the asymmetry of the nearby Quattrovalli Farmstead'layout, three blocks on two levels (classrooms, canteen, gymnasium) are staggered but interconnected along the longitudinal axis and the receded block of the multifunctional hall (auditorium, bridge-library, administration), to form an extended green forecourt and, in the rear court, the open-air sports field. The forecourt symbolically projects the raised solid of the library towards the town and the existing school complex in order to design a comprehensive landscape—i.e. "the education field". The hall is a public multifunctional square where a peculiar system of ramp ways starting in the hall and surrounding the auditorium, provides the library with an independent access from the learning spaces. The peripatetic device meets the design for all, stimulating spatial experience for all users.

Through an interdisciplinary approach, the design achieves a logical coherence between formal, technical and material choices, with particular reference to the selection of eco-compatible materials and technologies and to the reduction of net primary energy demand, GHG emissions, and energy costs. The synergic interaction between technological solutions has obtained a Nearly Zero Energy and Emission Building (NZEB): good thermal inertia; favorable orientation, brise-soleil, adjustable sun screens; radiant floor emission assisted by an air handling system to control relative humidity; air-to-water electric heat pump; photovoltaic system on the roof (peak power of 80 kWp.) caverning equal to approximately the 70% of the energy needed for HVAC purposes; thermal storage. The dynamic energy modelling of the building was carried out using the (BEST-Polimi energy simulation tool based on the EnergyPlus engine) expected annual consumption of electricity for heating, cooling and DHW production is estimated at about 130,000 kWh, corresponding to 27 kWh/m2 (Figs. 13, 14 and 15).

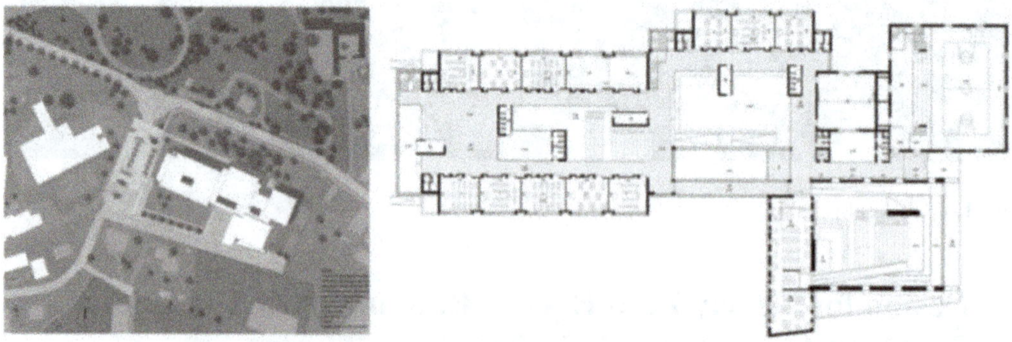

Fig. 13 Site plan and first floor plan

Fig. 14 View of the frontcourt, classrooms and library blocks from the entrance

Fig. 15 Interior views of the canteen and "knowledge deck", classroom nave and multifunctional hall

7 Competition to Design the "Primary and Secondary School" in Bolzano, 2018[6]

The big garden is the heart of the school which looks onto all of its spaces: the classrooms, study and play areas, laboratories, the canteen and collective areas. The garden, open to the landscape, is surrounded by a full-height pergola which gives unity to the different parts that make up the school, and it also protects its interiors from the sun.

Study areas are located on two floors in the two buildings facing each other across the garden, connected by a glass gallery. The canteen, which occupies the ground floor on the north-east side, looks out directly onto the garden.

The classrooms are surrounded by spaces for students' free activities. They are configured like luminous islands attachable through large openings, according to different configurations.

Collective activities included in the school, gym, auditorium-theatre and the library, are located along the north-west side of the garden, at the end of the axis that leads to the river. Each one is identified by a different volume which makes them recognisable. Connected between them and to the classrooms (albeit with independent entrances), they are higher in order to affirm their urban relevance and to appropriately conclude the long open view towards the river.

The courtyard in the middle of the project, a genuine "rain garden," has a strong micro-climatic impact.

The natural relationship with the courtyard, with its considerable pedagogic value, is accentuated by the presence of large, full-height glass facades which favour the use of natural ventilation during the summer season, also through the thermo-mitigating effects of the garden. The height-adjustable pergolas shield the facades from solar radiation.

The natural ventilation is supported by the direct air reaction which is activated in the classroom's cross-connection pass-through areas, interacting with them (Figs. 16 and 17).

Fig. 16 Raingarden courtyard view

[6]Team work: Tomaso Monestiroli, Raffaella Neri, Paolo Oliaro, Sergio Croce, Luca Cardani, Giovanni Uboldi.

Fig. 17 General plan

8 New Academy of Music in Krakow 2016[7]

The proximity of the Vistula River and the large green area in front of the project area make the place an extraordinary opportunity to build the new Academy of Music in a context dominated by nature. We believe that the issue of the relationship with nature is the main theme of our times and should influence the shape of the city and the architecture that determines it.

That is to say that the occasion of the competition is not only to design a school, but to define a growth hypothesis of the city which includes natural places. From the project area there is a view of the river, the woods that flank it and, farther away, the historic city of Krakow. An extraordinary site, in short, meant for public activities such as the school of music, a future large auditorium and various sporting activities in the green areas along the river.

The first choice made was to not place the new Academy of Music in only one building but to spread it across five separate buildings, aligned together and following the lope of the land facing west towards the future urban park and in the direction of Krakow old town. The relationship with the green areas is the foundation of the type of buildings that are arranged and oriented according to this general objective. The two buildings with classrooms are arranged at the centre of the linear system at the point where the project area changes direction. Even these buildings are adapted to

[7]Team work: Tomaso Monestiroli, Raffaella Neri, Luca Cardani, Maurizio Acito, M. Guazzotti, Giovanni Uboldi.

the land's rotation, creating a central place from which there are two wings, defined by the two aligned buildings that house the school areas and the premises of the rectorate, facing towards the landscape. Two small tower buildings will house 15 apartments required for guests and students.

The music school building located further north is the first to be reached by road. It is a linear building that is spread over three floors with the classrooms specified in the notice on both sides. To the west towards the park, there are four large classrooms for rehearsals, the orchestra, the lyric theatre, the choir and the organ. They are especially high classrooms measuring 16 m to ensure maximum acoustic efficiency, with an internal gallery measuring a height of 8 m, which allows for a steel external gallery to be reached, from which the landscape can be seen. Directly connected to the linear body of the school, where it is accessed from, these classrooms are enclosed in exposed brickwalls, respecting a building tradition that is highly palpable in the historic centre of Krakow.

The auditorium is the most important building of the entire facility and is located next to the library in the centre. In this way, a formal hierarchy is created between the two buildings open to the public and those intended for the school. The auditorium will also have a perimeter of pillars clad in exposed brickwork to emphasise its public purpose and its civic value. The interior of the building is a large rectangular-shape hall with a variable height appropriate to its acoustic performance where different areas are identifiable.

The library is constructed in a similar way to the auditorium with a perimeter of pillars clad with bricks. The building is arranged over four floors of reading rooms revolving around a central space lit by a large skylight which diffuses the light from above to all four levels. Each of the reading tables are placed near a large window opening out onto the landscape.

Near the library is a small residential building with 8 floors containing 15 apartments of about 40 m^2 each, consisting of a living room with kitchenette, a twin bedroom and a bathroom. Similarly, in this case, the building is clad in exposed brickwork.

The rectorate building concludes the series of the school buildings. This has the same linear type and the same construction form as the music school (Figs. 18 and 19).

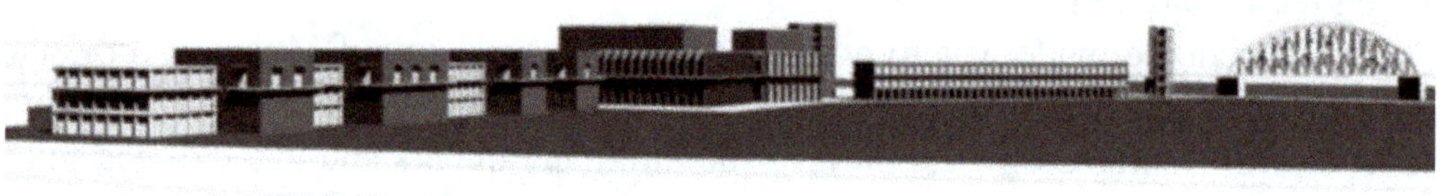

Fig. 18 General view

Fig. 19 General plan

9 Competition for a Kindergarten in Dolzago (LC)[8]

The urban island on which the project area is located—between the streets Montecuccoli, Parini-Piazza della Repubblica, Corsica and Donatori di Sangue—is made up of structures with different functions: the Bonomelli's factory; the large urban Alpini park; the Town Hall building and the new public car park area. Once the quantities relating to the different equipment are defined, the architectural layout is organized favouring some features regarding physical aspects, location and topography. The distribution of the required activities favours the placement of the spaces for the classrooms in the southern part, while the collective ones are in the northern part, towards the Alpini park.

The difference in height between the street level and that of the project is assumed as a resource, arranging, in correspondence with the secondary access on via Montecuccoli, the service spaces, the kitchen and, finally, in the existing underground floor, suitably transformed, the mechanical and heating systems. The layout of the spaces gives priority to the southern-west location of the classrooms, divided into three blocks around two intermediate garden patios, designed in continuity with the covered interior space. The layout of the classrooms is organized with different equipment: a space for free activity and one for planned activities. This space is not physically divided, but can be separated with the chance of being easily set up with

[8] Team work: Domenico Chizzoniti, Gloria Asnago, Emanuela Margione, Alessandra Rossi.

mobile furnishings, depending on the programming of different educational activities. The arrangement of the three sections on the south-west side in a radial pattern has allowed the construction of a versatile autonomous system in direct contact with the open space for free activities in the school garden.

A more protected intermediate space, the patio, is open to the interior space, through large openings in correspondence with the spaces destined for free activities. Each of the three sections is set up with different equipment, while the spring one, located close to the atrium of the main entrance, is equipped with a resting room for the younger children. The design of the outdoor space is carried out in continuity with the internal one. A suitable space is reserved for each classroom as a free outdoor area that serves as a living room, in continuity with the internal environment and as a link between open and closed spaces. These recesses, called "patios" have the function of connecting the spaces of the classrooms and provide sufficient natural light for indoor activities. Particular attention is paid to the differentiation of façade materials in order to support the conception of architectural bodies with the use of large windows, shielding elements, wooden doors and windows. Therefore, different colour choices are adopted in the façade cladding, giving to the materials the connotation of the different functions (Figs. 20, 21 and 22).

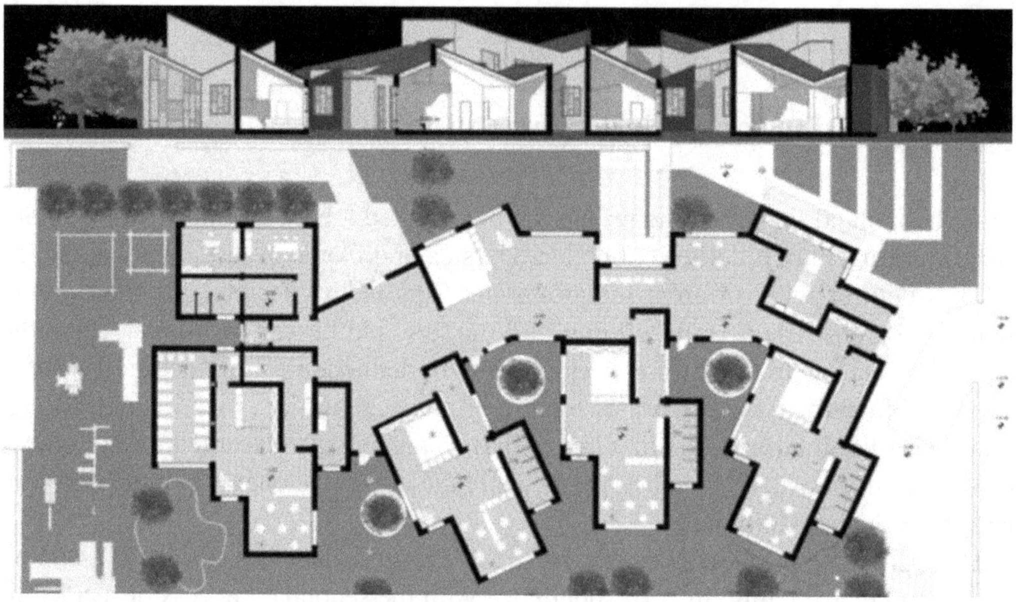

Fig. 20 General plan and longitudinal section

Fig. 21 Physical model

Fig. 22 Courtyard view; physical model view; elevation view

10 Kindergarten in Lurago D'Erba (Co), 2018[9]

The topography of the place and the beauty of the hilly landscape evoke the nursery's features, its location and its relationship with the surroundings.

The nursery is set on the gentle slope of the hill sloping towards the west: two masonry walls enclose the green spaces of the schools connected to each other but separated by walls of various heights. As a result they will define a large lawn all around the existing primary school, making future outdoor sports possible, and a more protected green terrace, belonging to the new nursery. The terrace, facing the countryside sloping down into the valley, is at the heart of the nursery and onto which all of its interior spaces are overlooked. These are dug into the sloping ground, forming terraces that do not obstruct the view of the houses above.

[9]Team work: Raffaella Neri, Sergio Croce, Elsa Garavaglia, Claudia Angarano, Alessia Cerri, Marvin Cukaj.

All spaces face old Durini factory and the lake's mountains, the true heart of the school complex and the settlement that surrounds it. The Leonardesque sweetness of the Brianza foothill landscape becomes the schools' quality element.

The nursery is set in terracing built at the level of the school, on the lower part of the slope, so as not to hinder the open views from the Via Madonnina and the houses above. Another level jump separates and protects the gardens of the two schools from the surrounding countryside: the valleys are bordered by a wall, which internally is, has the height of a parapet in order to leave the view open.

More than anything else, the nursery school is a metaphor for a house. It is the place where the child feels protected and develops the first forms of socialisation and learning through play.

As in a big house, it is accessed through the school garden, skirts the wall of the viewing point arrives at, you reach the terrace and through the porch, a modern portico for games and parking, which leads into the large common glass area. A wall that serves as a stage for theatrical performances and games reveals the canteen; the furniture can be changed around.

Sections are organised around five cloisters which allow the light to enter. Each one has an L-shape around its patio, a cosier place for outdoor games; through the glass walls, the common space can be spotted and, beyond this, the garden and the mountains. Outside to the south-east, the masonry wall borders a shady and cool space for games with different flooring. On the opposite side, on the wall facing south, there is a sunny area intended for educational gardens.

The nursery's green roof establishes continuity with the hill's slope.

The main entrance is from the car park to the north of the area; a secondary entrance is located to the south-west, linked to the cycle path. A final service entrance is located to the south-east and provides access to the technical rooms and kitchens across a service road that runs parallel to the cycle path (Fig. 23 a, b).

Fig. 23 Perspective views from the inside

The Paths to Innovation: Tools, Models and Processes

Massimiliano Bocciarelli, Laura Daglio and Raffaella Neri

The following Section focuses on the development and application of tools and approaches aimed at introducing original solutions to the design of schools, in order to improve the final quality of the buildings through a control of the construction process and of the relations among the diverse actors involved and to better respond to the changing demands of education and learning models.

Accordingly, the presented chapters highlight innovation trends in the design of educational facilities, including the construction of new buildings as well as the renovation of the existing heritage. In particular, the case studies collected reveal three main environments where to trace possible paths of enhancement.

First, the potentials emerging from the adoption of Building Information Modeling (BIM) softwares are disclosed, triggering a new approch to the overall construction process, able to improve the collaboration among the different stakeholders in charge. In fact, the building/renovation of a school belongs to the public works typology, thus involving the compliance with a complex set of national regulations aimed at costs and quality control from the design to the bidding, to the construction and also to the operation and the maintenance phases. The sharing of information and knowledge among the designers, contractors and the client is considered a crucial element to improve the efficiency of the process in terms of final results and scheduling, as well as to avoid possible unexpected interferences and over-budget problems.

The case study of the School of Melzo reports the positive experience of the application of the BIM process to control the information workflows during the design phase and the public procurement, including all graphic and performance information, organised in a database associated to the objects of the model. Moreover, the bidding evaluation and the construction phases were managed through the BIM tool providing a successful outcome, together with raising the awareness of the need for a necessary upgrading of the existing regulations to embed the advantages offered by the softwares.

In the experience of the School of Liscate the BIM is an instrument for the experimentation of a new contract typology for the delivery of public works within the Italian national regulatory context—the Framework Alliance Contract—which introduces a collaborative approach in the management of the process, bridging the gap between design and construction. A mutual agreement among parties (design teams, contractors and suppliers) is established, allowing for the improved sharing of information to align the interests of the different economic operators, to reduce inefficiencies in the supply chain, mistakes and misunderstandings among the professionals, to increase transparency and responsibility towards both the client and the other collaboration components.

The case study of the Progetto Iscol@ presents a more predictive than managerial application of the BIM tool: the optimisation of the school spaces quality in terms of fruition and use performances. Following the evolution of the learning methodologies, the interactions of the users in the educational enviroments is acquiring a significant role, thus requiring a special attention on the comfort and characters of space and on the school layout and organisation. Through the merging of different softwares currently available on the market, a solution to carry out a pre-occupancy simulation is developed, experimentated on three schools and consequently implemented to remove overcrowding phenomena, improve safety and simultaneosly reduce inefficient areas where low interactions occur.

In the design for the San Severino Marche High school and the Inveruno facility a second approach is discussed, concerning the multidisciplinary development of the project since the initial phase, again involving the BIM as a useful management tool and to share data. The information is exchanged horizontally, among the professionals with different specialisations and vertically, through the detailing design, in order to reduce the process duration and to enhance the performances in spite of the high complexity of the buildings. The public services have been conceived considering their respective symbolic role for the reference community. The San Severino Marche Technical College experiments a research collaboration with the goal of supporting design activities related to emergency situations. The new building, in fact, replaces the one destroyed after the 2016 earthquake in Central Italy and showcases a possible well timed intervention after the disastrous events, in addition to innovative constructive strategies. The collaborative design strategy adopted for the Inveruno case study aims at responding to the representative character, not only of the school as a public facility for the entire community, but also as a pivotal urban space and new civic centre of the town.

Finally, the impact of the rapidly evolving learning and teaching models on the design and organisation of the school spaces are addressed and examined. The project for the prefabricated "*Carro di Tespi*" pavilion offers a possible solution to add innovative learning spaces to the existing traditional school complexes, whose layouts are still based on outdated educational programmes. The proposed system can also be assembled as a temporary structure to provide shelter for education activities in areas affected by natural disasters.

The last theoretical contribution argues on the issue of space in the era of 2.0–3.0 schools; a significant question as already introduced by the Progetto Iscol@. In fact, according to the most recent learning and teaching methodologies, space detains a specific educational character and can be considered as a third teacher. Organizing space thus means organizing the metaphor of knowledge. New forms of teaching and learning, involving individuals or organized in groups and workshops, require layouts of greater spatial complexity and richness, yet to be flexible and multi-functional, and to reproduce the existing rapidly evolving situation.

In a rapidly changing society and culture, school design aims at providing the best learning environment implementing an adaptive heritage to be offered for the education of the future generations.

A BIM-Based Process from Building Design to Construction: A Case Study, the School of Melzo

Giuseppe Martino Di Giuda, Paolo Ettore Giana, Francesco Paleari, Marco Schievano, Elena Seghezzi and Valentina Villa

Abstract The digital transition in the construction industry is characterized by data and information management as central to a BIM process. The structuring of the design performance and data shared among the various stakeholders was designed to follow the entire building process: from the client requirements, up to the design, the construction phase and the final handover. The information workflow was drafted and validated on a new building school of Melzo case study.

Keywords School building design · Façade optimization · Building information management · Tender process · Construction management

1 Introduction

The new primary school of Melzo is designed to host 500 scholars, and is part of a wider programme for the school heritage set by the Municipality. The aim of this programme is the reorganization of existing building stock of the city. The final project was drawn up by the Municipality's UTC, as part of a research project in collaboration with the Politecnico di Milano for the development and implementation of the BIM processes for the management of the integrated contract, the tender with the most economically advantageous offer and support to works supervision for the controls in progress.

The project's aim was to create a morphological articulation capable of developing significant relations with the context, with specific attention to the internal-external relationship of the building. Particular attention was also paid to the study of an internal functional articulation that could be adapted to the needs of the contemporary school and tomorrow through a functional, flexible, unitary internal space.

G. M. Di Giuda (✉) · P. E. Giana · F. Paleari · M. Schievano · E. Seghezzi
Architecture, Built Environment and Construction Engineering—ABC Department, Politecnico di Milano, Milan, Italy
e-mail: giuseppe.digiuda@polimi.it

V. Villa
Department of Structural, Geotechnical and Building Engineering—DISEG, Politecnico di Torino, Turin, Italy

S. Della Torre et al. (eds.), *Buildings for Education*, Research for Development,
https://doi.org/10.1007/978-3-030-33687-5_14

The building is composed of three functional units, characterized by different construction techniques and technologies: a central plug in reinforced concrete that connects all the rooms as well as the didactic direction containing the offices, the library and the auditorium; on the central corridor there are three elements with wooden structure, containing the classrooms and the laboratories; finally, a double-height body in reinforced concrete contains the canteen, the gym and the technical rooms for the systems.

Particularly strategic appeared the option of an optimal use of the open space, characterized by a good exposure and practicability: it is in fact possible to think of it as a sort of expansion outside the school. The treatment of the facades enhances this aspect: the large glass surfaces, which replace the conventional concept of window and are intended to enhance the continuity of perception with the natural environment outside, symbolically express the desire to open to the context (Tagliabue and Villa 2017).

The process was managed with a BIM-based approach; this paper presents its application in the different stages of the process.

2 BIM Use During Tender Phases

The research work aimed at defining criteria, sub-criteria, evaluation procedures, methods, and formulas to allocate points for a Most Economically Advantageous Tender (MEAT). The use of MEAT is limited in Italy, as it implies the use of precise and clear explanation of the points allocation, a proper use of this procurement method (meaning that the organizational conditions to use it should be made clear). Nonetheless, if properly used, MEAT processes result in advantages and transparency (Di Giuda and Villa 2015).

In this sense, the BIM methodology was used in this case both in the project design phase, and in the tender process drafting.

The proposed work tries to have as goal the definition of the criteria, the sub-criteria, the evaluation procedures, the methods and the formulas to allocate points for a Most Economically Advantageous Tender. To the current state, in the tender notices is possible to often find requests of improvement of elements which are not included in the project, formulas of too generic or illegitimate bids which often to not report the explanation of the points allocation. Those factors necessarily entail a non-transparency in the work of the commission, which has to decide the procedures of points allocation during the appropriate session of evaluation. Therefore, incoherent data do not protect the contracting authority but, on the contrary, exposes it to a high litigation risk and to downwards negotiations during the execution of the work.

The tender process was managed through the use of the BIM project, as it included all graphic and performance information, organized in the database associated to the objects in the model. The model can be used to directly extract all the documents of the project and the tender information (graphic tables, and also documents, as Bill of Quantities, and performance specifications). The model was also used to obtain the

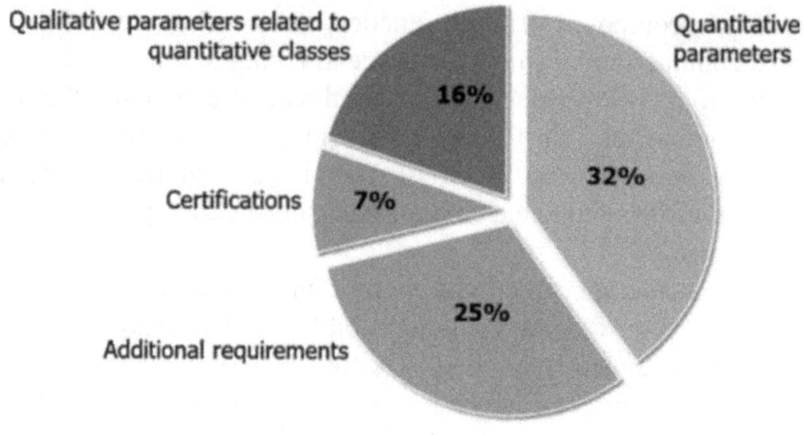

Fig. 1 Weight of proposed categories of parameters on the total score (image by authors)

bid sheets, useful to compose the guidelines for the compilation of the bid, attached to tender documentation (Di Giuda et al. 2015) (Fig. 1).

2.1 *Parameters for Bid Evaluation*

The integrated approach is outlined by innovation management system of the tender phase. In this sense, the methodology consisted in several steps. The first step is the parametric modeling of the building by the contracting authority. The model created is a virtual prototype of the project, consisting of virtual objects completely equivalent to the technical elements that constitute the building who will be assigned properties and attributes stored and managed by the database software. The model produced during the design phased was used to extract a coherent technical requirement document that contain the same information required by the client at the beginning of the design phase. This document constitutes the foundation of the "Allegato I" of the technical offer. Within the software criteria and sub-criteria of evaluation will be linked to objects on the form of parameters, which may already be present in the software or can be implemented manually. The contracting authority will draw in tabular form the latest information entered in the parameters, which are precisely the areas of activity of companies with respect to the project placed in the tender.

Analyzing the project was possible to identify the parts that could have been the subject of improvements. In particular, during the phase of use of the building, the focus was on the performance of the building envelope, on the performances of the plant machinery, on the hygiene and the managing of the resources. Another aspect on which the attention has been focusing on is the maintenance of the materials and the technological solution offered and the programmed maintenance of the architectural and plant design elements.

Each sub-parameter have been described in a thorough and punctual way the procedures of points' assignment, the formulas, the parameters of award, in order to the respect of the general principles related to transparency, non-discrimination and equal treatment. The points given to each part due to not alter the object of the award. In addition, the bids do not entail additional burdens for the contracting authority, but they have to guarantee improvements of the management, quality, or reliability of the contracting authority.

The parameters for bids evaluation were divided in four main categories:

- Quantitative parameters, that are valuable exclusively with mathematical formulas that have to be explained in the procedural guideline of the tender. Quantitative parameters include reduction of energy consumption, performances (thermal, acoustic, of resistance, etc.) of the materials or of technical solutions proposed, amount of waste and their management (reuse, recycling, disposal), cost of use and maintenance. Environmental parameters have been chosen in accordance to European Directive 2014/24/EU, stressing the importance of sustainable development and the potential contribution of contracting authorities to introduce environmental factors in tender notices. The calculation methods, the formulas and the pondering's parameters of the score have to be described in the tender and in the tender's procedural guideline, and the contacting authority has to provide all the information necessary to make the comparison competitive and equal.
- Qualitative parameters related to quantitative classes: every parameter in this category cannot be immediately identifiable through a numeric data, but can be related to "quality classes". These classes are defined in relation to technical requirements (for example, the maintenance of the expected solutions for the finishing on the pavement, siding, delivery conditions, services after the sell, technical assistance, etc.). The classes of quality can be defined in relation to the requirements of the contracting authority. Evaluation procedures, quality classes, and related points must be explained in the procedural guidelines of the tender.
- Qualitative requirements of subjective matters: this category regard all the cases where it is not possible to define objective qualitative classes. In these cases, it is mandatory to define precisely the object of the evaluation. It is also necessary to define the procedures for the evaluation. This category includes the evaluations related to the technical merit, the aesthetic and functional characteristics, the social characteristics, the organization of the personnel and everything not included in the points (a) (b) and (d).
- Additional requirements: this category includes certifications, and presentation of requirements additional to the ones which are strictly necessary for the participation in the tender. Examples of these aspects include: the rating of legality, the possession of environmental protection certifications (for example UNI ISO 14001) and certifications of management of health and security on the job (for example OHSAS 18001), etc., the possession of an ecological quality label of the European Union (Ecolabel EU) in relation to the goods and services listed in the agreement. It must be specified if the request of possession of one or more certificate, is made for the contracting company, for the entire group, or it is required from the suppliers (for example, for the materials with higher environmental impact)

and the construction companies. In addition, it must enlist the relative scores and how the calculation procedures are made.

The parameters and sub-parameters are here presented (Table 1), with the assignment of the relative points of the tender for the new school in Melzo. The last column to the right illustrates the category of each sub-parameter.

A great support has been given with the use of "guidelines for the compilation of the technical offers" flanking the tender and the tender procedural guidelines. In it is described:

- The documents that constitutes the technical offer;
- The explanation of each parameter and sub-parameter;
- The formulas used for the definition of the weight;
- The list of the technical elements, which have been the object of the evaluation;
- The procedures of evaluation of each element;
- The definition table of the classes of quality;
- The formulas for the re-setting of the points.

2.2 Scoring

For each sub-criterion (Table 1), the procedures for awarding scores, formulas and award criteria have been described exhaustively and precisely, in order to comply with the general principles of transparency, non-discrimination and equal treatment. The scores attributed to the individual parts are such as not to alter the object of the award. Furthermore, bids must not entail any additional costs for the contracting authority, but must guarantee improved management, quality or reliability for the contracting authority.

In order to determine the ranking of the bids, it was decided to use the aggregative compensatory method. This method makes it possible to compare the bids with the project on which the tender was based and to evaluate the relative improvements. With regard to quantitative evaluation elements, or those related to quantitative elements, the score is evaluated through linear interpolation between the coefficient equal to one, attributed to the values of the elements offered most convenient for the contracting authority, and coefficient equal to zero, attributed to the values of the elements offered equal to those placed on the basis of the tender.

Automatic checks ensure that: if the bid is worsening or identical, in terms of quality or performance, to the project on which the tender is based, the score paid to the bid will be zero for the sub-criterion; this indicates that the solution offered is inadequate and, therefore, not assessable and not acceptable to the commission.

Table 1 Criteria, sub-criteria, and scores of the evaluation

Cod.	Criteri di valutazione	Peso	Cod. sub.	Subcriteri	Peso subcrit.	
A.1	Trasmittanza involucro	10	A.1.1	Serramenti	5,9	(a)
			A.1.2	Chiusura verticale	1,3	(a)
			A.1.3	Chiusura verticale pref. (palestra/mensa)	1,7	(a)
			A.1.4	Copertura palestra/mensa	1,1	(a)
A.2	Requisiti materiali offerti	10	A.2.1	Distanza Località di prod. materiali offerti	3	(b)
			A.2.2	Limitazione ingombro delle partizioni interne	3	(a)
			A.2.3	Grado di manutenibilità dei materiali offerti	4	(b)
A.3	Requisiti ambientali	5	A.3.1	Cert. UNI EN ISO 14001 impresa affidataria	2	(d)
			A.3.2	Cert. UNI EN ISO 14001 prodotti prevalenti	3	(d)
B.1	Funzionamento impianto	12	B.1.1	Pompe di Calore (PdC1, PdC2)	6	(a)
			B.1.2	Unità di Trattamento Aria	6	(a)
B.2	Componenti impianto	2	B.2.1	Distanza centro manutenzione (PdC)	2	(b)
B.3	Incremento energie rinnovabili	4	B.3.1	Pannelli solari fotovoltaici (S x ε)	4	(a)
B.4	Utilizzo intelligente delle risorse	7	B.4.1	Sistemi di gestione e riduzione del consumo di energia elettrica (solo illuminazione)	4	(b)
			B.4.2	Sistemi di gestione e riduzione del consumo di acqua	3	(b)
C.1	Sicurezza	2	C.1.1	Certificazione OHSAS 18001 impresa	2	(d)

(continued)

Table 1 (continued)

Cod.	Criteri di valutazione	Peso	Cod. sub.	Subcriteri	Peso subcrit.	
C.2	Soluzioni costruttive, Gestione cantiere	13	C.2.1	Dettagli costruttivi	6	(c)
			C.2.2	Layout di cantiere (Fasi scavi e strutture)	4	(c)
			C.2.3	Gestione rifiuti/Sfridi (D.lgs. 152/06) con part. riferimento allo smaltimento rifiuti speciali	3	(a)
D.1	Man. parte Edile	5	D.1.1	Man. programmata parte edile	5	(c)
D.2	Man. parte Impianti	10	D.2.1	Man. programmata parte impianti	10	(c)

3 Use of BIM During Tender

An improvement of the procedure is represented by the integration of BIM methodology in the management of public procurement. In fact, if the proposed new method overcomes the discretion of the assessment by the Commission and makes objective evaluation criteria, a more advanced system management can leverage the BIM process to further expedite the procedure. In addition, the use of a single parametric model ensures completeness, consistency and homogeneity of information on the technical bids, reducing all the problems arising by the incongruity of the information.

During the publication of the tender, the contracting authority organized meetings to clarify potential questions, to explain the BIM methodology, to present criteria and sub criteria of the process. The use of guidelines for the compilation of technical offers was particularly helpful, as they described the required documents for the technical offer, the explanation of each parameter and sub-parameter, the formulas used for the definition of weights, the list of technical elements object of the evaluation, the procedures for the evaluation, the tables for quality classes definition, and the formulas for points resetting. Ten companies took part to the tender, but one did not deliver the whole documentation. Technical offers were filled in a thorough and detailed way, smartly combining parameters and sub-criteria to propose an improved offer. The values resulting from the offers demonstrate the truthfulness of the offers presented by the companies; the data were congruent, complete, and unequivocal.

4 Use of BIM During Construction Phases

During construction phases, the Construction Manager was able to verify that all materials and products supplied on site corresponded exactly to the offer provided in the tender, in terms of performance and of brands proposed. This possibility stems from the request of the technical data sheets of all the materials and components during the tendering phase. The BIM model plays a role also in construction phase, as the quantitative and geometric control is carried out using the model, verifying not only the correct positioning of the elements installed on site, but also the dimensional correspondence and the fixing methods (Fig. 2). The checks carried out on the model during the design phase and the controls during the execution phase guarantee the quality of the work carried out in terms of interference between structural, plant engineering or architectural elements, guarantee the performance of the individual elements, already verified as a whole, and reduce possible variations or rework to a minimum (Eastman et al. 2016).

The advantages of the BIM methodology can be found in all the phases illustrated, starting from the selection phase of the contractor where the economic discount of 5% of the winning offer stands out. The scores, mainly calculated using mathematical formulas or quality classes, derived directly from the services provided in the technical sheets attached to the offers. A check was made of the correspondence and congruence between the elements making up the offer, with values close to 100%. This eliminated any uncertainty in terms of design and technological choices and, consequently, the choice of brands and products to be implemented, greatly facilitating the process of verification and control in the execution phase.

Fig. 2 Comparison between façade in the model and façade during building works (image by authors)

5 BIM for Advanced Project Activities

A relevant element of the project is the curved façade cladded with coloured panels. The evaluation of the panels, in terms of quantity, dimensions, and pattern, was carried out through a script developed in Dynamo, and then linked to the BIM model (Fig. 3). This phase was carried out through a Design Optioneering approach, based on the setting of requirements conveying the needs of actors of the process, expressed in a quantitative form; each of them was then assigned a weight, to evaluate their mutual relevance in a multicriteria perspective.

The curved ventilated façade is composed of a metal structure, linked by fixing brackets to the concrete wall. The metal structure, composed of t-shaped steel profiles, acts as support for the cladding, made of high-pressure compact laminated panels. To create a pattern, the script included four inputs: the face of the geometry, width of panels, height of the coloured central panels, and RGB colours for the random pattern. The use of an algorithm allowed the creation of an adaptive family, embedded with the required parameters; the adaptive components were then placed on located points. The extraction of quantities from the BIM model allowed some considerations related to shape, costs, number of modules, and waste materials. A multi-criteria approach, based on the needs of the contractor, was used to evaluate the most suitable solution. This approach is particularly useful as it implies a tailored solution, based on stated needs and on quantitative evaluations, enhancing control over decision processes.

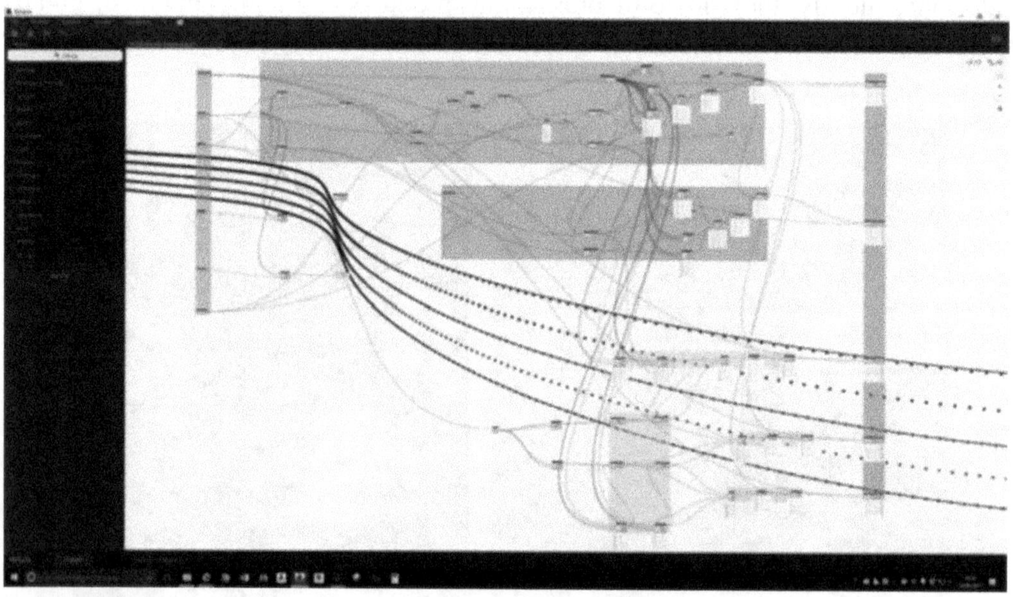

Fig. 3 Dynamo script of the ventilated façade (image by authors)

6 Limitations of the Proposed Method

Since the use of BIM in the tender phase was not required, the entire process was conducted in parallel with the traditional management of documentation both in the tender phase and in the executive phase. The objective was not to overload the company and the designers, allowing them to concentrate only on the project and the works, using the model only as a verification and feedback in terms of performance for research activities on the subject. Since the use of an information exchange platform has not been contractually agreed upon, the procedure for acceptance and approval of materials by the Works Management is carried out in the traditional way by means of certified minutes and e-mails.

In parallel, the same documents are placed on an experimental platform connected to the modelled elements, in order to deliver to the client an as-built BIM-based model, containing all the information, documents, and photographs of construction.

7 Conclusions and Further Outcomes

Besides the presented limitations, Melzo school is an example of the use of BIM methodology to improve the quality of the process and of the final result. In this sense, the application of BIM methodologies to other parts (and phases) of the projects could result in an even higher level of quality.

The use of a BIM model for the building management could assure effective maintenance (both preventive and corrective), and successful use of the building. The integration of sensors, combined with the BIM model of the building, provides a complete and immediate state of the asset to technicians and facility managers. Sensors and actuators installed on the school will in fact provide data for the monitoring phase. Data collected could be used to analyse the real use of the building, resulting in an optimization of resources, both in terms of personnel and of energy and resources.

Considering the entire process, the BIM model could be joined by a DMS (Document Management System), in which the processes of control of the works can be structured: acceptance controls of materials, checks in progress, and archiving of documentation. The collaborative process could contribute to create a structured supply chain within the construction market, considering production and delivery times and guaranteeing on the reliability of products and teams in charge of the installation. This process has as goal an improvement in quality of the final product.

The proposed approach demonstrated to be a valuable method for the application of BIM-based methodologies in several phases of construction projects, thanks to its flexibility.

References

Di Giuda GM, Villa V (2015) Verifica dei progetti e metodologia BIM, Ingegneri, nuove tecnologie, materiali, sistemi, processi, Maggioli Editore

Di Giuda GM, Villa V, Loreti L (2015) Il BIM per la gestione di una gara con il criterio dell'offerta economicamente più vantaggiosa-BIM to manage public procurement with award criterion Most Economically Advantageous Tender, Istea 2015: Sostenibilità ambientale e produzione Edilizia, Milano, 24–25 Sept

Eastman C, Teicholz P, Sack R, Liston K (2016) Il BIM Guida completa al Building Information Modeling per Committenti, Architetti, Ingegneri, Gestori immobiliari e imprese, Di Giuda GM, Villa V (eds). Hoepli

Tagliabue LC, Villa V (2017) Il BIM per le scuole. Analisi del patrimonio scolastico e strategie di intervento. A cura di Di Giuda, G M. Hoepli

A Collaborative Approach for AEC Industry Digital Transformation: A Case Study, the School of Liscate

Giuseppe Martino Di Giuda, Paolo Ettore Giana, Marco Schievano and Francesco Paleari

Abstract In the digital transformation process of the construction sector, the Client has a crucial role involving all the stakeholders and organizing workflows with a digital platform that ensure the flow of information in a fluid and coherent way. Collaboration becomes a key factor in making the process work the best way and supports BIM throughout the building's lifecycle. For this reason, the Framework Alliance Contract has been applied, for the first time in Italy, on a case study (the new School of Liscate). The construction process has been supported by a collaborative platform linked to the BIM model that allows the traceability and real-time control of the consistency and accuracy of project information.

Keywords Framework agreement · Information management · DMS · LoIN · Transparency

1 Introduction

The possibility of developing a perfect mechanism of competition and the allocation of resources in a market is subject to the occurrence of a condition of symmetrical information. To achieve this objective, the information must respect the characteristics of completeness and accessibility without any costs among parties. Unlike the hypothesis mentioned above, information is commonly considered as an economic good and for this reason, they are not accessible at no cost (Halac et al. 2012). Moreover, it is probable that not all the necessary information is available during economic transactions. It is, therefore, possible to define the existence of asymmetric information when, within an economic process, information is not entirely shared among parties in the same project (Sultan et al. 2008). This attitude allows a few agents involved to possess a greater amount of information compared to the competitors, on whom they would have a competitive advantage.

G. M. Di Giuda (✉) · P. E. Giana · M. Schievano · F. Paleari
Architecture, Built Environment and Construction Engineering—ABC Department, Politecnico di Milano, Milan, Italy
e-mail: giuseppe.digiuda@polimi.it

S. Della Torre et al. (eds.), *Buildings for Education*, Research for Development,
https://doi.org/10.1007/978-3-030-33687-5_15

The presented theories can be adapted also to the real estate sector and to the AEC sector. In these, both parties require fair compensation for the service they intend to offer. In this way, collaborative contract theories lead to a review of the process structure (Lahdenperä et al. 2012; Alwash et al. 2017; AIACC 2014; McKinsey 2017) and the inclusion of multiple contractual parties leads to a different view of the project (Serpa et al. 2016). However, in many cases, individuals tend to apply opportunistic attitude, without considering the benefit to the project. This approach produces neither a global gain nor the team and the project is damaged by selfish behaviours (Lenferink et al. 2013).

2 State of the Art of Collaborative Approaches

In recent decades, some project delivery systems (AIA 2007) have claimed to bridge the gap between design and construction. In this context, some collaboration contracts have been developed in many countries (e.g. United States, United Kingdom and Italy), they have mainly the same characteristics although born in different legislation (Mosey 2019). Due to their structure and composition, traditional contracts inevitably create (i) a conflict of interest that cannot be resolved and (ii) impose a strict division among stakeholders. These new working frameworks, on the contrary, allow achieving the final through a mutual agreement among parties. These relational approaches are based on several theories (Smith 2011; Sacks 2016), which have concluded that the optimal method of project implementation is an integrated approach that applies Lean principles (Suttie 2013). Collaborative relational contracts (RPDAs) apply these theories to align the interest of different economic operators allowing the optimization of the final result (Jalaei 2014). In such a revolutionary context, the use of information modelling allows applying what the contractual, sociological, psychological and economic theories support. Without a contractual change, the full application of information modelling and collaboration results difficult, even more, when BIM methodology is applied erroneously to opportunistic contracts (Raisbeck et al. 2010; Sacks et al. 2010; Singh et al. 2011).

The researchers investigated different contractual typology in a worldwide scenario (Di Giuda et al. 2017), among them researchers individuated the one closest to Italian legislation. This approach provided a solid base to adopt RPDAs in Italian legislation. The Framework Alliance Contract FAC-1 is a flexible meta-contractual model, in which parties are given the opportunity to consider efficiencies in the supply chain that make the flow of information more transparent and reduce the overall cost of performance. The Client could use the standard to create a collaboration, legally valid, among their sub-consultants and/or sub-contractors. The higher level of transparency and increased responsibility, towards both the Client and other Collaboration Components, required by each private operator in the pursuit of collaboration, are counterbalanced by the economic compensation provided. FAC-1 is a contract that regulates and manages the inter-relations among different contracts and, namely, the relation among parties, which are not directly associated over a contract. In addition,

FAC-1 is intended to build a solid legal framework for the BIM use in the construction sector. It allows to develop a positive interaction between the different design teams and to connect the different phases of the work construction. FAC-1 approach invites the participants to submit their Collaboration Proposals and enhances the professional expertise, thereby exploiting economies of scale and achieving cash or other benefits.

The research aims at understanding the practical benefit of a digital approach managed through a Relational Project Delivery Agreement in a real case study. The project allows the researchers to discover the limitations and difficulties of the approach.

3 Research Methodology

The proposed methodology starts analysing, according to ISO19650-1 (ISO 2018), the need of the public administration. The further step of the process consists in extracting the objectives and specific needs of the organization. According to the organization workflows, the researchers prepared both a contract procedure to fulfil the common procurement and a platform to manage the portfolio information.

These two generic tools need to be customised for the specific project. Namely, all the information, which need to be stored in Information Models and Databases, are breakdown and codified to have a consistency in the data even though they are produced by different stakeholders. Asset information are included in the technical sheet attached to the meta-contract. Best practice guidelines are developed and improved by iterative tests to better perform and fulfil the organization needs and expectation. This approach allows to customise information workflow among parties in a data-driven process (Fig. 1).

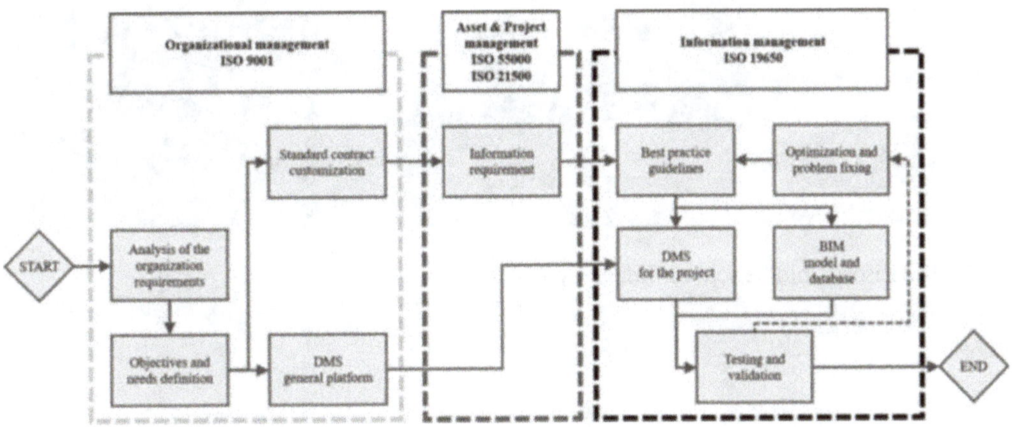

Fig. 1 Methodological approach (image by authors)

4 Case Study: The Liscate Approach

The "Adda Martesana" Municipality applies the Framework Alliance Contract (FAC-1) as part of the project to build a middle school in Liscate (Italy). It is a middle school project for 150 students and 5 M € of construction costs. The school was developed through a BIM approach, showing a high level of complexity (Fig. 2).

The collaborative contractual standard has provided the legal basis for optimising relations among parties and obtaining added value. The client, helped by the scientific consultant, has therefore drawn up a series of annexes which, according to the functions and schemes that support FAC-1, customise the standard contractual model to the specifics of the case. The client aims at including the most important parties of the construction process in order to allow better information exchange.

The agreement aims at including not only the general contractor, but also the Tier 1 of the supply chain according to the specific need of the project. The customization process started defining the objective of the collaboration, set by the client as "monitoring of the time and cost provided for in the Programme Contract and its annexes". All the features of the agreement were based on that assumption.

The agreement itself was used as the legal foundation to assure and control the information workflow that is essential in a data-driven process. To achieve this result, guidelines have been defined for the use of the platform for sharing information and the information required for the delivery of the model has been established. The DMS and BIM guidelines are annexes of the alliance agreement.

Fig. 2 The Liscate middle school (image by authors)

4.1 *Document Management System (DMS) for Construction Site*

The DMS platform adopted for the information management allows to have under control, in real time, data on supplies arriving at the site, documents to be approved, materials to be accepted. In order to streamline procedures and control the information flow between the parties involved, a DMS system is set up to serve as a basis for the information exchange. Through this approach, the documentation is automatically sent in digital format to people in charge following the regulatory flow. In the event that the verification should turn out to be negative, the information flow would be followed in the event of non-acceptance of the document. The system adopted in this way ensures a timely inspection of the control procedures on site. The data stored in this way allow the costs management and the monitoring of materials quantities delivered and to be delivered, so as to have a computational knowledge of the state of work consistency (Fig. 3).

A document platform has been set up to control and facilitate the management of the project through a structured information exchange between the parties. In this way, together with weekly coordination meetings, interests are aligned and information asymmetries between the parties are removed. This approach allows the use of always updated and consistent documentation according to the actual needs of the contracting parties. The folders are visible according to the roles assigned to the users, therefore the writing rules vary according to the permissions assigned. Administrators have access and editing power in all folders, as well as site managers. If a folder is not to be made visible only to some users, the inherited privileges must be removed (privileges automatically granted to all subfolders of a directory

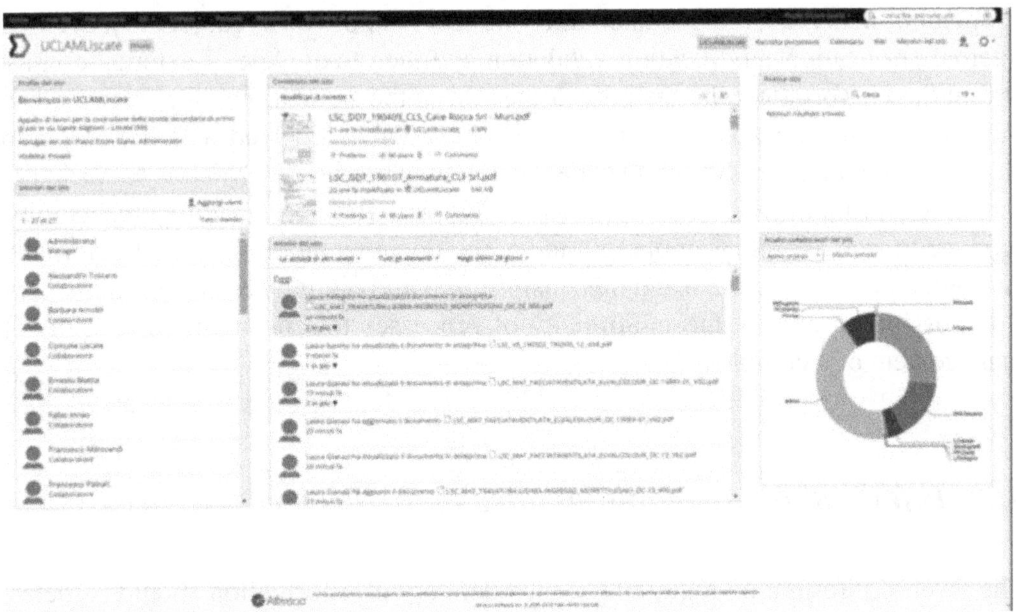

Fig. 3 The DMS platform (image by authors)

depending on the permissions of the directory itself) and locally you can customize the permissions outlined in the previous paragraphs.

For example, a contributor in a folder may have the privileges of a manager, while other contributors may not even have the privilege of seeing such a space. It is better to create user groups with the same permissions, rather than assigning individual permissions to users, so that if you add one or more users at a later time it, being part of a group, will have the same privileges as the other users in the group. The responsibilities in question concern the correct nomenclature of the files inserted and the correct procedures for using the platform.

The scientific consultants set up the DMS with roles and responsibilities for supervising the works on site and then defined a series of procedures to be used before, during and after the construction phase. All these activities must be carried out in accordance with solid procedures and roles defined in the platform management guidelines, where procedures (workflow) are exemplified for each decision-making operation established by current legislation. The solid definition of roles and workflows allows the efficient implementation of the DMS, i.e. the loading, updating, sharing and consultation of documents and information. In particular, the scientific consultants who set up the platform have established, according to the needs of the client, the rules of filing, nomenclature and responsibility, as well as additional information (metadata) related to each type of document.

The process digitization and the use of the DMS guarantee the possibility of managing complex projects, which otherwise would generate a loss of information and a lack of systematic control of processes. This provides a solid basis for obtaining valid, timely and consistent information, not only between the parties but also between different processes. The document management as-built, thus developed, provides for separate procedures and information depending on the type of information. In the case of processes that require verification and validation, as shown below for the approval of materials, the documentation subject to approval is subject to a cycle of internal verification-modification-validation and then to a cycle of external control within the platform to generate only at the end of the same a final report related. If the verification is negative, the non-approved version is archived with the proposed changes and then a new version of the document is produced, which is subjected to further verification cycles until approval is obtained. In the case of documents to be produced (such as, for example, minutes), the documentation is shared, internally verified, approved, recorded and then historicized and digitally archived. In this way it is possible to control the multiplicity of processes that occur in parallel in the management of a complex site.

4.2 *Information Modelling Applied to the Project*

The level of information required for each delivery of information should be determined according to the responses to which the contracting authority must respond in the design phase or project milestone. These requirements should include the

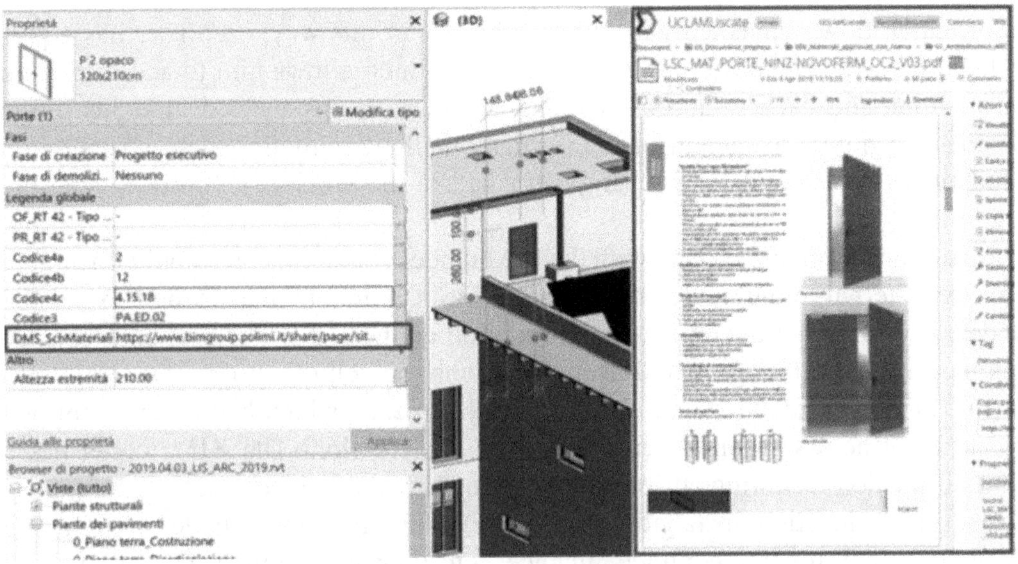

Fig. 4 Consistency and accuracy of information guaranteed by the link between BIM and DMS (image by authors)

appropriate determination of the quality, quantity and granularity of the graphic and alphanumeric information. Once these measures have been defined, they must be adopted to determine the level of information required throughout the whole project or asset. Establishing these requirements across projects, it allows for structured management of the data contained in the models and information databases of the entire real estate portfolio. The level of information required should be determined by the minimum amount of information required to meet the main information requirements, including information requested by other service providers and according to ISO19650-1:2018 standards. For this reason, the client establishes the levels of information per project phase of both technical elements and environmental units (Fig. 4).

This platform, the Document Management System (DMS), is a digital tool applicable to the information management of the site, which allows to improve and streamline the control of the information flow, structuring processes and information which are subject to validation, correction and archiving. The prepared DMS provides a collection of interconnected and persistent data and a series of applications used to access, update and manage the data, which is the data management system. This platform is a tool created and structured to digitally archive the documents of a construction site and monitor the flows of approvals. In particular, it can be assimilated to a database with a (web-based) interface, therefore accessible everywhere, with specific characteristics that allow users to: (i) manage documents; (ii) add metadata to documents; (iii) assign and execute tasks; (iv) manage users and their roles; (v) process data to create benchmarks.

Through the use of the DMS in relation to the methodology and the BIM model, each user can enter data, share, modify, manipulate and display them (depending on the privileges granted to them) in the database so that they follow a pre-established information flow connected with digital objects. In this way, each user has a predefined task, depending on his role in the job, which allows him to access certain information and data, to accept or request changes to instances/documents in the database. A digital management of information allows activities such as versioning (updating of documents with changes tracked in the versions), approval, sharing, tagging.

The connection of the DMS with the BIM model guarantees an as-built document archive, external to the model, but connected to it. This link makes the BIM model updated with all the documentary information connected with the model objects. In this way, guidelines provide stakeholders with the definition of robust workflows for the creation, archiving and updating of documents, which had to be performed by different actors before, during and after construction. The BIM model has a central function in storing attributes related to individual objects (e.g. performance, costs, timing, location, installation dates, etc.), while it is not suitable for archiving documents (e.g. reports, contracts, invoices, etc.). Data is a representation of facts, concepts or instructions in a formalized manner and suitable for communication, interpretation or processing by means of automated processes. The data contained in the model is made available according to the needs and roles within the company, with different reading and writing privileges. The process is completely transparent and traceable. The ability to access data from the web is essential for collaboration, data sharing and for uploading documents directly from the site, properly digitized. The document versioning function helps considerably in the analysis of possible inconsistencies or changes to the documents, to provide the customer (who has permission to view during the construction phase) with the most recent documents and information.

5 Discussion and Further Development

This meta-contractual form permits to increase the coordination of different subjects' activities with greater guarantees of results and with a reduction of unexpected interference, possible over-budget and overrun of time. Especially in complex project, this approach allows an efficient management with multiple subjects' contributions. Collaboration, in such a way, provides an added value in terms of work or service sustainability, site organization and working conditions efficiency, collaboration with the supply chain, reducing re-work. The early involvement of all professionals allows to prevent and/or reduce the mistakes, which must be reported to Alliance Manager that improves project final quality. The alliance members promote transparency in relation to the specific aims and objectives of collaboration. This standard provides the ability of team members to rely on the exchanges of BIM data and setting among different call-off contract the same rule in order to provide data consistently among parties. The use of FAC-1 has not been perceived by the parties as a mere bureaucratic aggravating factor, but as an element to improve collaboration and information exchanges, which aims to solve the relational frictions that traditionally occur.

In this sense, some shrewdness has been included as an activity of collaboration in the agreement to change the approach of the different contractual parties to the agreement itself.

This digital approach to information exchange benefits the control bodies (project management and security coordination office) which, on the basis of the categories and/or subcategories set and the different metadata used, can search for the desired documentation, speeding up the procedures and avoiding information loss in the life cycle of the asset. At the same time, the uniqueness of the information is always maintained regardless of how the research is carried out, which benefits the client in the management of the asset. In fact, the proposed approach facilitates document management by optimizing the archiving of documents and tracking the flows throughout the life of the asset. In conclusion, this approach ensures the traceability of information and information flows, increasing and maintaining the consistency of information that would otherwise be difficult to reach. Moreover, it allows the knowledge of the state of progress of the information flows that take place between the parties, simply querying the platform and the related information models.

At this point, the research provides an approach to the problem of the SMEs of the sector, in fact the collaboration that RPDA established among stakeholders is difficult to achieve and, most of the time, it is unattainable in a traditional process, although the promised success. The new contract improves the processes management decreasing the public administration burden often due to litigations caused by traditional contractual procedures.

References

AIACC (2014) Integrated project delivery: an updated working definition. AIA Calif Counc Sacramento, CA 3:1–18

Alwash A, Love PED, Olatunji O (2017) Impact and remedy of legal uncertainties in building information modeling. J Leg Aff Disput Resolut Eng Constr 1–7. https://doi.org/10.1061/(asce)la.1943-4170.0000219

American Institute of Architects (2007) Integrated project delivery : a guide

Di Giuda GM, Villa V, Giana PE (2017) Collaborative contract with building information modelling: comparison between USA and European approach. In: ISEC9

Halac M (2012) Relational contracts and the value of relationships. Am Econ Rev 102:750–779. https://doi.org/10.1257/aer.102.2.750

International Organization for Standardization (ISO) (2018) ISO 19650-1—Part 1: Concepts and Principles

Jalaei F, Jrade A (2014) Association between construction contracts and relational contract theory. In: Construction research congress 2014

Lahdenperä P (2012) Making sense of the multi-party contractual arrangements of project partnering, project alliancing and integrated project delivery. Constr Manag Econ 30:57–79. https://doi.org/10.1080/01446193.2011.648947

Lenferink S, Tillema T, Arts J (2013) Towards sustainable infrastructure development through integrated contracts: Experiences with inclusiveness in Dutch infrastructure projects. Int J Proj Manag 31:615–627. https://doi.org/10.1016/j.ijproman.2012.09.014

McKinsey Global Institute (2017) Reinventing construction: a route to higher productivity

Mosey D (2019) Collaborative construction procurement and improved value. Wiley-Blackwell

Raisbeck P, Millie R, Maher A (2010) Assessing integrated project delivery: a comparative analysis of IPD and alliance contracting procurement routes. In: Association of researchers in construction management, ARCOM 2010—proceedings of the 26th annual conference

Sacks R (2016) What constitutes good production flow in construction? Constr Manag Econ 34:641–656. https://doi.org/10.1080/01446193.2016.1200733

Sacks R, Radosavljevic M, Barak R (2010) Requirements for building information modeling based lean production management systems for construction. Autom Constr 19:641–655. https://doi.org/10.1016/j.autcon.2010.02.010

Serpa JC, Krishnan H (2016) The strategic role of business insurance. Manag Sci mnsc.2015.2348. https://doi.org/10.1287/mnsc.2015.2348

Singh V, Gu N, Wang X (2011) A theoretical framework of a BIM-based multi-disciplinary collaboration platform. Autom Constr 20:134–144. https://doi.org/10.1016/j.autcon.2010.09.011

Smith RE, Mossman A, Emmitt S (2011) Lean and integrated project delivery. Lean Constr J 1–16

Sultan A (2008) Lemons hypothesis reconsidered: an empirical analysis. Econ Lett 99:541–544. https://doi.org/10.1016/j.econlet.2007.09.038

Suttie JB a (2013) The impacts and effects of integrated project delivery on participating organisations with a focus on organisational culture Iglc-21 1:267–276

Use of Predictive Analyses for BIM-Based Space Quality Optimization: A Case Study, Progetto Iscol@

Giuseppe Martino Di Giuda and Matteo Frate

Abstract Predictive analyses on future uses, developed through usage patterns, provide a solid basis for space quality check. This paper aims at setting a methodology for the use of predictive analyses for project quality and effectiveness of school buildings design. Crowd simulations and pre-occupancy simulations are applied on BIM models of school buildings: data related to users' interactions, comfort evaluation help in increasing space quality and avoiding overcrowding or ineffective space distribution. The proposed approach is iterative, allowing the optimization of design, based on educational approach. This method has been tested in the design of a new school building located in Sardinia, in the framework of Progetto Iscol@.

Keywords Pre-occupancy simulation · Crowd simulation · Design optimization · Educational approach · Usage patterns

1 Introduction

The traditional design process implies the check for compliance of design solution with standards and requirements in order to get administrative approvals, during the phase of definition of the Progetto Definitivo (corresponding to RIBA Plan of Work's Technical Design (Royal Institute of British Architects (RIBA) 2013)) Referring to dimensional verification, two main categories of checks are necessary:

- Fire prevention regulation checks: these checks regard compliance with the minimum requirements set out in the regulations, as the length and width of escape routes and stairways, number and width of emergency exits, and the total evacuation time (Italian Parliament 1998);

G. M. Di Giuda (✉)
Architecture, Built Environment and Construction Engineering—ABC Department, Politecnico di Milano, Milan, Italy
e-mail: giuseppe.digiuda@polimi.it

M. Frate
Unità di Progetto Iscol@, Regione Sardegna, Cagliari, Italy

S. Della Torre et al. (eds.), *Buildings for Education*, Research for Development,
https://doi.org/10.1007/978-3-030-33687-5_16

- Minimum dimensions of spaces: the minimum size of rooms depends on their functions, the building use, and the number of users attending each space.

Nonetheless, the observance of minimum requirements and standards for hallways and spaces does not guarantee space quality in terms of use and user comfort, that are key elements of the Client's requirements. In a traditional process, the management of these aspects mainly comes from intuition and previous experience of the designer, supported by standards and requirements; nonetheless, these intuitions cannot be verified until the designed building is built and used. In most cases, the building is a prototype of itself with regard to the real use (Zimmerman and Martin 2001). In addition, during the phase of definition of the Progetto Definitivo, most aesthetical and technological features, layout and functions of the design are defined. As a result, the evaluation of the effective use of the building should take into account a large amount of aspects. It is relevant to detect high crowding and discomfort, both in ordinary and emergency conditions, such as fire emergency evacuations. These evaluations are essential when considering hallways, meeting and common areas, as these spaces are hosting a large number of users.

In this context, effective uses of spaces and users' comfort levels are hardly assessed. As stated before, they are usually defined only once the building is built and in use (Schaumann et al. 2016). During the operational phase, discrepancies between the built asset and the users' actual need can provide high costs for changes of the environment (Schaumann et al. 2016). As a result, the investigation of effective uses of spaces during the design phase can be useful to anticipate potential changes, leading to the decrease of costs of future modifications and an increase of design quality.

2 Pre-occupancy Simulations and BIM Approach

This work aims to provide a method to evaluate real use of spaces in design stages, through an integrated BIM approach. Building Information Modelling (BIM) methodology provides, by means of models representing the design alternatives, the association of information, such as number of users, space features, costs, to elements and spaces composing the model (Di Giuda and Villa 2016). The Information Modelling method strongly reduces time spent to produce documents and drawings required for administrative approvals, increasing therefore available time and resources for preliminary decision-making processes (Ciribini 2016). As a result, design alternatives can contain a huge amount of data, which are valuable to perform analyses on the model.

The method provides the application of Pre-occupancy simulations to analyse and verify usage patterns of design spaces and the interactions among users.

Pre-occupancy simulations are based on Crowd Simulation systems, which are computerized analyses of the movement of crowds. The main goal of these simulations is fire regulation check, related to spaces and escape routes and they are generally used for emergency simulations (Almeida et al. 2013; Montella 2012; Tang and Ren 2008): in the case study provided, the simulation reproduces the actual use of spaces, to verify the spatial quality of designed spaces and users' comfort levels. Spatial quality is defined as the ability of spaces to meet the intended use.

There are two kinds of Pre-occupancy simulation: Agent-based and Narrative driven, described below. Microscopic Agent-based simulations are used in the current case study: the users keep their own features and can act independently, while their behaviour influences choices and movements of nearby occupants (Ijaz et al. 2015). There is, therefore, an effect of individualisation of the movement of crowds (Santos and Aguirre 2004). The user reacts to simple motion rules called Keep It Simple and Stupid (K.I.S.S.) (Axelrod 1997) and a widespread A.I. (Artificial Intelligence) drives its actions and ensures randomness of movements in space (Santos and Aguirre 2004). Pure Agent-based simulation has a main limitation: it is possible to populate the model of a building, but the movement of the occupants is extremely chaotic (Simeone 2015).

As a consequence, it is necessary to apply a second type of simulation, defined Narrative Driven, which formalizes the sequence of activities the user carries out. A Narrative Driven simulation also presents some limits: the need to formalize every single activity carried out inside the spaces causes a strong rigidity of the simulation. It provides a reproduction of occupants' behaviour along a sequence of fixed activities, without the possibility of varying users' own path. As a result, the analysis appears more as an animation rather than as a simulation, making it nearly impossible to analyse the usage patterns of a building (Simeone 2015).

The adopted solution involves the mixed use of the two systems, thus removing the above limitations. This kind of simulation is not currently implemented in a traditional design process, resulting in lower quality of the design solution (Simeone 2015).

Pre-occupancy simulations were adapted to the case study to perform ordinary conditions analysis. The output of the simulation is the usage pattern of the building. As a result, the actual use of the building by users is assessed at the design stage, optimizing the design solution according to the results.

3 Case Study: Pre-occupancy Simulations to Evaluate Interactions Among Users

The proposed methodology (cf. Fig. 1) was applied to the design of a primary and secondary school in Posada, Sardinia, in the framework of Progetto Iscola@ (Locatelli and Pellegrini 2017). Progetto Iscol@ is a regional programme for the refurbishment and new construction of Sardinia's school heritage. One of the main goal of Progetto

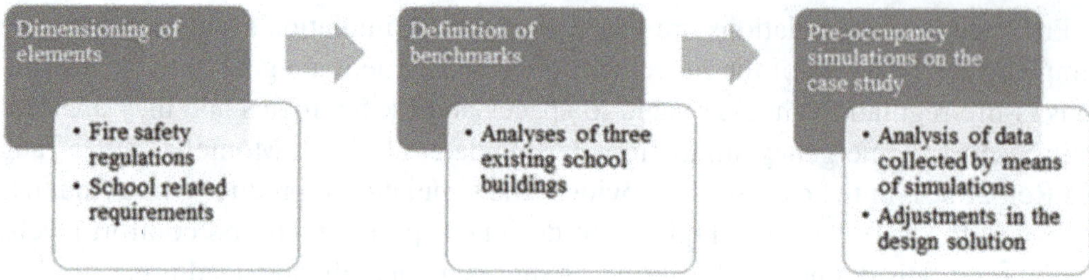

Fig. 1 Workflow of the analysis (image by authors)

Iscol@ is the integration between architecture and educational approach, also thanks to the introduction of innovative learning methodologies.

As mentioned above, a mixed approach based on Agent-based and Narrative Driven simulations was verified and allowed to measure the spatial quality. The model used is unable to replicate and simulate all aspects of human behaviour; it provides just predictive data on number and quality of interactions among users in spaces. The proposed model allows, therefore, to quantify, verify and evaluate the spatial configurations, according to the spatial quality. Spatial quality is defined as the number and variety of interactions, as well as the personal comfort of each user obtained by reducing the crowding phenomena (Locatelli and Pellegrini 2017). As a result, well-designed spaces can stimulate the spread of innovative learning methodologies, such as peer-tutoring and collaborative learning (Locatelli and Pellegrini 2017).

Pre-occupancy analyses allowed the evaluation of design alternatives in terms of layout, dimension, and shape to ensure quality and users' comfort, as well as compliance with educational requirements.

The workflow of the analysis was set as follows:

- dimensioning of hallways, stairs, common areas according to fire safety regulations and minimum size provided by school-related legislation (Italian Parliament 1975);
- definition of benchmarks to achieve, referring to interactions and usage patterns of school buildings;
- performing of Pre-occupancy simulations on the design solution;
- analysis and evaluation of Pre-occupancy simulations results.

The analysis was carried out choosing the time of day with higher probability of overcrowding and discomfort, i.e. school break, when the interactions among students and the number of activities performed are maximum.

The following table (cf. Table 1) shows the main features of the analysis carried out. Actually, the software used to perform the analyses is specific for emergency conditions analyses, therefore it was necessary to adapt simulation parameters in order to approximate ordinary conditions.

Shown parameters derive from the analysis of three existing school buildings, with similar features to the case study of the school in Posada: number of students, size and functions of spaces (Locatelli and Pellegrini 2017). The three schools were chosen as they represent positive examples of the application of innovative educational

Table 1 Simulation parameters

User speed	User speed was decreased in order to fit user behaviour in ordinary conditions, instead of panic conditions
	Primary school students move faster than secondary school ones. Indeed most children run, so their speed was set higher
Distance among users	The distance among users was decreased to simulate interaction and conversation
Sequence of activities	Flexibility in choosing the activity to be performed was increased to approximate the decision-making process of users in ordinary conditions. User priority is no longer just to reach the safe place as if in emergency simulations
	Users can choose the sequence and number of activities, according to defined probability functions and attractiveness percentages for each space
	The total percentage, as a sum of the above percentages, is equal to 80%, leaving students a 20% possibility not to choose any activity and stay in class
Duration of the analysis	The duration of the analysis has been set depending on the period of use of each space, related to its function

methodologies. The analysis of the schools by means of Pre-occupancy simulations allowed to measure the levels of crowding, number of interactions, maximum and minimum number of users in common spaces during the school break. As a result, it was possible to define benchmarks to compare the analysis results on the case study.

4 Results of the Case Study

The outputs of the Pre-occupancy simulation software are frequency and density maps (cf. Fig. 2), showing crowding data in any point in the space. Collected data are helpful to define main flows and space occupation indexes, allowing the evaluation of size and ability of the spaces to meet the future users' needs.

Density maps allow to investigate overcrowding, and represent a useful tool for designers to reduce discomfort. The value defined as "uncomfortable" is 3 people/m^2, which is the maximum acceptable value for dynamic crowds (Still 2000): over this value, people moving in a crowd can generate clashes that can be dangerous (Fruin 1993). As a result, it is possible to identify the most crowded areas and correct their size, shape and features.

As previously defined, the number of interactions provided with the analyses can be used to identify the capability of the building to facilitate interactions. For this purpose, spaces can be modified and optimised in order to both increase interactions and decrease overcrowding phenomena (Locatelli and Pellegrini 2017). Simulations on the information model, indeed, allow to anticipate the effects of the future occupants'

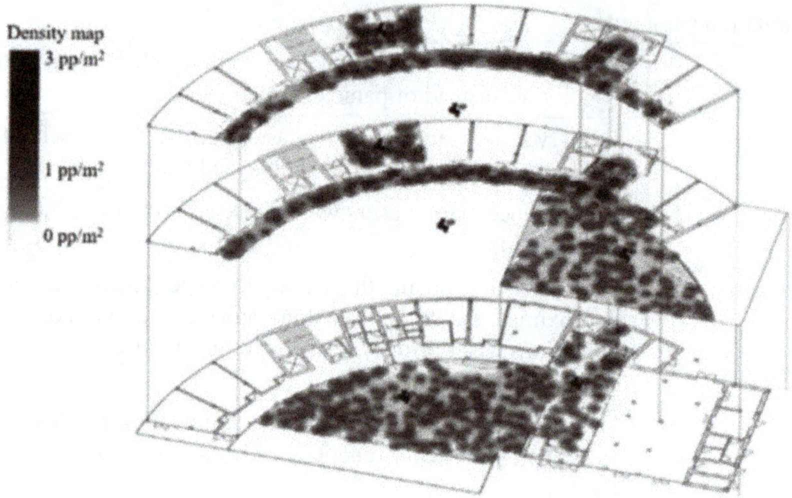

Fig. 2 Density maps showing overcrowding in some exits, stairs, hallways and spaces (Locatelli and Pellegrini 2017)

use of the building and their interactions (Shen et al. 2013), resulting in changes of designed spaces features.

Simulations on the case study identified high density values in some areas causing issues related to management of spaces (cf. Fig. 2). Analyses showed the need to review some aspects of the common spaces' design.

Simulation results highlighted the following issues:

- Overcrowding in entry and exit flows: this led to an increase in the number of exits to reduce the intensity flows.
- Dangerous values of overcrowding on the stairs: this result highlighted the need to review design and features of the stairs connecting the school levels, in order to avoid overcrowding phenomena. Stairs was widened to safely accommodate all the students.
- Spaces with low levels of interactions: the number of interactions in common areas was compared to optimum values defined through the three case studies. Spaces with few interactions were rearranged and their shape adjusted.
- Spaces with high levels of interactions: areas with high quantity of interactions were relocated in the layout, enlarged (when possible), or their exits were redesigned in order to match the expected large flows.

The most remarkable aspect of these results is that exits, hallways and stairs, adjusted on the basis of analysis data, were already sized according to fire regulations and other standards. As a result, the analyses revealed issues related to the actual use of the designed school and to the complexity of the planned users' flows. As stated before, fire prevention regulations define length and width of escape routes and stairways, number and width of emergency exits. These are static values: the legislation does not take into account the effects of possible funnel-shaped flows phenomena occurring in the case of large users' flows. This can cause both delays and discomforts in ordinary conditions, and dangerous situations in emergency conditions. The method proposed, by means of Pre-occupancy simulations, allowed to detect, assess and manage this kind of issues.

5 Conclusions and Further Developments

The proposed method and its application to the case study allowed the optimization of the design solution referring to usage patterns data during the early design stages. It is noticeable that spaces designed according to norms and requirements do not provide efficient levels of comfort, and do not avoid overcrowding, as shown in the simulations. Data collected from the analyses were useful to underline and solve issues related to overcrowding and interactions' level.

One of the main advantages is related to the possibility for designers to check and test their intuition in terms of layout. According to the simulation results, they can therefore modify and change the design solutions, selecting the best fitting one.

The evaluation of interactions among users also allowed to respond to the Client requirements in terms of educational approach. This kind of check would have been almost impossible without these simulations.

Another key aspect is the possibility to carry out Pre-occupancy simulations in the initial stages of the design process, by means of the information model. During these phases, variations on the design spaces provided minimum costs for changing and maximum impacts on the design quality.

The provided case study is a relatively simple building, but was helpful in defining a flexible and valuable methodology to optimise the design solution. Further developments include the application of the method to larger buildings, with several types of users and interactions patterns. The application of Pre-occupancy simulations may ensure the definition and evaluation of usage patterns, resulting in the optimisation of design solutions. Collected data may also allow the definition of a plan to manage user's flows depending on interaction patterns. This may avoid congestion, which can cause discomforts and delays in the operation of buildings such as airports or large facilities. This approach may therefore lead to greater benefits for complex buildings, where compliance with regulatory limits may not ensure quality and safety of spaces.

References

Almeida JE, Rosseti RJF, Coelho AL (2013) Crowd simulation modeling applied to emergency and evacuation simulations using multi-agent systems

Axelrod R (1997) the complexity of cooperation: agent-based models of competition and collaboration. Princeton University Press, Princeton

Ciribini ALC (2016) BIM e digitalizzazione dell'ambiente costruito. Grafill

Di Giuda GM, Villa V (2016) Il BIM. Guida completa al Building Information Modeling per committenti, architetti, ingegneri, gestori immobiliari e imprese. Hoepli

Fruin JJ (1993) The causes and prevention of crowd disasters

Ijaz K, Sohail S, Hashish S (2015) A survey of latest approaches for crowd simulation and modeling using hybrid techniques. In: 17th UKSIM-AMSS international conference on modelling and simulation, pp 111–116

Italian Parliament DM (1998) Criteri generali di sicurezza antincendio e per la gestione dell'emergenza nei luoghi di lavoro, 10 Mar 1998

Italian Parliament DM (1975) Norme tecniche aggiornate relative all'edilizia scolastica, ivi compresi gli indici di funzionalità didattica, edilizia ed urbanistica, da osservarsi nella esecuzione di opere di edilizia scolastica, 18 Dec 1975

Locatelli M, Pellegrini L (2017) La modellazione informativa per il Progetto Iscol@, il nuovo campus dell'istruzione a Posada. Politecnico di Milano

Montella DR (2012) Fire safety management

Royal Institute of British Architects (RIBA) (2013) Plan of Work, UK

Santos G, Aguirre BE (2004) A critical review of emergency evacuation simulations models

Schaumann D, Pilosof NP, Date K, Kalay YE (2016) A study of human behavior simulation in architectural design for healthcare facilities. Annali dell'Istituto Superiore di Sanità 52(2016):24–32

Shen W, Zhang, X, Qiping Shen G, Fernando T (2013) The user pre-occupancy evaluation method in designer–client communication in early design stage: a case study. Autom Constr 32:112–124. Elsevier

Simeone D (2015) Simulare il comportamento umano negli edifici. Un modello previsionale. Gangemi editore, Roma

Still GK (2000) Crowd dynamics. University of Warwick, Warwick

Tang F, Ren A (2008) Agent-Based evacuation model incorporating fire scene and building geometry. Tsinghua Sci Technol 13(5):708–714

Zimmerman A, Martin M (2001) Post-occupancy evaluation: benefits and barriers. Build Res Inf 29(2):168–174

Technical-Scientific Support for the Definition of the Project for the Reconstruction of School Buildings Involved in Seismic Events

Emilio Pizzi, Maurizio Acito, Claudio Del Pero, Elena Seghezzi, Valentina Villa and Enrico Sergio Mazzucchelli

Abstract This research regards the development of the project of the new school of San Severino Marche. The school is located in a region severely affected by the earthquake, and for this reason the design of the new building was based on a high degree of structural capacity, as well as strong innovations on typological and technological level, in accordance with specific educational needs. The research work is developed through a BIM approach that allowed the proper coordination of the disciplines involved.

Keywords Reconstruction · School building design · Information modelling · Seismic design · Design process

1 Introduction

This research work is part of the programme set by Government Commissioner for the reconstruction of areas affected by the earthquake of August, 24 2016, in the framework of the Memorandum of Understanding between the Commissioner and the Conference of Rectors of Italian Universities.

The research steps carried out aimed at facilitating all the project activities, in order to reduce times of project development, and to reduce construction times of heavily damaged buildings. At the same time, the goals of this research include the coordination of advanced disciplinary skills, leading to the design and construction of new buildings with a high degree of structural capacity, as well as strong innovations on typological and technological level, in accordance with specific educational needs. The remarkable research work is based on the use of a BIM approach

E. Pizzi (✉) · M. Acito · C. Del Pero · E. Seghezzi · E. S. Mazzucchelli
Architecture, Built Environment and Construction Engineering—ABC Department, Politecnico di Milano, Milan, Italy
e-mail: emilio.pizzi@polimi.it

V. Villa
Department of Structural, Geotechnical and Building Engineering—DISEG, Politecnico di Torino, Turin, Italy

S. Della Torre et al. (eds.), *Buildings for Education*, Research for Development,
https://doi.org/10.1007/978-3-030-33687-5_17

and methodology, in order to optimize the coordination of the various disciplines involved.

The project area assigned from the Conference of Rectors to the research group referring to ABC Department was the most challenging, due to its extensions and to project costs, among all the work programs defined in the Memorandum of Understanding. For this reason, the engagement of a multiplicity of available professionals with different skills was crucial to guarantee the quality of the intervention. The project activities were carried out with the collaboration of Area Tecnico Edilizia, the technical office of Politecnico di Milano.

The project activities, carried out in 2 months, required the definition of all the contents of concept and developed design. Flexibility was a primary aspect of the project development, allowing for potential future adaptation of functional and organizational layouts. Another relevant concern was seismic vulnerability, related to the special conditions of ground acceleration of the area, managed through innovative construction techniques. The use of dry construction systems helped in both these aspects, implying—in addition—high-performance levels in terms of energy consumptions reduction, in the context of optimization of circular economy concepts.

The Technical Institute E. Divini is a relevant institution in the region, as it provides advanced technical education to professionals in an active manufacturing environment, located in Macerata province. Before the earthquake, the institute hosted five specializations, divided in 32 classes, resulting in the total amount of 657 scholars, with a growing trend.

The amount of scholars used for the project development, defined by the Provincia di Macerata for the reconstruction, was set on 800 scholars. Considering this requirement, the project choices led to the definition of a building with a volume of 47.177 m^3. The previously existing building had a volume of 39.000 m^3.

The new designed building will stand, together with the laboratories that are currently under construction, in the linear part of the area, located near Viale Mazzini. The main part of the building will be located in the Southeast-corner, be-tween Viale Mazzini and Via Monte Conero, near the Sports Center.

Several aspects guided the project concept and building layout: the morphology and geometrical features of the area, the optimal solar orientation, and, most importantly, the opportunity to create a building that could act as a landmark, in the corner of the urban block. This space was previously marked by an indefinite cluster of volumes, lacking of organizing approach.

The integration and coordination of different competencies and skills was possible thanks to a BIM-based approach.

The presented project work can be seen as a relevant case study for future research collaborations, with the goal of supporting project activities related to emergency conditions. Disastrous events, such as earthquakes, require in fact well-timed interventions, but also the adoption of innovative strategies.

2 Information Modelling Approach

The information modelling strategy of the new buildings was guided by four main approaches:

- Compliance with regulatory and legislative consistency (DM 1975);
- Contextual modelling on various discipline's models, due to the connection among them;
- Coordination and management of clashes between elements and components of the various disciplines involved;
- Coordinated export of project drawings and documents.

The number of models created for the entire school building is eight:

- 1 architecture model for each building (A-School building and B-Gym);
- 1 structural model for each building;
- 1 building systems model for each building;
- 1 mass model for the third building (building C), hosting laboratories, that is already under construction. This building was modeled as a unique volume, as it is not part of the contract. This building was anyway taken into account during architectural design in terms of space distribution and volume management of the entire building site.
- 1 Master model, where all the models are linked. The Master model also includes the surroundings and the external areas of the buildings.

Considering technical systems, only HVAC systems are modeled, excluding therefore water related systems and electrical systems; this choice is due to the larger dimensions of HVAC ducts.

The three models of building A and building B have been linked before being inserted in the master model, in order to allow clash detections among the elements of the different disciplines. For this reason, the structural model is linked inside the architectural model, and then linked inside the building systems model (Fig. 1).

The information modelling operations of the buildings allows to take advantages of all the potentials of this approach from the early stages of design (Di Giuda and Villa 2016). The model has been gradually detailed during the process, from concept design to technical design.

In concept design, the project idea was translated from freehand sketches to models. The first step regarded the production of mass models representing the buildings' volumes integrated in the existing context, taking into account site layout and

Fig. 1 BIM model (architectural, structural and building systems models) of building A—School

altitudes. This operation was useful to develop the best-fitting shape of the buildings in relation with the surroundings, the solar path, and the shading patterns. A second step regarded an analysis of the relation between the new buildings and the existing one (building C), defining the distribution of internal spaces. The internal layout was carried out based on minimum surfaces, orientation, and connections with the central building hosting laboratories.

The second phase of the project regarded the progressive development of details in the models through the integration of technical elements (façades, roofs, vertical walls, and horizontal slabs) and spatial elements, such as rooms. Considering minimum dimensions of spaces, the specific requirements and legislative standards were taken into account.

During the project development, areas and volumes of the rooms were often changing due to the moving of internal walls and the changes in thickness of load-bearing walls. Nonetheless, the use of BIM approach and code-checking allowed to control rooms' areas, guaranteeing a continuous check of minimum surfaces (Tagliabue and Villa 2017). This control was carried out through tables in the BIM-based software, where calculation rules facilitate the checking of parameters and variations (Fig. 2).

Interoperability between Autodesk Revit and Midas Gen, a software for structure calculation, have been tested. Non-graphical documents, such as bills of quantities, have been exported through Quantity Take Off from the BIM model.

Render images were realized using 3ds Max, exporting from Autodesk Revit and choosing materials and textures for objects and surfaces (Fig. 3).

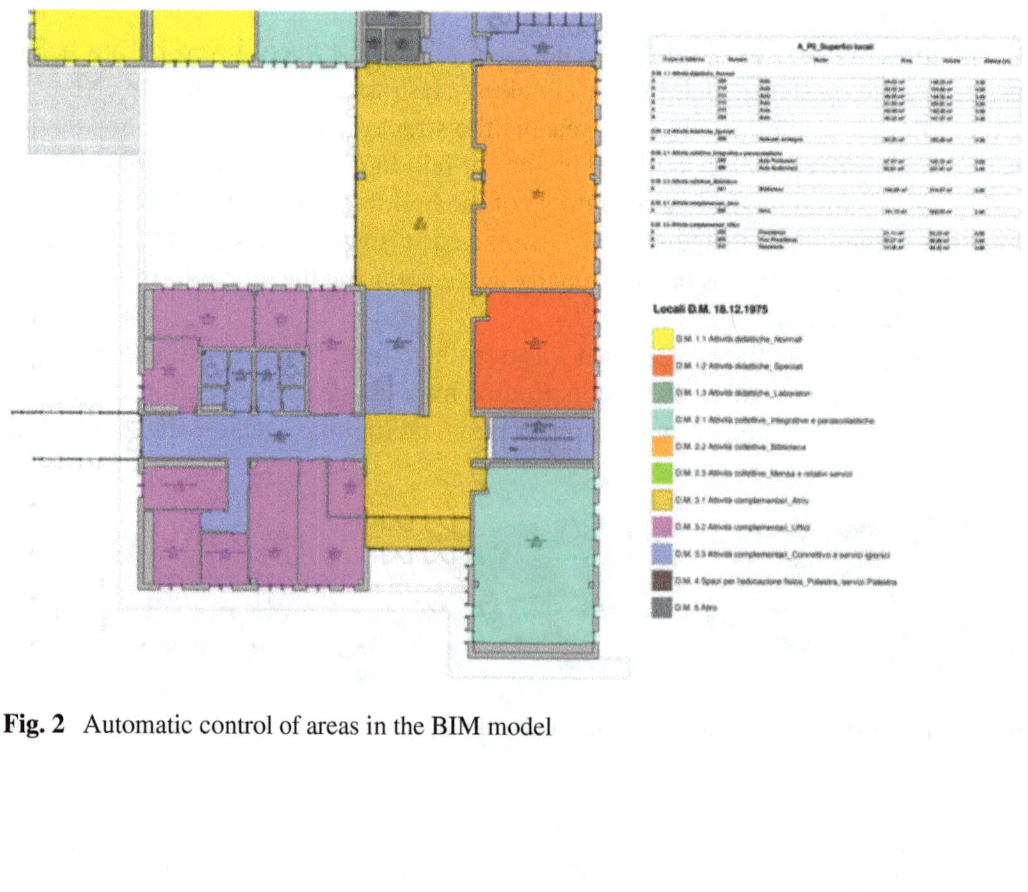

Fig. 2 Automatic control of areas in the BIM model

Fig. 3 Render image of the main entry

The use of BIM modelling resulted therefore in several advantages: the continuous control over the intervention's budget, the comparison of different layouts, the iterative approach to project design, and finally the fulfillment of a balance between the Client's requirements and the total cost of the building (Eastman et al. 2018).

3 Structural Design

3.1 Objectives and Design Criteria

The following are the guiding principles and the innovative techniques that inspired the design of the structures of the new ITIS Divini School, in San Severino Marche. The performance requirements considered in the structural design have concerned both the need to optimize construction times and the seismic safety requirements, dictated by the knowledge acquired on the local seismic hazard (Fig. 4) and by the will, given the frequency with which important seismic events are repeated in this area, to minimize the effects of damage to the structures, according to more rigorous multilevel performance, required by modern seismic design. The project involves the construction of two buildings. The first (main building), containing classrooms and laboratories, has a C shape in plan. The second, constituted by the gym with attached changing rooms and services. For brevity, we refer here to the main building only.

The San Severino school reconstruction project has been developed considering materials and techniques that optimize the fulfillment of the objectives of:

- rapid construction;
- recyclability of the components and basic materials;
- durability of the materials;
- maintenance costs minimization;
- low construction costs (contents within 1100–1200 €/m^2).

The design solution, in addition to fulfilling these objectives in the best possible way, was strongly conditioned by the following constraints:

- the connection with the project of mechanical Laboratories already in a construction phase;
- the availability of an underground level (given by the demolition of existing buildings).

The reference parameters for the seismic design are the following (NTC 2008):

- Seismic area: II
- Class of use of the structure: IV
- Nominal design life VN = 50 years
- Importance factor Cu = 2 (buildings with strategic functions)
- Reference seismic life (VR = VN Cu = 100 years)
- Ground type: C
- Topographic category: T1.

The definition of the design seismic actions to be adopted was refined through the analysis of local seismic response, carried out on the project stratigraphy. This analysis led to the definition of the SLV response spectrum shown in Fig. 4.

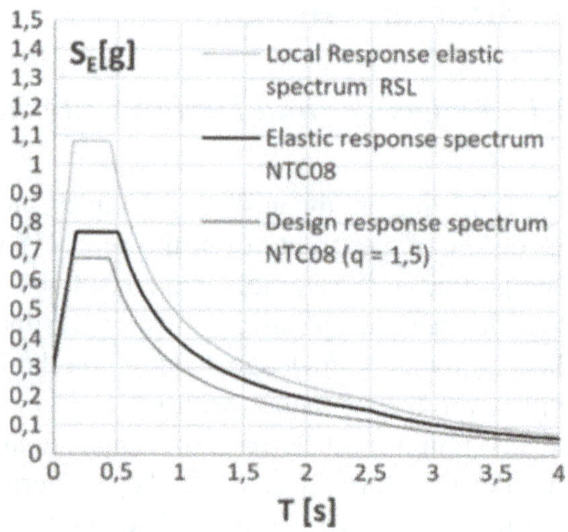

Fig. 4 Design response spectrum

3.2 *Description of the Structures of the Main Building*

The structure was based on a system that would allow to have the advantages of using reinforced concrete precast elements and assembling by components, similar to metal structures (Acito 2019).

The structural system was chosen in order to best optimize the objectives of points 1–5 and the constraints of points 6, 7 of the previous paragraph (Acito and Lavermicocca 2019).

The assumed system uses as a seismic-resistant system (Fig. 5), a classic structural system with post-tension pre-stressed reinforced concrete (p.r.c.) shear walls (Fig. 5).

The structural system is completed with the use of linear structural elements and panels, partially or totally pre-casted, which in the assembly phase guarantee an adequate self-supporting capacity for the construction phases, designed for vertical loads only (Acito and Jain 2019).

In particular, the structural system assumed for the structure is currently provided by prefabrication companies in the sector, with different commercial names, for which the structural elements differ only in some detail aspects. The innovation introduced in this project, perhaps for the first time in Italy, is linked to the idea of resisting the seismic actions with the only p.r.c shear walls, designed for the realization of stairwells, whose geometric and mechanical consistency was found to be adequate only thanks to the introduction of post-tension (Acito and Chesi 2019). In the specific case, the assumed system uses the following structural components:

- seismic-resistant walls in p.r.c. (Fig. 5);
- precast columns in c.a.;
- alternatively, partially precast columns could be used, made with reinforced concrete filled steel tubes columns (RCFST). In which the concrete reinforcement is designed to withstand the vertical design actions and the steel tube designed for the construction phase actions and for the fire resistance (Circolare n.617 2009);
- composite steel-concrete beams, partially pre-casted with the designed reinforcement for self-supporting in the various construction phases;
- decks made with pre-stressed self-supporting hollow core panels on the entire length, for all the construction phases (Fig. 5).

The design was developed in the final design following the procedures of the integrated design in the BIM environment. The structures were calculated using the Midas Gen. software (Fig. 6).

4 Building Services

One of the main design purposes was the limitation of the building energy consumption through the use of high performance envelopes, high-efficiency HVAC (Heating Ventilation and Air Conditioning) systems and by the exploitation of renewable

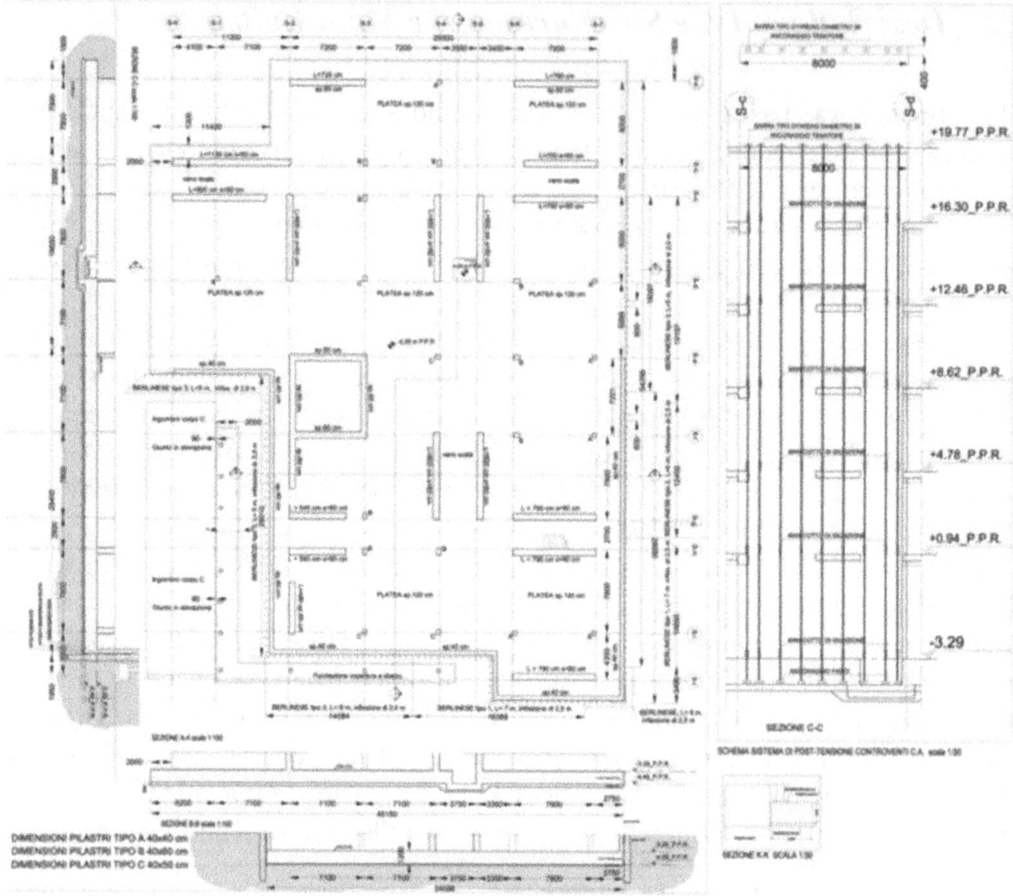

Fig. 5 Structural system: **a** Foundation plan; **b** Post-tension scheme of the shear walls

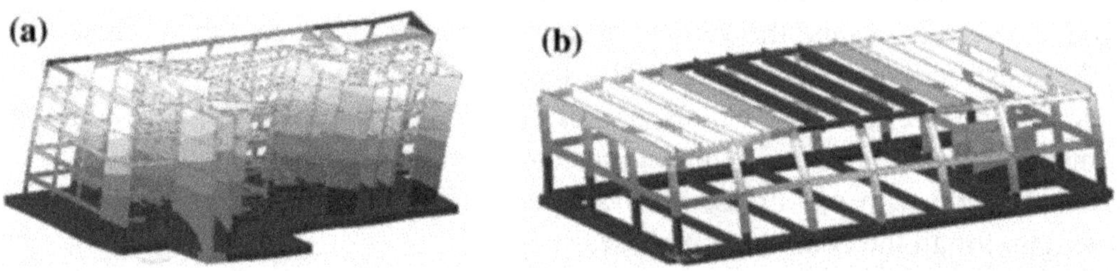

Fig. 6 Numerical models: **a** Numerical model of the main building; **b** Numerical model of the gym

energy sources. In this regard, the building services design of the Divini Technical Institute was oriented towards integrated solutions able to guarantee a high environmental comfort with a minimum amount of non-renewable primary energy, providing an efficient use of energy resources and optimal indoor conditions according to outdoor climate, crowding, equipment use, etc.

In detail, the generation system consists of two high efficiency hybrid air/water compression heat pumps, equipped with a total heat recovery system. One of such heat pump is fully dedicated to the air conditioning system, while the second one is used for heating purposes but with priority to domestic hot water (DHW) production. These units are equipped with a primary air exchanger and a secondary water exchanger, that can be connected to a ground water source or to solar thermal collectors if necessary, further increasing the overall system efficiency. Furthermore, a considerable part of the electricity demand of the heat pumps is generated by a photovoltaic system installed on the building.

Classrooms and offices are equipped with low-temperature and low-inertia radiant panels. In this way, it is possible to easily obtain both indoor comfort conditions and high energy efficiency, with a particularly low environmental impact. This is achieved also through a dedicated building services control and management system, monitoring internal conditions, external climatic variables and actual occupation of the spaces in real time.

According to the most advanced guidelines, the HVAC system has been designed to ensure the highest flexibility, with solutions that allow to obtain optimal thermo-hygrometric conditions with low installation costs. The AHUs (Air Handling Unit) are characterized by a high-efficiency heat recovery system, both static and thermodynamic, and by an adiabatic-type cooling (and humidification) system. They can operate at variable flow rates by modulating the fan rotation speed, in order to maintain a constant pressure/depression in the supply/return ducts according to the number of the served classrooms. This because the air conditioning of each classroom can be activated or deactivated at any time according to the effective need. This solution introduces a significant energy saving since, typically, the occupation profile of classrooms can be different.

The supply and return ducts layout has been integrated in the BIM model, in order to identify and eliminate any interference with structural and technological systems. This because the building systems' integration had to achieve two main goals:

- the functional coordination of the various technical subsystems in order to optimize the overall performances and their control;
- the spatial coordination of the various subsystems, to simplify their installation, accessibility and maintenance, in coherence with the needs of the periodic total or partial replacement of their components. In this regard the use of BIM becomes fundamental.

Furthermore, all the mechanical systems are controlled by a supervision system. The objectives of the centralized regulation and supervision system are the following:

- reduce the building services management costs;
- ensure a continuous building services monitoring;
- increase the overall efficiency and service life of the system, supporting a scheduled and preventive maintenance (thus reducing to a minimum the possibility of failure).

Therefore, the choice of all the mechanical systems was carried out with the aim of optimizing the energy performance of the building, but also of making the systems extremely reliable, competitive and monitorable, as required by the particular building use. Finally yet importantly, the use of advanced and modern technologies, their integration into the building, as well as the use of renewable energy sources, take on a significant educational and application function, especially in a technical school.

Acknowledgements The authors would like to mention and acknowledge Eng. Francesco Paleari and Eng. Marco Schievano for their valuable contribution to the project development.

References

Acito M (2019) Thermal and axial compression behaviour of full-scale RCFST columns exposed to fire: experimental study. Insights and innovations in structural engineering, mechanics and computation. Taylor & Francis Group, London

Acito M, Chesi C (2019) RCFST columns exposed to fire: non-standard thermal test and "hot" compression test results discussion. Insights and innovations in structural engineering, mechanics and computation. Taylor & Francis Group, London

Acito M, Jain A (2019) RCFST columns exposed to fire: residual compression test results discussion. Insights and innovations in structural engineering, mechanics and computation. Taylor & Francis Group, London

Acito M, Lavermicocca V (2019) RCFST columns exposed to fire: standard thermal test results discussion. Insights and innovations in structural engineering, mechanics and computation. Taylor & Francis Group, London

Circolare n. 617 (2009) Istruzioni per l'applicazione delle nuove norme tecniche per le costruzioni di cui al decreto ministeriale 14 gennaio; 2008

Decreto Ministeriale (14 January 2008) Norme Tecniche per le Costruzioni (NTC)

Decreto Ministeriale (18 December 1975) Norme tecniche aggiornate relative all'edilizia scolastica, ivi compresi gli indici di funzionalità didattica, edilizia ed urbanistica, da osservarsi nella esecuzione di opere di edilizia scolastica

Di Giuda G, Villa V (2016) Guida completa al Building Information Modeling per committenti, architetti, ingegneri, gestori immobiliari e imprese. Hoepli

Eastman C, Teicholz P, Sacks R, Liston K (2018) BIM handbook: a guide to building information modeling for owners, designers, engineers, contractors and facility managers. Wiley

Tagliabue LC, Villa V (2017) Analisi del patrimonio scolastico e strategie di intervento. Hoepli, Milano

"A Factory for the Future": Inveruno New School

Tomaso Monestiroli, Francesco Menegatti, Maurizio Acito, Giuseppe Martino Di Giuda, Franco Guzzetti and Paolo Oliaro

Abstract School is a rather complex issue that involves a range of different disciplines—the technical disciplines of architectural, system and structural design, along with the disciplines of training that define the educational project, as well as the disciplines of sociology and urban studies. Given such complexity, the relationship among public institutions such as City, Province and Regional administrations and the seats of scientific research such as University Departments where the above-mentioned specific expertise is developed, becomes fundamental for an innovative school project.

Keywords School · Architecture · Inter-disciplinary · Sustainability · Nzeb · Anti-seismic · Public authority collaboration · BIM

1 Introduction

Here, school is intended as a relational space open to the territory and designed to host public activities accessible to the entire urban community. The new school of Inveruno is, in this sense, the new civic center of the city, a representative building and a place of cultural integration. The school is designed to offer flexible and permeable spaces (sliding walls, movable furniture, glazed rooms, covered and open-air collective areas, etc.) where innovative education becomes the key principle to provide students with adequate skills. Therefore, the school offers in equal measure individual spaces for education and study, spaces for exploration where students may experiment and as a group practice the skills they have acquired (cross-disciplinary workshops), and group spaces where they may present and discuss the results of their work with the school and city community. Just as important are the open space between the buildings and the central square designed to host open-air events that involve the entire school community.

T. Monestiroli (✉) · F. Menegatti · M. Acito · G. M. Di Giuda · F. Guzzetti · P. Oliaro
Architecture, Built Environment and Construction Engineering—ABC Department, Politecnico di Milano, Milan, Italy
e-mail: tomaso.monestiroli@polimi.it

S. Della Torre et al. (eds.), *Buildings for Education*, Research for Development,
https://doi.org/10.1007/978-3-030-33687-5_18

2 Urban Context

Inveruno, a town with 8600 residents, is part of the metropolitan city of Milan in the Lombardy region.

Its urban structure mainly resulted from the prevailing agricultural activity that experienced a renewed impulse after the opening of the Canale Villoresi in the second half of the nineteenth century.

The modification of the territory after the construction of the canal resulted in a new system of smaller canals and in the innovation of the agricultural activity and brought remarkable transformations in the history of the town. During the modern age, Inveruno owed its urban expansion to the creation of industrial activities such as the Muggiani textile mill, the Officine Elettriche Colombini and more recently the Belloli oil-mill.

The development of industrial activities resulted in a gradual decline of agriculture with remarkable consequences on the urban structure of Inveruno. Later on, the decommissioning of most industrial facilities has left several sites now requiring adequate redesign and redevelopment as they play a strategic role in the urban structure.

The urban fabric of Inveruno comprises a first and oldest core connected to the establishment of the rural hamlet and featuring linear buildings that follow the morphology of the territory. Their layout defines closed blocks with interior courtyards. This oldest core is complemented by sections of residential fabric comprising low houses within which the landmarks of the town community, starting with piazza San Martino bordered by the parish complex, emerged. A third evolution followed this second development, which appears more relevant in terms of quantities rather than for settlement reasons. This is a disjointed and patchy fabric made of one- or two-family houses built during the phase of industrial development. This expansion resulted in an uncontrolled sprawl within which, however, the old structure defined by orthogonal hydrographic canals that shape a road network that still organizes the actual expansion of the town's urban boundaries is still recognizable.

Within this heterogeneous context, the industrial site of the decommissioned Belloli oil-mill, now undergoing a design rehabilitation, is located in the northeastern part of the town and the urban void that defines it plays the role of a cornerstone between two kinds of road tracing of the urban expansion. Therefore, the rehabilitation of this decommissioned site, in terms of its location and contextual features, is strategically important for the urban transformation of Inveruno.

3 The Site

The decommissioned Belloli site is a large urban void bordered by via Brera, via Fratelli Bandiera, via IV Novembre, and the provincial road 129. Built in 1919, the Belloli industrial facility increased its activity in the post-WW2 period and closed down in 1979. The large void of the decommissioned factory within the town was surrounded along the perimeter by industrial facilities demolished in 2009 for safety reasons and still features a towering reinforced concrete silo used to store seeds. Built in the 1960s, the silo is a reinforced concrete structure supported by "V"-shaped pillars that emerges as an actual landmark in the town and over the years has almost become a historic monument in the life of its residents. The site is currently in a state of disrepair and heavy deterioration. Its rehabilitation is necessary because it occupies a central spot along a green axis established by the park across from the Town Villa overlooked by the town library that extends along viale Piemonte, connects the areas of the Inveruno Sports Union and further on, the green areas of the Luigino Garavaglia Town Stadium. This green axis is particularly important for the Town of Inveruno as it connects the areas that traditionally accommodate the old San Martino Fair, the main agricultural fair of the region for over four centuries.

4 Architectural Design

The principle underlying the design of the new school complex results from the belief that the current buildings fail to meet some fundamental requirements as a good quality educational facility. The two primary schools have small classrooms and lack collective spaces and workshops, without mentioning the fact that their cafeterias are in the basement. The secondary school building is hardly functional in terms of the standards of a modern secondary school: the building is not properly insulated and therefore underperforming on an energy level and the implementation of safety measures from the static point of view would require a significant investment.

The project results from the belief that the school represents a place of primary importance and recognition for the town community within the urban structure. For this reason, the project addresses multiple levels in the rehabilitation of the decommissioned Belloli oil-mill site. The core of the project is an open public space created for the gathering of the entire community. The two schools with their sports facilities and a small auditorium that closes the perspective from via IV Novembre overlook this new green square. The school complex is conceived as a small campus where open space is prevailing and the layout of the individual buildings acquires a particular importance. Indeed, in order to respect the vocation of this large urban void, the buildings are recessed from the boundary of the roads so that they create widenings and resting spaces for the town.

Both buildings, a primary school and a middle school, feature an open courtyard overlooking a central square with a slight rotation that follows the layout of the context in order to create articulated and differentiated volumes around the central square as well as to distinguish their sites and accesses.

The middle school complex lies to the north-west from via IV Novembre and comprises the classrooms and the sports buildings, while the open-air sports facilities and the cafeteria directly connected to the school's building lie in the back from the square.

At the southeast, there is the primary school complex comprising, from the entrance, the indoor sports facilities followed by the actual school building in a recessed position from the road. Like the first building, this complex has its green open spaces and the cafeteria at the back. The buildings are autonomous and independent volumes also in terms of their potential of use by the citizenship. The school offers sections others than the educational facilities that can be used in different ways as well.

5 The Single Buildings and Their Interior Spaces

The middle school building includes four classes with attached workshops and special classrooms, while the primary school includes three classes and an additional Montessori Method program for one class of the school of Furato, a hamlet of Inveruno.

The middle school has a courtyard layout with a system of load-bearing columns that define its inner and outer perimeter. This solution allows treating the elevations in a differentiated manner by opting for either glazed walls or opaque infills according to the needs. This construction system also offers additional advantages in terms of transparency and visual openness as it optimizes the potential of natural lighting. The school is accessible through the green courtyard overlooking the square. At the ground floor, a large lobby connects the interior courtyard and the garden in the back, a large collective space designed to host temporary exhibitions of the students' works. All the classrooms at the ground and first floors along with the associated workshops overlook the large central courtyard according to a layout that benefits from the best sun exposure. The generous distributive system becomes an informal space (a fundamental element in the guidelines of innovative educational facilities) where alternative education- or study-related activities may be organized. The staircase cores and services are located at the sides of the courtyard so that they are immediately visible and accessible. One of such cores provides access to the locker rooms of the gym through an underground passage.

The class-B gym is designed to host junior league provincial and regional sports games. The facility relies on load-bearing walls and features two entirely glazed and shielded elevations overlooking the courtyard. The gym's main elevation directly overlooks the new green square so that the building may be accessible independently from the school. Spectators and users may reach the gym directly from a lobby in order to get either to the stands on the field's sides or to the locker rooms and ancillary services at the underground level. From the northern elevation, instead, it is possible to access directly the outdoor sports courts in the school's large garden.

One last small building accommodates the cafeteria directly connected with the school building from one of the distribution cores through a covered and heated passage. The cafeteria is entirely glazed and openable towards the exterior.

The primary school has the same features as the middle school except for the necessary distinctions related to the latter facility. Classrooms and workshops are similar in terms of size to those of the middle school, except for the selection of specific furniture that guarantees high flexibility in the subdivision of spaces. The gym is smaller than the one of the middle school and designed to accommodate motor activities for children. It is likewise independent and directly accessible from the green square even for after-school programs such as sports activities that require smaller courts—martial arts, yoga or dance classes.

The cafeteria is larger in this case and divided into smaller "rooms" in order to avoid overcrowding strongly discouraged by the scientific community of educators.

A civic hall completes the square in the terminal part as a facility designed to operate independently from the school and host activities for both students and the entire town community.

6 Technical Design Choices

The technical choices reflect the compliance to the following main criteria: reduction of energy consumption, reduction of the building's environmental impact, reduction of construction times, construction and use flexibility, simplified maintenance and management.

7 Building Life Cycle

The project relies on the use of construction technologies based on the dry assembly of single components. Vertical prefabricated concrete structures are mounted on a concrete basement for the construction of the load-bearing frame and are completed by floors and interior laminated wood walls designed to guarantee a fast and efficient construction on the one hand, and dimensional precision and flexibility of spaces on the other hand. Prefabrication allows for a high quality level due to the possibility of selective dismantling and replacement of parts in case of maintenance. The absence of the seasoning times required by concrete and the installation of completed components allow for a shorter construction phase. A particular care was devoted to the distributive flexibility of classrooms, which have no structural elements dividing them and therefore may be repurposed in case of changed use requirements simply by dismantling and moving the wood dividing walls. The two courtyard layouts allow for an optimal distribution network of systems, the connection ridges of which are in the readily serviceable false ceiling and in the raised floors in order to guarantee flexibility of use and an easy maintenance.

8 Materials, Safety and Comfort

The wooden elements of the floors allow for a perfect reaction in case of earthquake as they perform as a monolithic plate unbounded to the concrete frames. The now consolidated use of both elastomeric and sliding seismic isolators guarantees a high resistance to earthquakes. Aside from technological aspects, we tried to provide the building with features that would guarantee maximum efficiency to emergency escape routes in case of danger.

The insulation materials we chose are of natural origin and recyclable; they provide a high stability of performance across time and maximum fire resistance. The wooden floors are completed by cork suspended ceilings, which are highly efficient in reducing noise reflection.

In order to ensure the natural ventilation of spaces, the pivoted windows are equipped with vertical opening leaves. Low-emissivity glass and rolling blinds complete the equipment of windows in order to guarantee high levels of comfort.

The geometry of the façade, which is recessed from the floors, allows for a satisfying control of solar radiation, while the stringcourse beams perform as brise-soleils in order to protect the elevations.

The walls are enclosed in ventilated walls equipped with interior insulation and a particular care was devoted to the elimination of cold bridges.

Green roofs concur to the control of solar radiation. If sown with low-water requirement plants such as succulents, they will guarantee a further thermal insulation to the surface.

The rooms are equipped with underfloor radiant heating panels; each room is equipped with temperature control calibrated on crowding and solar radiation. A ventilation system channels filtered, hot and humidified air into the rooms through airflow vents and high induction diffusers with air intake vents and grills in toilets and corridors.

9 Environmental and Energy Sustainability

The widespread reliance on prefabrication, the careful design of the shell and the use of natural materials, efficient glazing and solar shields complement energy-efficient air conditioning systems.

The production of domestic hot water is centralized and fueled by heat pumps. Winter/summer air conditioning exclusively relies on renewable power sources.

The geothermal system fueled by reversible heat pumps is the only source of thermo-refrigeration. The system we propose relies on a station with two heat pumps, one of which functioning in a polyvalent reversible mode. Both concur to cover the winter energy supply, while the polyvalent heat pump covers the summer energy supply of the building, as well as the yearly production of domestic hot water (DHW) with a total recovery of heat during the summer period.

A system of photovoltaic panels would be installed on the roof of the gym.

The school is as a relational space open to the territory designed to host public activities for the entire town community. In this sense, the new school of Inveruno is intended as the new civic center of the town, a representative building and a place of cultural integration. The school provides flexible and permeable spaces (sliding walls, movable furniture, glazed classrooms, covered and outdoor collective spaces, etc.) where innovative learning methods become the key element in helping the students acquire the right skills. For this reason, the school equally offers individual spaces for learning and study, exploration spaces where students may experiment and practice as a group the skills they have acquired (cross-disciplinary workshops), and group spaces where they may present and discuss their works with the school and town community. Just as important are the open space between the buildings and the central square where open-air events that involve the entire school community may be organized (Figs. 1, 2, 3 4 and 5).

Fig. 1 Aerial photography photomontage

Fig. 2 View of the main courtyard garden

Fig. 3 View from the high school courtyard

Fig. 4 View from the high school courtyard garden

Fig. 5 General plan

Field of Education and "*Corpus Socialis*"

Riccardo Canella and Micaela Bordin

Abstract The Italian school system is still affected by the "Berlinguer Reform"—it never entered into force but it became the basis of every subsequent reform that was implemented—for which the legislator intended to compensate for the imbalances of the Italian school, in the relationship with mass-studies, with the transposition of European directives that have substantially changed the schools of every order and degree and introduced a "3 + 2" structure in the university educational system, stiffening the entire school cycle and causing further fragmentation. The essay presents a pilot project of a reversible wooden pavilion as the primary nucleus of (the) experimental teaching, for the recovery of degraded and typologically insufficient public schools in Milan, but also for the reuse of the "mother houses", the farmhouses in Lombardy and also for the restoration of the "*forum*" in the Italian places damaged by the earthquake.

Keywords Architectural composition (architectural design) · Architectural theory · Italian architecture · Typology · School complex project · Prefabrication

---

Riccardo Canella, Micaela Bordin with Alessandro Piacentini, Camilla Laura Pietrasanta. Translation by Laura Canella and Guido jr. Canella.

R. Canella (✉)
Architecture, Built Environment and Construction Engineering—ABC Department, Politecnico di Milano, Milan, Italy
e-mail: riccardo.canella@polimi.it

M. Bordin
Milan, Italy

S. Della Torre et al. (eds.), *Buildings for Education*, Research for Development,
https://doi.org/10.1007/978-3-030-33687-5_19

1 Public Compulsory School in Municipal Complexes for a New Culture-Civilization: Project for the "Carro di Tespi" Pavilion[1]

"*Test suckers and roots to build a new ethnicity*".[2] If we consider how it is possible that the origin (*the poleogenesis*) of Italian cities (whether large, medium or small) is to be attributed fundamentally to "facts of structure" (in the meaning given by Ferdinand de Saussure)—as the historian Henri Pirenne seems to suggest, paraphrasing the well-known thesis about the paleogenetic dualism of the medieval city (fortified *nucleus* and mercantile town) that became functional to the proto-capitalistic structure (Henri Pirenne 1927)—we could therefore argue that the interaction between the resources and the endowments of the city-center and resources and endowments of the peripheral area and of the metropolitan concentric structure becomes necessary to the understanding of the *phenomena* of transformation into an inalienable coherent destiny.

If after all it is considered possible that the most careful and in-depth analyzed urbanistic critique now tends to favor this hypothesis, it is also true that, in the not so disorderly growth of the same Italian cities, hazarding a generalization, there are entirely original characters that distinguish, in adherence or in transgression, the destiny of some of the most representative and emblematic among them.

It would seem that these characters may depend precisely on the "suckers" and on the "roots", that is the propensity of the humanized environment and of the work culture that takes place in the relationship with the "*longue durée*" (like *Annales*), underlining once again the relative autonomy of architecture and of composition (Guido Canella 1969), but also suggesting the way to complete its "knowledge". An architecture of the city that is capable, precisely, of bringing that "knowledge", when it is considered "behavioral" architecture, which we believe must be the basic philosophy and ultimate goal filtered into the project and into the construction.

[1]The Carro di Tespi (or Pavilion) were mobile theaters built with covered wooden structures used by the comedians of the popular Italian nomad theater for their street theater, starting from the late Nineteenth century. They were mounted "on the town square" and remained set up for 40/50 days during which the companies of the "guitti" wanderers recited a different script night after night, exhausting completely their repertoire. They owe their name to the mythical figure of the theatrical actor Tespi d'Icaria, described by Horace in the *Ars poetica* and were anchored to the idea of a mass theater with a strong emotional impact and capable of conveying theatrical culture to forgotten sections of the population. The fascist regime used this experience to build an outdoor traveling theater in 1929.

[2]For a constructive intervention strategy by parts in the polycentric city: "*Saggiare polloni e radici fino a costruircene nuova etnia*". Title of the famous essay by Lucio Stellario d'Angiolini published in "Hinterland" n. 4, *For a metropolitan museum*, monographic number dedicated to the museum, July–August 1978, pp. 50–54.

And it would not seem on the other hand, still generalizing, to contradict this hypothesis with the appropriate differentiation between capital cities, military cities, trade fair cities, ration cities, etc. These non-ordinary, genuine characteristics of Italian cities, superimposed on the "structural facts" that have conditioned their development, would seem to be the expression of a whole culture, even if specific point to point. What then could be a *minimum* common denominator that can confirm the belonging of these cities to that level of merit that is attributed to the "boroughs of Italy" and that can condition the project of the "modern"?

It could be, for instance, a geographical factor, which is declined at first in the great Italian cities[3] divided, for example, by climatic bands (north, south, but also coast, countryside, mountain); or for medium and small cities it could be the effect of irradiating the characters of the same major cities of reference on the territory and the other way around. Characters that combine and recombine with different degrees of intensity and elaboration to create a "skein" whose in-depth and punctual deciphering is decisive for understanding the true nature of every Italian city.

However, it seems to be the so-called second and third order poles, precisely the "boroughs", that contain in their genetic heritage—typological and figurative but also urbanistic-morphological—that clarity and transparency of "behavioral" intentions that it seemed to have been reached with the medieval construction of the primary space (the town square) of the compact city in a system however "polycentric", that of the "boroughs of Italy".[4]

We could therefore be led to suppose that the structuring factor for excellence is represented by the *agorà*, understood as an assembly[5] and by its permanent surrogate, the public school, considering the epistemological question and the academic political-cultural distinction.

---

[3]Among the "big" names, the cases of Venice, Rome and Florence are memorable, but also those of Milan, Turin, Genoa, Naples, Palermo and of the others, which have become such since the realization of central quarters surrounding the square of the government, of the ritual and of the exchange, generally in the Middle Ages, up to the construction of the so-called "historical periphery" (this is the term-concept with which Guido Canella used to define the first suburb of Milan, the productive one, which seems to have been able to express its own original "character" since its formation), at times capable of relating to the countryside in a fruitful relationship of reciprocal regeneration, prevailing a physiocratic conception *ante litteram*, but at times also a reservoir of that workforce capable of sustaining and reviving the fortunes of the city itself, this, on the other hand, belonging more specifically to the modern era.

[4]Pisa, Siena, Lucca, Verona, etc.—considering its consolidated historical centers with only the adjacencies of the "historical periphery", omitting the opportunistic and troubled expansion of the second half of the twentieth century—, but also for example Syracuse, naturally considering only the island of Ortigia with the adjacency of the neighborhood and the port on the inland sea.

[5]Ultimately by the "school of Athens", understanding as a representation of the seven liberal arts: grammar, arithmetic, music, geometry, astronomy, rhetoric, dialectic.

For a didactic offer that is coherent in a renewed global course of studies that are truly "of the *futuribles*"[6]—in the context of the Italian public school of every order and degree—a role that seems to us to be decisive should be covered by the instruction given by the universities, as well as by the research that is carried out within them in the name of them. But we are obliged to acknowledge that the didactic offer presented to university students doesn't follow a coherent academic organizational program, even if we consider a new faculty and the implicit epistemological and methodological assumptions.

Therefore, a possible new direction for a degree course today should inevitably be placed in coherence with a general reference assumption (perhaps incorporated as *stigmata* of the same faculty), with the sense of belonging to a critical thought.

Nevertheless, belonging to that partisanship that is consistent with the assumption, we are trying to introduce into the debate to circumscribe and define a "problematic and operational"[7] approach to knowledge, aimed at forming critical intellectuals and not just specialists or professionals with a trade.[8]

The coordinated professors within this new direction, beyond possible differences for cultural positions, would necessarily be united by the same "holistic" conception of reality, by virtue of which the approach to knowledge can be global, dialectical and historical, in total antithesis with that of ontological and methodological individualism, or "Robinsonian".[9]

---

[6]"Città dei futuribili", an architectural column curated by Guido Canella, which appeared from 1968 to 1970 on "Il Confronto", a magazine on politics and culture, in which appear the first critical writings by Guido Canella and his friends M. Achilli, G. Polesello, A. Rossi, F. Tentori and others.

[7]This term-concept defines the modalities of teaching developed by the research group "Architecture and City" coordinated by Guido Canella and Lucio Stellario D'Angiolini in the Faculty of Architecture of the Polytechnic of Milan from the Sixties of the last century.

[8]The experience of "field research" therefore seems to be the only cognitive approach that, on the one hand, allows an authentic contact with reality and an accelerated scientific education—creating among students interest in an unprejudiced study of the chosen problems and the need to corroborate it through direct relationships with the operators involved—and, on the other hand, allows a partialization of theories without necessarily renouncing organic conceptual relationships. The cultural project would be proposed as an "activity project"—in view of a critically assumed structure framework, in function of a policy of interventions capable of affecting the nature of the development of cultural and productive forces, their organization throughout the territory and the expectation generated by society, in the search of a new culture-civilization—and not as a mere expression of a good "scholarly" attitude and an involvement in the standards of the discipline. And never less as a tool of neoliberalism that is dominant today, functional to the needs of the market, guaranteeing an operational flexibility that the timely satisfaction of particular interests would demand from time to time.

[9]Here understood as "individualism" in the sense given by Marx in the following essay: Karl Marx, *Formen, die der kapitalistischen Produktion vorhergehen* (1858), Dietz Verlag, Berlin, 1952.

Learning in this way could express itself in maximum awareness as a dialectic expression of a historically determined civilization. By virtue of an adequate ability to interpret the needs of society, it would be able to stand out on the identity of the European (and of the world) city and on those of the historical and problematic essence of its disciplinary heritage, escaping from a notion of cultural project which today is increasingly equated with the pursuit of the vogue too often claimed in the global market of postmodernist culture.

If we become aware of the underlying gnoseological and epistemological discriminating factor, perhaps the spaces intended for education should be reformulated in reverse order: from the configuration of a university to the possible configuration of a school complex that includes high school, passing through middle school, to end with primary school and kindergarten, where the configuration of a middle school would have a dominant role, as Giuseppe Samonà had already underlined back in the 1960s: "*It is likely that in the future there will be large localizations of educational establishments of middle schools that will be much more significant than* (*the*) *universities, because in them the intelligences will mature and a very lively social life will be formed.*

So it could be said that, by filiation, the school, considered as a functional and figural device, should present that same typological "icasticity" and that predisposition towards the central role of the "behaviors" of the space of life, if not universal, of the center—church-palace-square—of the "boroughs of Italy". Those same behaviors that are necessary for the learner to build their own critical intelligence corroborated by the juxtaposition of preparatory spaces delegated to their formation.

The research for a new way of child and of adolescent education based on "doing", able to put the student at the center as an actor and not just as a user of their development, a new "Montessorian spring"[10] (Fig. 1) seems to be a viable way within a *scenario* that appears to be completely fluid. To achieve these goals today the school should radically transform and renew itself, with an "ontological-social"[11] attitude, into a "school-laboratory" made up of different *ateliers*, special rooms aggregated around a space that we could define as a "library" or as a town square "forum *scentiam*-forum of knowledge", where children can carry out appropriate activities and become aware of the cognitive problems to be deepened through the aid of books

---

[10]According to Maria Montessori—and according to Friedrich Fröbel, Rosa and Carolina Agazzi, but also according to Giuseppina Pizzigoni, Rudolf Steiner and others—school education should have overcome the division between theory and practice and favor a critical learning method based on direct and concrete experience.

[11]See the essays on the subject by Georg Lukács and Costanzo Preve.

Fig. 1 Phototypesetting for the presentation of the project references. E. Beaudouin, M. Lods, *École en plein air*, Suresnes, Paris 1932–35; «Hinterland» directed by G. Canella, n. 17, 1981 and n. 3, 1978, dedicated to the subject of education; G. Folli, *Open air school in the Trotter*, Milan, 1918–1927; R. Steiner, *First Goetheanum*, Dornach, 1908–25; «Casabella-Continuità» directed by E. N. Rogers, n. 249, 1961 and n. 245, 1960, dedicated to the subject of school; T. Crosby, *Shakespeare's Globe Theatre*, London 1997; G. Canella, P. Bonaretti, *Technical Institute Giambattista Bodoni*, Parma 1985; G. Canella, *Service Center Piazza Monte d'Ago Quarter*, Passo di Varano, Ancona 1984; A. Belloni, *Primary School Rinnovata Pizzigoni*, Milan 1924–27

and, above all, learn the art of permanent assembly as a form of culture-civilization.[12]

The project[13] involves the prefabrication of a medium-sized structure, a "special" classroom to be placed in the courtyards of public schools of every order and degree that requests it, typologically preordained for those special operating activities that the teaching requires, when it wants to have the features and characteristics of a "problematic and operating" approach.

We are naturally favoring the atavistic distinction that there is between the work of industrial design—dominated by practicality as a form of knowledge induced by the dominant traction of ergonomics applied to the "object of use"—and the work of architecture—pervaded by practicability as a form of knowledge induced by the dominant traction of the typology applied to the public building for the city. It has been difficult for us to find away to identify "knowledge" in the structuring "composition" process that has characterized this research on prefabrication as an architectural product of a work that otherwise could be attributable to the design of the "object of use". Thus it is uprooted from any "allocation context", while still abstracting from the practice of reconstruction, imitation or anastylosis and evoking instead, in the construction of this prefabricated humanized environment, a tendential "approach by figures".

The theme is therefore the search for the possible conformation of a reversible pavilion that can be inserted in the courtyards of the degraded and typologically insufficient public school complexes, but also for the reuse of the "mother houses", typical of the irrigated countryside farmhouses, and also for the restoration of town squares in places damaged by the earthquakes, as the primary nucleus of experimental teaching.

The project requires the constructive completion of a school complex in the outskirts of Milan,[14] located at the intersection of two waterways, the Naviglio Grande canal and the Lambro river, characteristics that make it a microcosm, but unfortunately "wounded boroughs" hit by degradation and neglect, and characterized by instances of superfetation that have over time altered and mutilated their practicability (Fig. 2).

We therefore chose a planning strategy that envisaged the re-triggering of the endemic territorial polycentrism which, as always, also involved the small Italian cities, the so-called "boroughs", and thus, by extension, the foundation cities, the additions of parts of cities, the quarters and "formally completed parts of cities".[15]

[12]The content of teaching has always been the study of human activities placed critically within a new mode of development—alternative to that of globalization or of generalized, polarized and financialized oligopolies, as Samir Amin calls them (in a recent essay later formalized in 2012 in the text *The Crisis*)—in the hypothesis of a coexistence of the capitalist system and the socialist system, trying to overcome the seasons of capitalism that have alternated as follows: competitive capitalism, oligopolies and monopolies, organization of the transnational network.

[13]Developed in collaboration with the M2B Medie Montessori Association and conceived by its co-founder Andrea Perugini.

[14]The Ilaria Alpi State Comprehensive Institute in Via San Colombano.

[15]Carlo Aymonino used to denounce the relationship that his architecture establishes with the city.

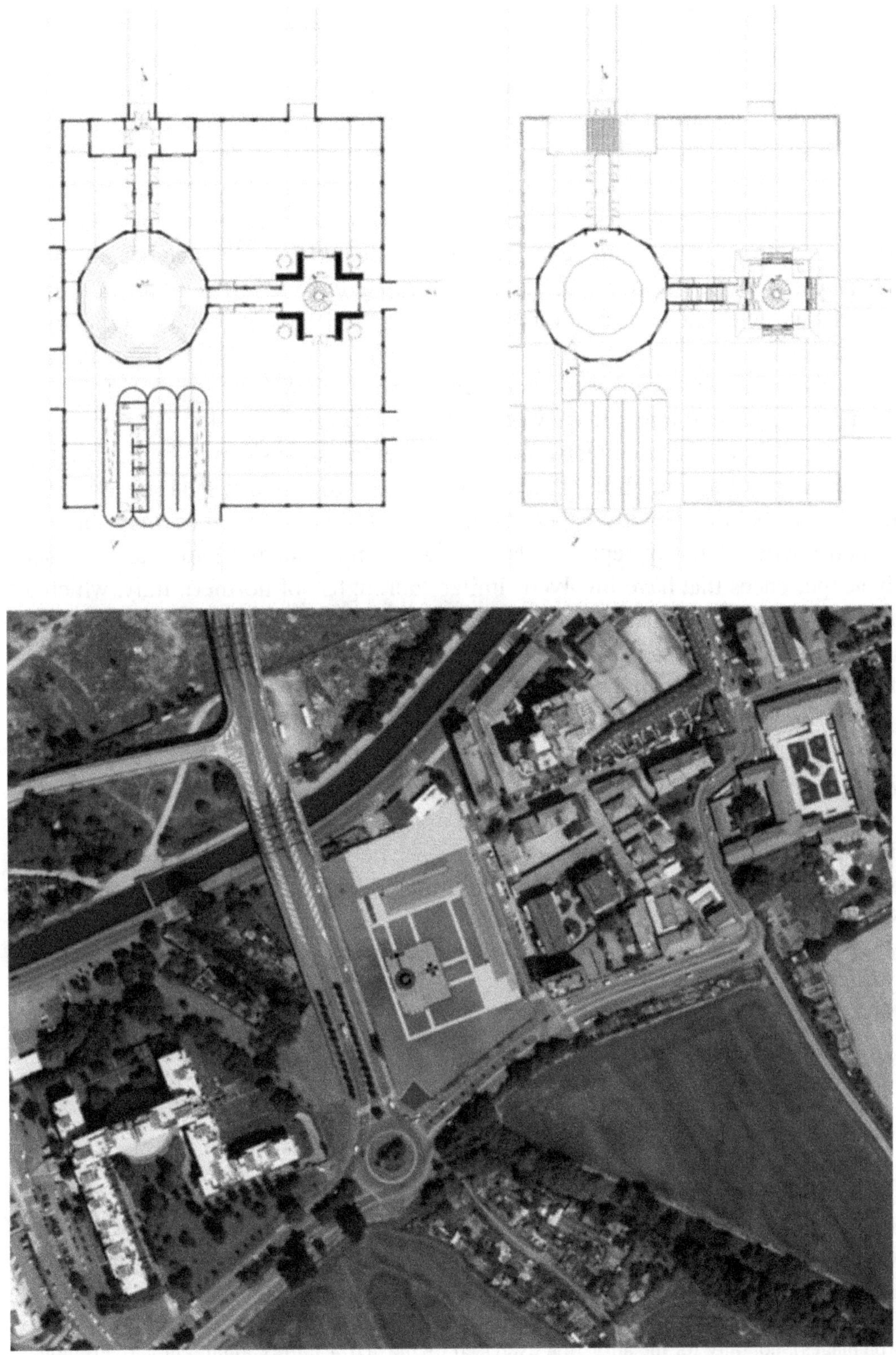

Fig. 2 The pavilion "Carro di Tespi". Ground floor plan, First floor plan, Masterplan

The pavilion, "Carro di Tespi", is envisaged as prefabricated with laminated wooden structures and floors, and vertical infill walls in sandwich panels complete with building services and insulations, such as x-lam system. The panels are uniform, modular, square of three meters side, self-supporting and interchangeable, with three finishing solutions, so as to allow the construction of different sequences of "reversible" aggregation spaces, so to speak "in palimpsest". For the reconstruction hypothesis, therefore, the research has provided for the maintenance of the original building, a work that would seem part of that eclecticism "of manner" of Milan ("Novecento" style without frills and tinsel) which, in the complicity of some architects dedicated to the construction of the city of the 1930s, it prefers a relationship with the context, not directly from a geographical point of view, nor from an exclusively historical point of view and not even from a purely linguistic point of view, but from a more general point of view of "evocation" (Figs. 3 and 4).[16] The concept of "evocation" that these architects of the "Novecento" style seem to be transplanting for a criterion of assimilability that can be defined "of distance in absence" and "of temporal detachment" with respect to the chosen models, and it seems sublimated in their poetics. This concept can also be, for example, an alternative to other decisive experiences that have involved similar "boroughs" of northern Italy, which are prerogative of enlightened entrepreneurs (like Adriano Olivetti in Ivrea) who were able to operate in those same years through the wise planning and construction of the already industrial suburbs.

This practice of "evocation", in the case of the school on Via San Colombano, seems to be an appropriate path to conform to, in the act of composing "by figures", not so differently, on the other hand, from what Leonardo da Vinci undertook proposing for Milan—city for which the most original proposals were made over time, even though they have almost always been betrayed—the project of placement of his "giant" equestrian sculpture, which should have been allocated either in the Sforza internal courtyard (the *Rocchetta*) of the *Castello Sforzesco* or in the transplantation of the "Square of Central Italian tradition" newly formed in front of the Castle itself.

[16]This attitude is more appropriately reminiscent of Central Italy, where it would seem that these same architects received assignments from various institutions, even "total" (Army, Ministry of Education but also the Vatican, for the construction of prisons, schools, asylums, hospitals, churches, orphanages), not only for the attraction exercised by the "*Urbe*", the capital, but perhaps more as compensation to those common "rebels" and "secessionists" of the belt of Rome, because of the mutilated regional membership (the "secession" from the Lazio Region). These same architects would seem to operate through a series of institutional "grafts", according to a practice of "evocation" of the ancient place of the decentralized government out of town, which brought "municipality" into the "*ager centuriato*", countryside.

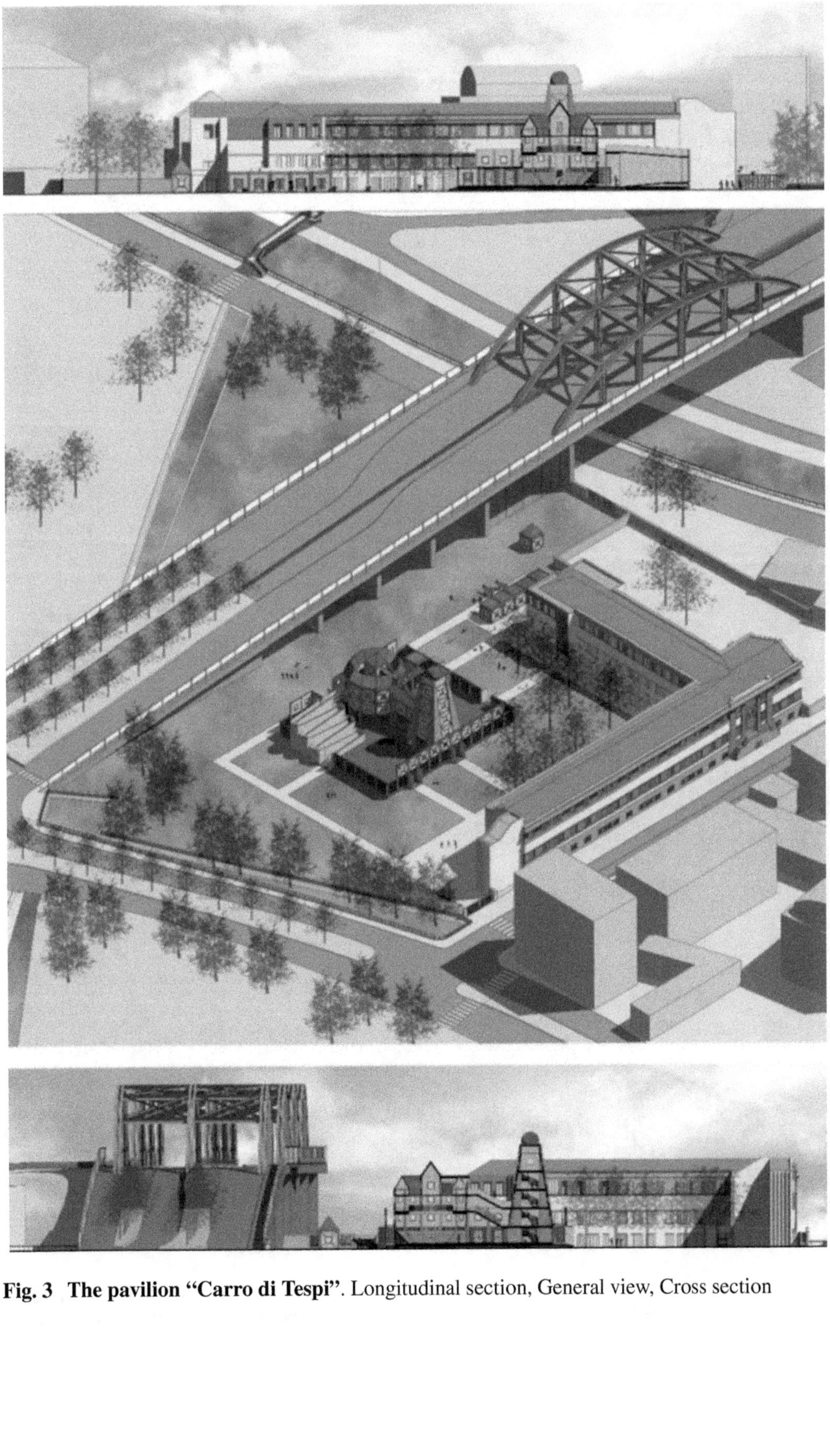

Fig. 3 The pavilion "Carro di Tespi". Longitudinal section, General view, Cross section

Fig. 4 The pavilion "Carro di Tespi". Front elevation east, Photographic insertion of the project in the Ilaria Alpi school, Front elevation north–west. (For all the images, rights are reserved for authors.)

Reference

Giuseppe Samonà in Vv.Aa., La città territorio, Leonardo da Vinci Editrice, Bari 1964

Guido Canella, Un ruolo per l'architettura, Clup, Milan 1969

Henri Pirenne, Les villes du Moyen Âge, essai d'histoire économique et sociale, Lamertin, Bruxelles 1927

Karl Marx, Formen, die der kapitalistischen Produktion vorhergehen, (1858), Dietz Verlag, Berlin 1952

Lucio Stellario d'Angiolini, Saggiare polloni e radici fino a costruircene nuova etnia in "Hinterland" n. 4, For a metropolitan museum, July-August 1978, pp. 50-54

Samir Amin, La crisi. Uscire dalla crisi del capitalismo o uscire dal capitalismo in crisi?, Punto Rosso, Milan 2009

Space-Places and Third Teacher: The Issue of Architectural Space in the Age of Knowledge Cities and Schools 3.0

Laura Anna Pezzetti

Abstract Information and knowledge are not synonyms; rather they are quite distinct facts. As a form of knowing never separated from the critical processing of subjects, knowledge is sensitive to space. The city as a *knowledge hub* demands a dense exchange of context where urban morphologies cannot be replaced by dispersed relations allowed by ICT networks and smart efficiency. Symmetrically in school buildings, that is, the basis of knowledge infrastructures, *learning architecture* is not replaceable by an unstructured environment, mechanically derived from a new flexibility allowed by digital technologies and specific mainstream views on "innovative teaching". This paper critically explores the role of architectural space in the age of 2.0–3.0 schools, discussing the relationship between transformations introduced by the unstructured classroom upgraded by digital technologies and new necessary experimentations on architectural space, the *third teacher*. Architectural space is not only an active player in influencing learning and development but is also a constitutive element in the formation of thought and a specific tool of critical, cultural and imaginative knowledge of reality. Organising space means organising the metaphor of knowledge.

Keywords Knowledge cities · School buildings architecture · Learning spaces · Third teacher

1 The City and School Building as *Learning Spaces*

In the century defined as the "century of knowledge" (Drucker 1968, 1997), the development and sharing of knowledge is said to be the driving force of Europe's economic competitiveness. Within the new economic, social and urban paradigm, the concept of cities as *knowledge hubs* has been consolidated as a desirable future for European cities (Lisbon Strategy 2000; European Year of Creativity and Innovation

L. A. Pezzetti (✉)
Architecture, Built Environment and Construction Engineering—ABC Department, Politecnico di Milano, Milan, Italy
e-mail: laura.pezzetti@polimi.it

S. Della Torre et al. (eds.), *Buildings for Education*, Research for Development,
https://doi.org/10.1007/978-3-030-33687-5_20

2009).[1] While in the 1990s many economists and a number of planners considered the city a residue of the industrial era, surpassed by the "erasing geography through technology" (Koolhaas 1994) or "the annihilation of space through time" (Harvey 1988), that is, by the indifference to locations favoured by networks, the knowledge economy considers the city as a physical place and the privileged context of exchange.

It is within the complexity of urban systems that knowledge is processed and disseminated; innovations, culture and creativity are produced (Florida 2002); talents are developed and accessibility to global economies is focussed on (van Winden 2010; Yigitcanlar 2007; Franz cir. in Yigitcanlar et al. 2008). It is in cities as well that universities, the true infrastructures of knowledge, find consolidation and development.

Information and knowledge are definitely not synonyms, but quite distinct facts. As a form of knowing never separated from critical elaboration, knowledge is sensitive to physical distance and demands a physical and dense exchange context: space, which is not replaceable by dispersed relations of ICT networks and Smart City technologies. The objective of a knowledge city is to promote the development, sharing and dissemination of knowledge by enhancing relationships, intersections and porosity between the various functional urban systems and, within these, between different fields and competences (Pezzetti 2012a, b). Specialisation and self-segregation do not correspond to the cross-sectoral and creative contamination on which innovation is based, as well as stiff boundaries between disciplines; they instead, today, constitute an obstacle to scientific research.

Cities themselves must be redesigned as *knowledge hubs*—that is "purposefully designed to encourage the nurturing of knowledge" according to Leif Edvinsson (2003 cit. in Dvir and Pasher 2004), a leading expert on intellectual capital—distinguishing not only for their ability in *clustering* innovation, creativity and research, but also for the opportunity of self-realisation and development offered to their citizens and the creative potential and resilience of the communities established.

Since the *century of knowledge* is not reduced to the "century of information", it will be necessary to also include in this horizon the context of exchange, critically rethinking the efficiency of Smart Cities (so free of hindrances and managed by big technological companies) and the role and culture belonging to physical, built and urban space (Pezzetti 2012a, b).

Aspects that are the basis of creative innovation such as *serendipity*, culture and community are also at the basis of urban life and challenge the rational deterministic approach to innovation itself.

The reorganisation of the city and its functional systems should entail innovative concepts of urban design and experimentation on new prototypes of learning spaces, which are key tools to achieve this aim.

---

[1] Starting from the Lisbon Strategy (2000) and more recently with the European Year of Creativity and Innovation (2009), the European Council has recognised knowledge and culture as catalysts of innovation and creativity. Cultural activities and the *creative industries* were thus recognised as potential for economic innovation and for individual and social development.

Demanding creative connections between the functional systems and the assets present in a given context, the characteristics of a knowledge society could finally reverse the tendency to build the city through mono-functional enclosures and isolated objects, boosting the role of urban morphologies as a continuous system of urban and spatial relationships. On the other hand, learning building needs to extend the regime of collective activities encouraging their integration and contamination with the urban structure.

The role of the city as the place, where knowledge concentrates and is transmitted, is inherent to the very origin of the European city, from the *gymnasia* of the Hellenistic cities to the cathedral schools of the revived medieval urban communities, and to the emancipation of *universitas* and the institutionalisation of compulsory and higher education as an indispensable guarantor of modern industrial development. Just as a certain degree of concentration and integration of multiple functions is inherent to the early places devoted to education, or the leading role they played in questioning consolidated schemes of knowledge and mentality.

The early type of Hellenistic *gymnasium*, for instance, showed a dynamic relationship between the introvert microcosm of its colonnaded peristyle and its opening to the whole city, as well as the integration of different activities: from the study plan that put literary, philosophical and grammatical knowledge on the same plane as physical education—which was the original purpose, to the simultaneous presence of lessons, conferences and banquets. The integration of the baths, in Roman times, juxtaposed the simple *xystus* system—the covered portico of the Hellenistic *gymnasium*—with an elaborate series of rooms, giving thus rise to the idea of inner space (Pezzetti 2016).

The challenge of the knowledge society evidently invests schools of every grade, which are "knowledge infrastructures" themselves since they form, pass down and certify knowledge, and because in turn they employ knowledge workers while integrating innovative forms of transmitting knowledge, by working in network with other schools and other social players.

Only in recent years, however, supporting the knowledge culture has been understood as investing in education at all levels, starting from the founding substrate of primary schools, while also stressing that only a country of educated people can move towards sustainable development (De Maio 2011).

Spread countrywide, school buildings constitute not only poles of education but also the resources around which the civic dimension of society condenses. Hence, it becomes important to consolidate their "second life", extending their fruition beyond the teaching time and turning them into civic centres, poles of reference for local neighbourhoods and in a network with the territory, no longer dedicated exclusively to ordinary learning but also educational, recreational, cultural and cooperative activities targeting a wider audience.

Creatively circulating the culture of places together with that of knowledge institutions cannot overlook the design of these places as a *constructed idea*. It is not merely about regulating and making school buildings safe, adapting them to educational changes or improving their "performance". It is about integrally transforming improper buildings or banal containers that are the fruit of prefabrication logic, entirely redesigning the built environment or, if impossible, converting it to other functions and replacing it with real knowledge architecture.[2]

Schools need to become *hubs* for lifelong learning, stimulating places not only for students but also for adults, teachers and staff, hosting hybrid spaces to produce, exchange and disseminate culture, while opening up to the outside, to the neighbourhood and the city. The innovation of the concept of classroom and collective spaces themselves must be able to encourage this change. The architectural space must return to be a further pedagogical tool. As Rogers wrote back in Rogers (1947) "If a sacrifice is imposed, no budget item is better justified" (Rogers 1953).

The current crisis places European cities and their economies in a state of transition. A condition which, according to Hall (2010), already proved to be favourable to cultural creativity and to the transition towards new and unexplored ways of organisation.[3]

2 Space-Places and Commonplaces. The Issue of Architectural Space in the Era of 2.0–3.0 Schools

If we aim to foster settlement systems as *knowledge hubs* according to a culturally broad sense, one that is also contextually specified, then instead of following a-critically up-to-date planning clichés (*creative cities, science cities, smart cities,* etc.) or programmes delegating innovation essentially to the collaborative sharing of the 2.0–3.0 web information, we should launch an approach that is in itself creative, experimental and multidisciplinary. A renewed attention to the educational character of space itself is needed, since space is the *third teacher* (Malaguzzi)[4]: organising space means organising the metaphor of knowledge.

Just as the knowledge city is not a spatially neutral phenomenon, knowledge buildings cannot disregard the knowledge intrinsic in architecture as a specific way of critical, cultural and imaginative knowledge of reality.

---

[2]According to the data provided by the Italian Ministry of Public Education, authorised school buildings in Italy amount to more than 42,000 (42,292), of which 33,825 are operational. Of these, 55% were built before 1976, and only 70% were built specifically for use in the education sector.

[3]"*An entirely different, historically-based approach came from the present author in a study of six 'creative cities' in Hall (1998): Athens in the fifth century BC; Renaissance Florence; Shakespearean London; Vienna in the eighteenth and nineteenth centuries; Paris between 1870 and 1910; and Berlin in the 1920s. The first three of these cities became culturally creative long before they proved very adept either at technological advance, or in managing themselves effectively. All enjoyed golden ages even while the majority of their citizens laboured in abject poverty, and even while most people lived in conditions of abject squalor—at least, by today's standards*" (Hall 2010).

[4]According to the educationalist Loris Malaguzzi, who in the WWII aftermath was the founder of the Reggio Emilia educational system, known throughout the world as the "Reggio Approach", children have three educators: adults, other children and the physical environment they are immersed in. The concept of the "third teacher" was recently resumed in O'Donnell WP, Peterson BM et al., *The Third Teacher* (2010).

With regards to school buildings, the issue was already clear to Ernesto Nathan Rogers back in 1947 when, complaining about the continual cuts in education's budgets, in an editorial for the magazine *Casabella* he declared that progressive pedagogy could not ignore its architectural dimension, since "the problems of education cannot be accomplished without a learning architecture" (Rogers 1947).

Architectural space is not only an active player in influencing learning and development. Since architecture is a language—while the vague and indeterminate concept of "environment" is not, architectural space is a constitutive element of the formation of thought: "Beautiful schools are good schools" (Rogers 1953).

New forms of teaching geared towards working individually, in groups or in workshops, to supplement traditional face-to-face teaching, require layouts of greater spatial complexity in school buildings, while at the same time allowing their evolution over time.

To stimulate conditions for multiple forms of learning, integrating the classroom with the school and the school with the community—while also tackling the challenge of a multicultural population and transformations in employment—the new schools neither will be confined to simply unstructuring the organisation of previous types nor stifling the challenges linked to the emergence of digital technologies by unstructuring teaching into modular multipurpose environments equipped with multimedia support while remaining substantially conventional in the poverty of their spatial qualities.[5]

Although early innovative cooperative theories and accomplishments were established after World War II by the Italian experience of the "Reggio Approach" led by Loris Malaguzzi, which have been and still are considered worldwide (Gandini and Gambretti 1997; Edwards et al. 2011) to be a leading reference[6] for the emphasis they place on creating beautiful environments to support children's emotional, cognitive and social development (the "challenging" and "creative" child, i.e. the future active citizen of the city), the difficulties of the educational system in understanding the need to give innovative teaching a consistent architecture emerge clearly.

Shifting from *learning environment* to *learning architecture*, some challenges, simplifications and contradictions appears.

In an interview I carried out in 2015, Herman Hertzberger while discussing the school built at Romanina outlined the distance between innovative educational approaches and the acceptance of real spatial innovations, which affected the design.

---

[5]The flexible learning spaces promoted by "Future Classroom Labs" by European Schoolnet are in fact substantially devoid of formal and architectural connotations, focusing solely on functional flexible aggregation of environments, modular furnishing and introduction 2.0–3.0 equipment. See Bannister (2017), *Guidelines on Exploring and Adapting Learning Spaces in Schools*; Mosa (2013), "Nuovi spazi per l'apprendimento", https://www.insegnantiduepuntozero.files.wordpress.com/2013/07/quandolospazioinsegna.pdf; the teaching model "without partitions and classrooms" like Vittra Telefonplan (2011) in Stockholm, assumed as a model in http://www.indire.it/quandolospazioinsegna/scuole/vittra.

[6]In 1991, the "*Newsweek*" devoted an article to the "Reggio Approach" describing it as the best educational system in the world.

Despite Italy's "New Guidelines" (MIUR 2013) have substituted quantitative regulations with performance standards, foreseeing five space-types (agorà, classroom, workshop, individual and informal space), the programme "Scuole Innovative" (2016) has failed to coordinate coherently the competition programme since most pilot projects required incoherent dimensioning of learning spaces based on previous legislation standards and related educational models.

Innovative programmes for school building design are an international trend but few EU countries have promoted it in a strategic and coordinated way, among which are the UK (BSF), Netherlands, Portugal (Parque Escolar) and Denmark (SKUB), although with different outcomes in architectural quality. The Danish SKUB "The School of the Future" (1998–2010), developed in Gentofte's municipality—where the milestone Munkegård School, designed by Jacobsen in 1956 was built—brought to the construction of the celebrated Hellerup School (2002), a "*school-container*" that repeats some themes developed by Hertzberger's schools although inserting them in an open plan. Even more extreme is the open plan of Ørestad Gymnasium (3XN 2007), a mega-block allowing for teaching and learning decks that overlap and interact with no distinct borders, the latter being perceived as obstacles to the fluidity of multidisciplinary interaction.

If those cases still rely on design, the overcoming of the classroom-based layout mostly relies on modular environments (e.g. Epping View Primary, Melbourne, 2009) then obviating the banality of the "*container*" by the upgraded design of sectional furnishing. Modularity and open space appear as easy shortcuts compared to questioning how to overcome the classroom/corridor opposition by granting the former a more permeable configuration and to the connecting spaces the value as a *collective classroom*, *square* or *Learning Street*, thus enabling students to explore different degrees of self-responsibility and self-learning possibilities.

Flexibility and multifunctional spaces in architecture—for example the schools by Canella and Hertzberger shown in different ways—are not necessarily synonymous with modularity and uniformity, namely the absence of architectural character as these concepts, already explored by manuals from the 1960s, and which subsequently failed, usually entail.

Should the school be just an "explicit rendition of the latest theoretical theories" as Dudek's book (2000) maintains?

Flipped classroom, digital teaching, IWB or BYOD are some of the practices derived from the UK and the United States that are taking foot in the EU, which see the alliance between the use of new technologies and active education focussed on the learner rather than on the teacher and on competences rather than on transmitted knowledge, not without raising some actual questioning on the theoretical weakness of their fundamentals.[7]

The challenges of digital technologies have produced the new cliché of 2.0–3.0 schools (soon 4.0), where the instance of the active participation of students seems to focus on interaction with the technological fetish-object, whether it be an IWB

[7]A petition against competency-based teaching was launched in 2018 in Italy, gathering a large number of people, professors and intellectuals.

or a *tablet* (BYOD), almost as if the content is guaranteed to the learner by the collaborative use of the *medium*—"the *medium* is the message", as the essay by McLuhan back in 1967.[8] Or as if knowledge could coincide with information, or the experience of space with that of a nomad who wanders from workshop to workshop with a *tablet* underarm.

The tactile and dynamic exploration of physical space, the manner in which the activities are structured, and the boundaries between individual and collective space seem to give way to the seduction (and overestimation) of a virtual reality that is simulated, and soon, perhaps, augmented; in any case dislocated and surrogate, to be experienced in a "halted motionless" vision inside a "container" that is spatially indifferent and undifferentiated.

Cooperative learning, it is said, "unstructures" the entire school, thus justifying the informal space with the revolution brought about by the ICT technological devices. However, if the various areas lose their distinctive character and everything becomes unstructured and indefinite, there is no longer much left to explore, exchange or recognise (Hertzberger 2008).

Form and thought are linked. The open plan, where all spaces and boundaries become blurred and flow into one another, as an ideological choice seems to reflect the informal and dissolutive character of contemporary socio-technical-scientific knowledge, which is the informal character of the *liquid society* and its *non-places.*

Yet, a learning space should not content itself with reproducing the *status quo* by becoming a mere description of the present. Architectural language does not merely transcribe the existing world but, like other artistic languages, produces it via the language itself (Pezzetti 2010, 2015).

The type linked to an architectural theme provides a structuring principle to which the spatial organisation of the parts and the identification of the various space-units that make up the school layout are subordinates, expressing a sense of unity, identity and construction for the community.

Holistic learning is based on the integration of knowledge and disciplines, not on their dissolution. Instead of dissolving the architectural space into modular, undifferentiated or blurred environments, a true integration of knowledge and creativity involves the ability to experiment on new types and their articulation in space-places, each devoted to proposing *centres of attention* through spatial themes, in a dialectic balance between individual freedom and a sense of belonging to the community.

The place where learning is developed jointly because of relationships with others is much more than the environment made up generically of open classrooms, workshops, shapes, colours, furnishings, yards and gardens. The school building is the first place where the learner experiences an architectural space that has the analogy and complexity of a small city and landscape, in its full richness of space-places, meanings and symbols, metaphors and metonymies attributed to forms; in the play of

[8]The reflection of M. McLuhan, in *The Medium is the Message* (1967), grasped early effects of the pervasive technological medium on the collective imagination regardless of the contents of the information conveyed.

different scales, heights and layouts which predispose and stimulate different kinds of behaviour; and in the expression of tactile and aesthetic-perceptive values.

A space rich in educational space-places can stimulate active learning, thus allowing students to pick and choose what captures their attention. Children are guided by what they see and by gaining experience of differentiated spatial units or strolling through the school via a sort of architectural *pedagogical promenade* they discover the possibilities of relations or learning offered to them.

Learning architecture and active education can therefore learn from museums, fostering the arrangement of building in multiple space-places, tasked with proposing as many possible *centres of attention* and stimulating emotional and aesthetic qualities.

In association with the classroom, conceived as a *home base* providing a feeling of identity, social spaces, informal gathering spaces, and single or group workspaces can form a continuous fabric made of different depths of field and heights, degrees of partition and sharing; rooms, habitable recesses, *squares* or multifunctional theatres; shaded patios, ramps and paths; and gardens. Those spaces also serve the school as a community centre, already emphasised by the Italian experience of schools related to typological criticism (Tafuri 1968), namely in the projects by Aymonino at Pesaro and Canella in the Milan hinterland.

They support the aesthetic emotion of knowledge and are necessary for the school to work as a community centre as well.

We build as we dwell, and reciprocally "only if we are capable of dwelling, can we build" (Heidegger 1951). Architecture as the *third teacher*, therefore, always plays a decisive role in influencing attention, learning, development and the initiation into the aesthetic and knowledge experience.

Architecture is the art of slowness and permanence. Design innovation and the ordering of space should never depend on any specific view on education, which is just a starting point for design. Architects should instead explore spatial conditions that favour and widen the possibilities for learning within a general framework that is flexible enough to respond to continuous changes in educational pathways (Hertzberger 2008) while being characterised around durable themes and spaces.

Significantly, the UK's BSF programme launched in 2003, aiming for high-quality school design,[9] commissioned by a number of pilot projects to a number of architectural firms selected expressly because they did not specialise in school buildings, asking them to rethink educational architecture from its fundamentals (DfES 2004).

There is a need and potential urge for a more in-depth fertile interaction between educational theories and architectural space.

The design of "open-air schools" between the 1920s and 1960s, Van Eyck's multifunction halls and Sharoun's *learning streets*, Jacobsen's morphologies and Quaroni's urban and mix-use composition, Canella and Rossi's typological *montages*, up to the latest prototypes by Equipo Mazzanti in Colombia, just to quote a few, turns

[9]Before 2011's cuts, the BSF programme involved the construction of around 706 of over 1400 of the new schools planned.

the simple transcription of transient educational theories into architectural themes and a quest for spatial resources (Pezzetti 2012a, b, 2016).

Louis Kahn's metaphor of the "school before school"—a group of people sitting under a tree, intent on exchanging their knowledge without even knowing that they are, respectively, teacher and students—leads the relationship between form and design to the need to rethink not only the building but even before that the institution, right from its statutes and founding principles (Kahn 1961a).

The spatial arrangement of a work of architecture does not in fact originate from the functional programme, which rarely includes the "problem", that is, the translation into spaces of a given institution's "will of being". Architecture is exactly what is lacking in the programme but which the architect offers to the aspirations of mankind:

> This is why I think it so important that the architect never follows the program given but simply uses it as a point of departure in terms of quantity, never of quality. For the very reason that the program is not architecture, it is simply an indication, like a prescription for the pharmacist. Because in the program, when it says *atrium* the architect must transform this into a place for entering. The corridors must become galleries. The *budgets* must become economies and the areas spaces (Kahn 1961b).

Moreover, as Ernesto N. Rogers stated (Rogers et al. 1965):

> The matter is to activate the concept of utopia: to think pragmatically of a better society […] There is no better place than the school to deal with such an issue […] If you think how necessary it is to forge the tools to overcome the difficulties of the world rather than to comply with the current conditions - with the illusion of a guarantee-, we must not only accept but also promote the use of criticism and imagination. Which are the cornerstones of architectural research.

The school, as the public building *par excellence*, must aspire to be a *learning urban architecture*.

References

Bannister D (2017) Guidelines on exploring and adapting learning spaces in schools. Bruxelles: European Schoolnet. http://www.indire.it/wp-content/uploads/2018/04/Learning_spaces_guidelines_ENG.pdf. Accessed June 2018

De Maio A (2011) L'innovazione vincente. Brioschi Editore, Milan

Department for Education and Skills (DfES) (2004) Exemplar designs, concepts and ideas. Department for Education and Skills, London

Drucker PF (1968) In the age of discontinuity. Harper e Row, New York

Drucker PF (1997) From capitalism to knowledge society. In: Neef D (ed) The knowledge economy. Butterworth-Heinemann, Boston

Dudek M (2000) Architecture of schools: the new learning environments. Architectural Press, Oxford

Dvir R, Pasher E (2004) Innovation engines for knowledge cities: an innovation ecology perspective. J Knowl Manag 8(5):16–27

Edwards C, Gandini L, Forman G (2011) The hundred languages of children: the Reggio Emilia experience in transformation. Praeger, Santa Monica

Florida R (2002) The rise of the creative class. Basic Books, New York

Gandini L, Gambetti A (1997) An inclusive system based on cooperation: the schools for young children in Reggio Emilia, Italy. New Directions for School Leadership (3)

Hall P (2010) Hard policy instruments and urban development. Bartlett School of Architecture and Planning—University College London, London

Harvey D (1988) Time-space and the postmodern condition. In: The condition of postmodernity: an enquiry into the origins of cultural change. Basil Black Well, Oxford

Heidegger M (1976) Costruire, Abitare, Pensare (1951). In: Gianni V (ed) Saggi e discorsi (tran: Gianni V). Mursia, Milan

Hertzberger H (2008) Space and learning: lessons in architecture. 010 Publisher, Rotterdam

Kahn LI (1991a) Form and design (1961). In: Latour A (ed) Louis I. Kahn: writings, lectures, interview. Rizzoli International, New York

Kahn LI (1991b) Statements on architecture (1961). In: Latour A (ed) Louis I. Kahn: writings, lectures, interview. Rizzoli International, New York

Koolhaas R (1994) Construire la ville à partir des infrastructures, entretien avec Rem Koolhaas. In: L'Architecture interieure créé. Histoire et enjeux urbains, Monographic number (262)

MIUR (2013). *Nuove line guida per l'edilizia scolastica.* http://hubmiur.pubblica.istruzione.it/web/ministero/cs110413. Accessed Dec 2014

Pezzetti LA (2010) Architettura senza composizione? (Architecture without composition?). A.C. (3), Pezzetti LA, Canella R, (eds) Architettura/composizione, Monographic number, pp 3–4

Pezzetti LA (2012a) Architetture per la scuola. Impianto, forma, idea (Architectures for Schools. Lay-out, Form, Idea). CLEAN, Naples

Pezzetti LA (2012b) Knowledge Hubs: la questione urbana. Città Europee e città cinesi. In: Corradi V, Tacchi EM (eds) Nuove società urbane. Trasformazioni della città tra Europa e Asia. FrancoAngeli, Milan

Pezzetti LA (2015) Architettura educatrice. Ideologia del reale o utopia della realtà? (Learning Architecture. Ideology of Reality or Utopia of Reality). In: Canella G, Manganaro E, Locatelli L (eds) Per una architettura realista. Maggioli Editore, Milan

Pezzetti LA (2016) Storia e progetto dell'edificio scolastico. Architettura educatrice nella città della conoscenza. In: Poli C (ed) Rivoluzione scuola. Valori, spazi, metodi. Overview Editore, Padua

O'Donnell WP, Peterson BM et al (2010) The third teacher. Abrams Books, New York

Rogers EN (1947) Architettura educatrice. Domus—La Casa dell'uomo, No 220, June

Rogers EN (1953–54) L'Italia è assente. Casabella-Continuità (199)

Rogers EN et al (1965) L'utopia della realtà: un esperimento didattico sulla tipologia della scuola primaria. Leonardo da Vinci, Bari

Tafuri M (1968) Teorie e storia dell'architettura. Laterza, Bari

Van Winden W (2010) Urban Hotspot 2.0: The challenge of integrating knowledge hubs in the city. http://www.urbact.eu. Accessed Dec 2011

Yigitcanlar T (2007) The making of urban spaces for the knowledge economy: global practices. In: Proceedings of the 2nd international symposium on knowledge cities: future of cities in the knowledge economy, Malaysia, pp 73–97

Yigitcanlar T, Velibeyoglu K, Baum S (eds) (2008) Knowledge-based urban development: planning and applications in the information era. IGI Global, New York

Management, Transformation and Enhancement of the Built Heritage

Massimiliano Bocciarelli, Laura Daglio and Raffaella Neri

Most part of the school building heritage in Italy is old (it dates back mainly to the 1960s and 1970s) and is not appropriate both to the modern teaching systems and to the current levels of safety and efficiency required by the current legislation on seismic vulnerability, energy saving and fire safety. For this reason, in Italy a vast program of reorganization, requalification and consolidation of existing school buildings has been recently launched and financed on a public basis.

Often the buildings that currently host schools consist of historic constructions belonging to the cultural heritage and this makes the reorganization, requalification and consolidation processes even more difficult in order to deal with and to respect the historical cultural value of the building.

This need of reorganization of the school buildings is often due also to a process concerning the transformation of educational and pedagogical approaches, aimed at improving the effectiveness of the learning models.

On a broader scale, all these needs offer the possibility of redesigning complex existing buildings and developing projects that play an important role also at the urban level by becoming reference places, opportunities for redevelopment of degraded parts of a city, cultural and civic centers.

These themes have long been a field of great interest, experimentation and application of researches aimed at developing projects, models and intervention strategies for the upgrading of existing educational buildings, where the synergetic combination of the different disciplines and skills involved has always been fundamental to the success of the intervention.

The possibility of giving old places a new identity, to update buildings according to the new educational and teaching models, to develop projects that take into account the needs of energy savings and structural safety, has been deeply investigated in the following chapters, describing the recent activities which have been carried out at the Politecnico di Milano (Department of Architecture, Built Environment and Construction engineering) in relation to the before mentioned topics.

The works presented turn out to be particularly innovative in view of the effort to propose interventions, not as a mere statement of rules or design methodologies, but with strategies able to deal with the current situation of the school institutes in terms of financial resources and the need to not interrupt the normal school activity.

On the national territory, the school building situation results to be rather diversified in several aspects, the most important of which are structural and fire safety and energy saving. With respect to this issue, the first chapter presents an effective control system and a decision-making process for the identification of the priorities in the intervention needs.

The reuse of existing historical buildings for the implementation of modern teaching systems is investigated in the two following Chapters, concerning the project for the transfer of some departments of the Brera Academy to the area of Scalo Farini located in the center of Milan. Particular relevance is given to the design of the new teaching spaces, in due respect of the conservation of the historical evidence of the existing construction, as a part of the urban regeneration project of the whole former marshalling yard area.

A similar aim is pursued in the contribution by Poggioli, which includes studies and projects for the architectural conservation and the reutilization of the monastery of San Sepolcro in Piacenza, by inserting spaces for a new university of medicine and nursing, within the overall regeneration project of the area of the Guglielmo da Saliceto Hospital.

Regarding structural safety, in Italy, numerous seismic evaluation programs of public buildings have been implemented with Ordinances 3274/2003 and 3362/2004, which highlighted the high vulnerability of school buildings, especially of the older ones, built without any seismic protection and, sometimes, in a state of degradation. The reduction of the seismic vulnerability of these buildings requires huge financial resources. Consequently, the possibility of opting for seismic improvement rather than for full adaptation must be carefully considered in order to intervene more quickly and on a greater number of buildings with the same budget, and by carrying out structural interventions without interfering with the normal school activity. With respect to this issue, the Chapters by Calabrese et al. and D'Antino et al. describe the use of fabric-reinforced cementitious matrix (FRCM) and composite-reinforced mortar (CRM) as externally bonded reinforcement of existing masonry structures (the former) and to strengthen different types of slabs (the latter), on the basis of the results obtained from an experimental campaign conducted at the Politecnico di Milano.

The final two chapters describe some research and consulting activities aimed at the energy and environmental requalification of different existing school buildings located in Italy. Particular relevance is given to the definition of the optimal energy performance targets as a compromise among different aspects: energy saving, life cycle cost of retrofitting works and environmental enhancement in terms of the use of ecological materials, the recycling of demolition materials or the use of renewable energy sources.

School Building Surveying: A Support Tool for School Building Registry Office

Angela S. Pavesi, Genny Cia, Cristiana Perego and Marzia Morena

Abstract On the national territory, the school building situation appears patchy, with very different situations in several respects, the most important of which is related to structural safety, to possessing the correct certification for fire prevention or hygiene. In fact, many schools were built before the 1970s and require very costly interventions; not only that, but in the meantime, teaching has changed, and it is not always possible or convenient to intervene on these buildings. From the ongoing dialogue with the Department of Education, Educational Policies and School Building of ANCI (Sabrina Gastaldi, Head of the ANCI Education School Department) and with the Department of School Education of Legambiente, (Vannessa Pallucchi, Vice President of Legambiente and National Head of Legambiente School and Training) what emerges is the importance of substantiating the survey on the quality of school buildings (*Ecosistema Scuola* (2018), now in its 19th edition, continues to show a clear gap between north and south of our country). School Building Registry Office, through the portal of the Ministry of Education, University and Research, started a transparency operation on the data concerning the health state of school buildings present on the national territory. However, it is necessary to define an effective control system and a decision-making process for the identification of priorities. This paper presents a due diligence tool developed with the aim of supporting the School Building Registry Office in the collection and processing of data.

Keywords School building survey · School building registry office · Due diligence tool

1 School Buildings in Italy: State-of-the-Art

The current state of public school buildings shows an extremely varied condition throughout the national territory with regard to the quality of buildings and the related supply of services for teaching. This is confirmed by the 19th Legambiente

---

A. S. Pavesi (✉) · G. Cia · C. Perego · M. Morena
Architecture, Built Environment and Construction Engineering—ABC Department,
Politecnico di Milano, Milan, Italy
e-mail: angela.pavesi@polimi.it

S. Della Torre et al. (eds.), *Buildings for Education*, Research for Development,
https://doi.org/10.1007/978-3-030-33687-5_21

Report, which states that "*school is not the same for everyone*" (Legambiente 2018). The *Ecosistema Scuola* survey aims to provide a database for the School Building Registry Office to quantitatively and qualitatively describe the current situation of Italian schools from different perspectives: defining the innovation rate and representative indicators of schools also in respect to the quality of the services it offers, in order to "*reposition the school at the territory centre as an educational and cultural agency and as a model of sustainability and well-being processes*". The data collected shows clear territorial inequalities, especially in the south and on the main islands compared to the north and central Italy, which have a heritage of school buildings that are in better safety and maintenance conditions. The most worrying phenomenon concerns the seismic fragility of territories which reveals that in the southern provincial capitals, three out of four schools are in seismic risk areas.[1] Legambiente supports the importance of "*directing funding and planning towards priority structural objectives such as new schools, requalification actions aimed at seismic upgrading and/or energy efficiency*" and of supporting "*the planning capacity and design quality of those administrations that are most lacking and inefficient*" (Legambiente 2018). Data provided by *#italiasicura.scuole* processed by Legambiente reveal that out of 2,787 construction sites started in recent years, for new schools or improvements, less than half have actually been completed. This is a consequence of the fact that investing in school construction is wrongly often considered only as a cost, not to mention the importance of developing an efficient and sustainable school project, which can in itself has a positive impact on costs, ensuring a non-negligible economic return.[2]

63.3% of school buildings were built before 1974 (the year in which regulations for construction in seismic areas were issued), thus interventions on these assets can be very expensive; moreover, teaching has changed, requiring different spaces and services and making the existing buildings inadequate. In Italy, only 42.2% of buildings have a fire prevention certificate, 60.4% are fit for use and 53.7% are static tested (Legambiente 2018).

This scenario describes the national territory in a condition characterized by an uneven resources allocation and more generally by a lack of "management": situation which requires the development of a model of best practice throughout Italy. In order to implement such a system, a precise knowledge of the state-of-the-art of public school buildings is necessary. For this reason MIUR[3] has launched an operation of transparency on the data concerning the state of health of school buildings across the national territory through the creation of the School Building Registry Office, accessible through an online platform. The Minister of Education Marco Bussetti underlines the importance of this tool to move quickly in identifying priorities for

[1]In Sicily 98.4% of schools are in seismic risk areas.

[2]In Bolzano, the energy efficiency of all schools has seen a 50% reduction in energy consumption.

[3]Ministry of Education, University and Research.

intervention working in close collaboration with local authorities to speed up maintenance works.[4] Another action aimed at promoting transparency was the convening and reopening of the Observatory for School Building, which provides a "control room" based in MIUR, involving representatives of the Ministry of Infrastructure and Transport, ANCI, UPI and individual regions.[5]

Legambiente Dossier suggests a number of future prospects for actions relating to school building: the completion of the Registry Office with an updated analysis of the static condition of schools in seismic risk areas to speed up any safety work; the establishment of a guarantee fund to support expenditure for energy and seismic upgrading of buildings accessible to local authorities; support to municipalities, trying to eliminate barriers to the involvement of private resources in interventions; and the re-launch of public policies in support of services for teaching (Legambiente 2018).

2 School Buildings in Italy: The ANCI Document and the ANCE Programme

During the Observatory for School Building, held on 23 January 2019, a very significant criticism was presented to the Minister of Education, Marco Bussetti. According to an estimate of ANCI[6] (2019), six billion of the resources allocated to school building have remained in the State coffers, waiting to be used, due to the long timeframe in the administrative process management (Scuola in Comune 2019). For this reason, at the Unified Conference held on 6 September 2018, ANCI and UPI delivered a joint document with some recommendations.[7] ANCI National Council document indicates an indicative figure of the overall need for interventions on school construction: "*it is possible that the figure is around 30 billion euros that can only be supported with a multiannual program and with a decisive turnaround on administrative procedures which currently intercept and retain for months and years all the resources allocated by the State to compete with public resources for a commitment that local authorities alone cannot support*" (ANCI 2019). There are many areas in which there is a lack of resources in relation to the real needs identified. For instance, total requests received from local authorities following the regional calls for loans from the EIB for 2018/2020, in fact, shows that the overall need is of around 10 billion euros compared to the 1.7 billion euros currently available. Concerning the ongoing seismic vulnerability investigations, the problem is split into the difficulty of finding resources for urgent strengthening interventions and the need to identify

[4]MIUR (2019), Edilizia Scolastica, Anagrafe, http://www.istruzione.it/edilizia_scolastica/anagrafe.shtml.

[5]MIUR (2019), Edilizia Scolastica, Osservatorio, http://www.istruzione.it/edilizia_scolastica/osservatorio.shtml.

[6]Associazione Nazionale Comuni Italiani—National Association of Italian Municipalities.

[7]At the moment only a small part of the recommendations has been implemented.

adequate arrangements for the continuation of educational activities. During the Unified Conference, it has been proposed that an *ad hoc* fund be set up to act on the most urgent cases and to define action priorities (ANCI 2019). In this fragmented scenario, an innovative proposal emerges from ANCE,[8] which through its operating company Ispredil s.p.a., since 2008, has initiated research into implementing a programme of redevelopment of public-school infrastructure in the country. The ANCE programme implements a model in public–private partnership (PPP)[9] with the intervention of private resources. ANCE developed this system, thanks to the participation of its own network of local associations and companies, in a constructive dialogue with numerous administrations and sharing the initiative with ANCI, UPI and the Chamber of Commerce System (Pavesi and Zanata 2013). The ANCE programme follows a set of fundamental principles, including building renovation for functional and energy-related aspects, scrapping and redevelopment of the school heritage with the logic of "efficient replacement", concentration and relocation of users in a view that is more consistent with demand and mobility networks, and the innovation of the management model, integrating school spaces with complementary functions and services. This programme is capable of activating investments of around 30 billion euros, involving SMEs from the public works and private construction sectors, service companies and local economic players in a widespread manner, intervening on about a third of the current surface areas. The ANCE school programme has all the potential to activate and trigger innovative processes, activating a new specialized market for management services, the integration of teaching with complementary functions, a source of profitability able to support systems such as those of collective transport, as well as a school polarity able to offer services to families and neighborhoods proposing itself as a "catalyst" for urban regeneration processes. It is a new system of services offered by the school, proposing a change of perspective in the orientation of public spending and favouring the first PPP programme launched with a central direction in our country (Pavesi and Zanata 2013). The strength of this model lies in its ability to meet a variety of needs of different stakeholders in the sector, both public and private, while also providing the administration with the ability to resolve quickly and definitively all critical issues relating to the current state of the existing school heritage. Concerning possible financial and corporate instruments to activate the model, it is possible to propose to the Government to jointly study financial mechanisms, involving Cassa Depositi e Prestiti (CDP), to support the financial exposure phase that operators would face for the construction of new school complexes, and the possible creation of a fee guarantee fund, managed by the CDP with strict technical criteria defined by itself, pursuing the objective of simplifying and accelerating access to credit and lowering the cost of money for operators involved in the interventions (Pavesi and Zanata 2013).

---

[8]Associazione Nazionale Costruttori Edili—National Association of Building Contractors.

[9]PPP is a new business that combines public and private operators, enhancing its features and optimizing results.

3 The requalification of School Buildings concerning Safety, Sustainability and Innovation Requirements

Analysing the state-of-the-art of public schools, shifting the attention from management to technical aspects in the requalification of school buildings, the guiding principles are safety, sustainability and innovation. Requalifying the existing properties and designing innovative interventions are strictly related to compliance with the existing legislation.

The D.M. "Nuove Norme Tecniche per le Costruzioni" of 14 January 2008 is the only technical reference standard, in force since 1 July 2009. It defines the rules for the design, execution and performance principles of the works but also the procedures for the qualification and acceptance of construction materials governed by EU Regulation no. 305/2011 of Construction Products (CPR).[10]

Italian school construction system highlights, among many different critical points, those relating to seismic safety and sustainability.

Since the Italian territory is characterized by a high level of seismic risk, school buildings play a fundamental role in safeguarding their users, also because their functional efficiency is strategic for emergency management. The building stock is lacking in terms of structural design, materials quality and its preservation state, so the knowledge of these assets is the starting point to define also the causes of seismic vulnerability and for planning interventions. It is important to underline that a large part of the national territory has been classified as seismic only in the last few years and a good percentage of the buildings was built without considering the seismic action, in areas that were later recognized as seismic zones (Pavesi and Zanata 2013).

It is possible to provide targeted interventions for the adaptation of the school building heritage in terms of seismic safety, but it is necessary to know the risk level of each school through the identification of its elements of vulnerability. Therefore, the state-of-the-art survey phase by an expert technician is fundamental. Subsequently, the planning interventions phase must pay attention to its convenience in terms of cost—often the requalification cost is similar to the cost of a new construction—and to the remaining period of use of the building.

The sustainability issue, instead, involves the conjugation of three inseparable recognized dimensions: environmental, economic and social sustainability.[11]

Thus, building sustainable schools means making the structures themselves an educational message for the new generations, increasing their didactic value.

Directive 2010/31/EU, in addition to dealing with new buildings, suggest a redevelopment of the existing buildings through a preliminary energy diagnosis, followed

[10]Construction Products Regulation. The seven essential requirements of products are: mechanical strength and stability, safety in case of fire, hygiene, health and the environment, safety and accessibility in use, protection against noise, energy saving and heat retention and sustainable use of natural resources.

[11]World Summit on Sustainable Development, 2002, Johannesburg.

by the systemization of the data collected in a proper software to elaborate a project proposal that increases energy performance and decreases management costs.

Applying the environmental impact reduction in the context of public buildings, the multitude of users can prove to be the vehicle for rooting the awareness of a well-being in harmony with the natural environment. Public schools can be the bearers of an educational message for a new generation design approach (Pavesi and Zanata 2013).

In the area of digital innovation, school buildings implement several tools oriented towards "digital education". The MIUR website hosts a section dedicated to the "National Digital School Plan" which aims to modify learning environments through the integration of teaching technologies to create a sense of citizenship and achieve "smart, sustainable and inclusive growth". The plan includes several actions including: the LIM (Multimedia Interactive Board) programme, the @urora action for minors and action beyond @urora, HSH@Network, to support hospitalized or in-therapy students, and the Pact for Scuol@2.0, to develop knowledge in collaborative and dynamic spaces.

For the European Digital Agenda[12] information and communication technologies are fundamental for achieving high levels of education, employment and revitalizing the competitiveness of the economic fabric and social growth (Pavesi and Zanata 2013).

4 Funding for the Safety of School Buildings

After analysing the requalification of school buildings from a technical point of view, it is necessary to examine the financial instruments to be implemented and the resources currently allocated by institutions to reach such an end. In fact, MIUR has set up a fund of 50 million euros[13] to finance the planning of safety measures for school buildings by the competent local authorities (Falconio 2019).

Applications for grants were accepted only the design of safety interventions of school buildings surveyed in the National School Building Registry Office.

The evaluation of the requests for the assignment of the contributions took place on the basis of various criteria, including the age of the buildings used for educational purposes, the co-financing quota, the seismic zone, the possibility that the building was included in the 2018/2020 three-year programming or beneficiary of other funding, the lack of the certificate of viability, an ordinance or provision for closing the building on a date prior to the public notice and the number of students in the school building (MIUR 2019).

---

[12]Europe 2020 Strategy Project.

[13]MIUR (2019), Edilizia Scolastica, Finanziamenti, http://www.istruzione.it/edilizia_scolastica/fin-progettazione-interventi-sicurezza.shtml.

5 Survey Check List and Regulatory Handbook: A Support Tool for School Building Registry Office

The awareness of the critical conditions currently facing the existing school buildings highlights the need to rethink an effective system of control and "management" of this heritage, which can guide the decision-making processes for identifying priorities and develop a due diligence tool that can support the School Building Registry in collecting and processing data.

Therefore, there is a need to initialize a system for managing and monitoring the status of school real estate assets by identifying a model that facilitates implementation. An innovative due diligence tool has been created through the revision of the registry file provided by the School Building Registry Office and the reorganization and implementation of the present fields, comparing data already in place with further requests deriving from tools such as the Technical Standards of school building, the Building Booklet and new guidelines issued by MIUR. Thus, it was possible to elaborate a complete model of references and normative excerpts, to be supplied to technicians in order to allow them to optimize survey operations and equip them with a checklist able to identify the parameters required, consistent with what is required by recent regulations. This is an updated tool for identifying, monitoring real estate assets and detecting deficiencies in the Italian school buildings stocks.

A survey questionnaire and a Regulatory Handbook were prepared to support the Scholastic Registry. The questionnaire aims to analyze each school from the general to the particular, as well as to provide a description of the maintenance status of the school building and the characterization, both quantitative and qualitative, of its interior spaces. Therefore, this tool results in a snapshot of the school, highlighting its shortcomings and the possibilities for improvement, also acting as a basis for planning the management and maintenance of its parts.

The methodology adopted for drafting the questionnaire followed several phases. The first phase focused on the analysis of data required by the survey form created in 1996 for the implementation of the registry and comparing it with the information contained in important tools such as Technical Standards, UNI standards and guidelines. A reorganization of the information required to create new sections and the formulation of an updated and more complete questionnaire was prepared, also considering the changes in the field of school building. The second phase identified, also in relation to the updated sections of the questionnaire, the relevant reference regulations alongside the fields to be filled in, to create a support for the survey work through a real regulatory handbook which, at fixed intervals, will need to be updated. The third phase has seen the creation of an internal checklist that identifies, among the requested data, those able to define schools according to the new regulatory requirements related to safety and innovation.

The questionnaire provides for a territorial contextualization moving into a morphological and dimensional identification of the building with priority given to compliance with anti-seismic criteria, and then "dissecting" and codifying the building on a spatial level (safety and well-being) and on the technological system level (state of conservation).

This tool has the prerequisites to promote the future design of an application programme able to manage and monitor school buildings, which will become increasingly flexible structures and as centers that will offer a space for growth to develop the integration and reception of the surrounding neighborhood. The potential of this innovative approach also lies in the possibility of implementing this tool, going deeper into technical and regulatory details, evolving into a useful tool for designing new school buildings. The periodic updating and the variation in the questionnaire in parallel with the normative evolutions will be of fundamental importance; as for the possibility of in-depth analysis concerning the definition of a maintenance programme over time.

Creating a tool able to be in control of this real estate asset is fundamental because schools are the structures within which future generations are educated.

Acknowledgements Thanks to Sabrina Gastaldi of ANCI and Vanessa Pallucchi of Legambiente for their contribution to the general research. Thanks to Veronica Oliva for her contribution to the advancement of the research.

References

ANCI, National Council, Rome, 31st Jan 2019. http://www.anci.lombardia.it/documenti/8339-Aggiornamenti%20edilizia%20scolastica.pdf

Falconio E (2019) ANCI, Miur public notice for the financing of safety measures in school buildings, 8th Mar. http://www.anci.it/avviso-pubblico-miur-per-finanziamento-interventi-di-messa-in-sicurezza-degli-edifici-scolastici/

Legambiente (2018) *Ecosistema Scuola*, 18th Oct. https://www.legambiente.it/ecosistema-scuola/

Legambiente, Dossier (2018) *Ecosistema Scuola*, https://www.legambiente.it/wp-content/uploads/ecosistema_scuola_2018.pdf

MIUR (2019) Public notice concerning the concession of grants to local authorities for the design of measures to improve the safety of school buildings. http://www.istruzione.it/edilizia_scolastica/allegati/Avviso%20di%20concessione%20contributi%20per%20la%20progettazione.pdf

Pavesi AS, Zanata G (2013) PUBLIC SCHOOL BUILDING Tools for the regeneration of school heritage in Italy, Maggioli Editore

Scuola in Comune (2019) ANCI document on school building, 14th Feb. http://scuolaincomune.it/index.php/2019/02/14/documento-anci-su-ediliziascolastica/

Extension for the Accademia di Brera at the Farini Marshalling Yard in Milan: The Architecture of the Campus and Spaces Frames for Teaching

Luca Monica, Luca Bergamaschi, Giovanni Luca Ferreri, Paola Galbiati and Massimiliano Nastri

Abstract A study of the architectural tradition of the types and spaces used to teach fine arts underlays the research to develop the Accademia di Brera project at the Farini marshalling yard, a brand-new campus for the arts as part of the urban regeneration of the whole of the former marshalling yard area. The system, which uses different varieties of newly built multiple teaching spaces, is inserted into the existing long galleries which feature frame construction.

Keywords University campus architecture · Academies of art · Architectural composition · Urban design · Building types

1 Among the Academies of Europe

There are many reasons why the Brera Academy of Fine Arts needs more room, in a new, more rational, separate site; a theme that has traveled its history, with various projects throughout the twentieth century in common with many other European academies.

Most certainly we can thank the beautiful book *Academies of Art. Past and Present* by Pevsner (1940) for a modern recognition of the European identity of a social institution which was long the cornerstone of the artistic and cultural formation and ferments in Western thought throughout the nineteenth century. Albeit in the complex variety of institutional roles and even in its controversial role as a guide, between "academy" and "anti-academy", Pevsner's text succeeded in building a profile that recognizes the common characteristics of the long course of the social history of art,

L. Monica (✉) · M. Nastri
Architecture, Built Environment and Construction Engineering—ABC Department, Politecnico di Milano, Milan, Italy
e-mail: luca.monica@polimi.it

L. Bergamaschi
Università IUAV, Venezia, Italy

G. L. Ferreri · P. Galbiati
Milan, Italy

S. Della Torre et al. (eds.), *Buildings for Education*, Research for Development,
https://doi.org/10.1007/978-3-030-33687-5_22

split among decidedly European contexts in the large capitals and the more marginal provinces.

Pevsner's story extends up to the threshold of the Modern Movement, to the "awakening of the industrial arts" at Gropius' Bauhaus, but could ultimately continue up to our own times, given that today many of these academies still play an active and identifiable role within a common European history, which, although arguably lacking a reference community policy, has encouraged local developments that are not always easily comparable.

Looking at it today, that network of exchanges which united the academies in Europe and across the world certainly ceased to exist as soon as the educational statutes evolved in line with the positions of higher university education in different ways from one country to the other.

These all tend to be found in historical contexts and buildings of maximum value, also thanks to their inaugural layouts and spaces (a school, workshop and museum all at the same time). Secondly, the academies possess and defend a historical-artistic heritage of enormous value still alive and well among their spaces (Cassese 2013).

2 The Farini Marshalling Yard and an Architectural Research Tradition

The Northwest trajectory was recognized by many as one of the most important constitutive elements and a vital component in the system of radial trajectories which formed the layout of modern Milan, based on real, formally complete "linear cities". Its high potential for transformation, as well as its extraordinary endowment of relations and infrastructure systems towards the productive "hinterland", resisted the slow process of welding the urban expansion fabric together (Canella 1981).

Boasting important architectural and urban projects in the best tradition of Milanese rationalism, this was precisely the terrain for the application of a number of research and teaching theories of the Faculty of Architecture of Milan.

With the decommissioning and consequent displacement of the Trade Fair quarter to the outlying zone of Rho-Pero, the propulsive sense of the Sempione-Fair axis was partially shut down, only a few years earlier still fixed in the collective imagination and in its functions as a management centre as an area perennially available for transformation, a sort of palimpsest open to the dynamics of economic territorial development, through its exceptional, strongly experimental architecture.

In the meantime, other research goals had found an important parallel in the Garibaldi-Bovisa axis, where, in addition to the great accessibility given by the historical trajectories of the railway lines, the dismantling of the production base was producing a gradual transformation of the area.

The first educational and research projects were developed back in the 1970s by Guido Canella's group (*Variations on the Wolkenbügel* (cloud-hangers) *of El*

Lissitsky-Mart Stam, 1974), for a new technology department of the Polytechnic at Bovisa, a first anticipation of the future detachment (Fiorese 1984).

Experiments which followed were by projects for the exhibition *Le città immaginate* at the 17th Milan Triennale (1987), with projects for the Bovisa Polytechnic by Guido Canella, or for a big service connector and park at the Farini marshalling yard by Gustav Peichl, and the first polytechnic masterplan for a new campus (1990) designed by Antonio Monestiroli, with the buildings designed by teachers of architecture from the Faculty itself.

On the Garibaldi–Farini–Bovisa axis, the ex-marshalling yard presents itself as an important large "*vuoto urbano*" (urban vacuum) to be given over to a metropolitan park with amenities. The first thesis came in 1978, *Milano-Bovisa: riconfigurazione dello storico porto in terra metropolitano dal recupero del Cimitero Monumentale* (supervised by A. Acuto, G. Canella), followed by the project by Vittoriano Viganò in 1979 for the Garibaldi Station area extended as far as the marshalling yard, part of a mega-structure with a linear frame.

In 2009, the School of *Architettura Civile* along with the Municipality of Milan, in view of the future *Accordo di Programma*, addressed these issues at a workshop entitled *Milano. Scali ferroviari e trasformazioni urbane*. If, on one hand, the various hypotheses tried to make the property imbalance indicated by the agreement compatible, other research groups tried to continue the hypothesis of the urban trajectory and the linear business district, extended to 10 km from Garibaldi to Fiera-Expo, recovering the monumental sense of its architecture and landscape, as the projects by the groups of Canella and Bordogna (Protasoni 2012), through a series of strategic functions. Following that, the research project La parte elementare della città. Progetti per Scalo Farini a Milano (A. Monestiroli, R. Neri, A. Dal Fabbro and others) concentrated on the quality of the urban architectural project for a residential district (Neri 2014).

With regard to the extension for the Accademia at the Farini marshalling yard, proposals were developed as alternatives to the original *Grande Brera* project which wanted to allocate the Accademia into the barracks in Via Mascheroni-Monti (Monica 2013, 2014; Bon Valsassina 2014). Among the first works, it is worth recalling the thesis entitled *Per Campus Brera un nuovo modello teresiano. Scalo Farini Milano* (supervised by G. Fiorese in 2010) and other academic works (Monica et al. 2015), later shown at a larger exhibition in the Accademia itself in 2016 (teachers M. Dezzi Bardeschi, G. Guarisco, L. Monica, A. Torricelli), which was followed by an in-house debate which had already begun at the Accademia (Bonini 2019, p. 8; Cusatelli 2019; Monica 2017; Bianchi and de Lillis 2017).

The aim of these proposals for the Farini marshalling yard, for a higher education establishment such as an academy of art, was to seek an urban regeneration process that was compatible with the assumed property imbalance envisaged in the *Accordo di Programma*, of which today only in the light of the first results of the design contest for the *Masterplan*, we could glimpse the concrete and predictably reductive directions compared to its premises and possibilities which the research had theorized (OMA and Laboratorio Permanente 2019).

Looking then at the long tradition of architectural research and projects around the theme of urban recovery in the Farini marshalling yard, the conviction remains today that the establishment of strategic functions on a metropolitan and territorial scale—public architecture for services—remains the necessary point of departure for this renewal.

3 The New Campus for the Artists' Community of Brera

The prominent aspect of the great goods depot at the Farini marshalling yard is its large overall dimension and its typological and structural uniqueness, constituted by two galleries of around 350 m in length, which the original installation was based on before 1914, made of reinforced concrete using a typical frame structure. The serial nature of the internal structure, clearly visible in the full length of the two galleries, endows the building with an unusual and modernly monumental scale.

The entire area of the Farini marshalling yard is part of the *Accordo di Programma* arranged by the Municipality of Milan which covers the part of the building of the former goods depot intended for the Accademia di Brera. In the future urban development of the area, it is planned to expand the "Campus of the Arts", which could actually be considered as a relatively autonomous part with respect to the overall design of the large park and the predictable residential district. An autonomy guaranteed especially by its strategic public function for the city (a university campus) is in itself able to act as a solvent in a hub interconnecting the neighborhoods and multiple amenities surrounding the former yard.

This first scheme therefore involves the entire yard when it comes to organizing the Brera Campus, which will proceed in phases with a first settlement in the former goods depot, taking advantage of the full occupation of the entire former depot, and considering its shared use with the public spaces for exhibitions, leisure activities and services for the area, thus preserving its architectural and monumental unity with respect to the entire area of the yard (Figs. 1 and 2).

A new building will be added which will be taller, to contain all the functions that the Accademia requires to complete the university complex (service spaces for studying, lecture rooms, departmental offices for teachers and ateliers), and different in type and spaces, forcing the empty spaces intended as an aesthetic reserve, perceptible from the large ex-depot now belonging to the Accademia (Fig. 3).

4 Frames, Urban Settings and Suspended Images

Following a hypothetical ideal route, visitors should reach the new spaces of the Accademia di Brera at the Farini marshalling yard by arriving from the Monumental Cemetery complex. On entering they would be standing in front of the monument built in 1946 by the BBPR group of Milanese architects in memory of those deported to concentration camps. Precisely this frame, resolved in a cube, marks the stopping point of an independent path of Italy's figurative culture; after this, the suspension of the search for a city moral expressed itself in addition to the certainties of the aesthetic consent "without a relationship with the taste of the majority" (Persico 1934, 1964, p. 255).

This structure at the cemetery appears as a fragment of what had already been tried out in 1934 by Edoardo Persico and Marcello Nizzoli for the Medaglie d'Oro room at the Esposizione Aeronautica Italiana in Milan's Palazzo dell'Arte, and that same year for an advertising structure erected at the crossing inside the Galleria Vittorio Emanuele II, using a framework of Mannesmann tubes.

The frame structure therefore becomes a device for fine-tuning the relationship between dimensional aspects and habitability.

These installations are "lyrical visions of construction" (Veronesi 1953), visions that continue the investigation into the city and its changes. This work can be traced back to the itinerary of pictorial research which has always related with the meeting between the urban dimension and temporality by tabulating symbols, allegories and formal abstractions. In fact, as far back as the fourteenth century—even before the awareness given by the science of perspective had arrived—the city was depicted by arranging human vicissitudes and urban sequences on superposed, occasionally ambivalent planes. In these representations, architecture was the limit condition between public experience and family life.

Perspective, seen as a representation of objective reality, was demonstrated through a mix of empiricism and esotericism. With the precision of the vanishing point, the works of architecture were conditioned by the choice of the viewpoint and had also to take immediately account of their constructive datum.

Fig. 1 Aerial view of the Farini marshalling yard, 1998. Photo S. Topuntoli

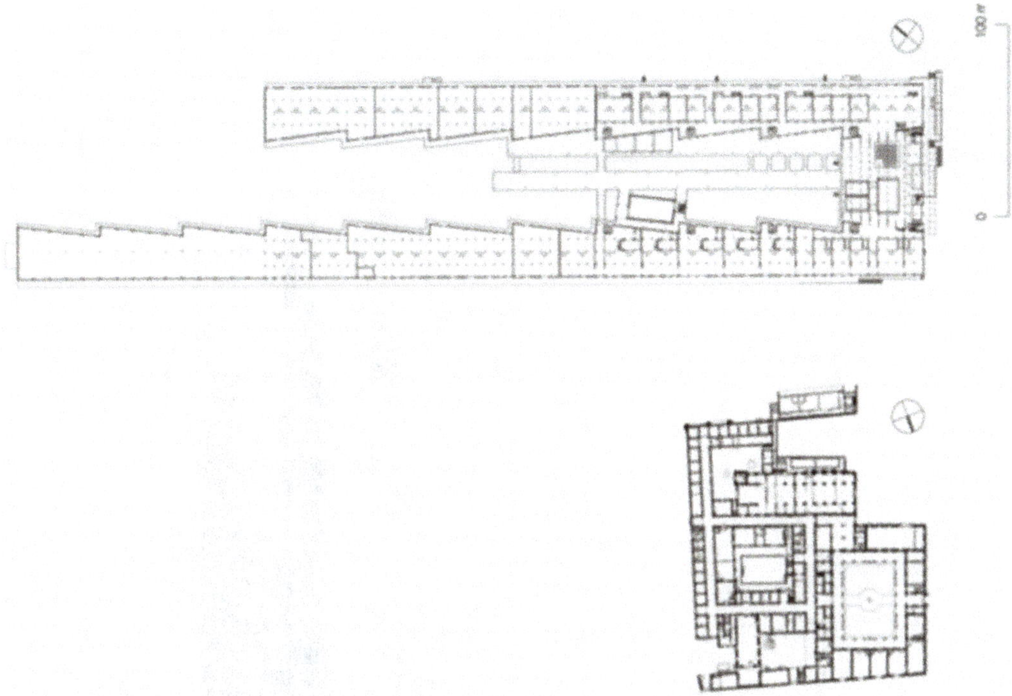

Fig. 2 Plans at the same scale of the Palazzo di Brera and the former goods depot at the Farini marshalling yard

For the group of rationalist architects who gravitate around the Milanese area, the frame is translated by urban civilization and almost no longer belongs to construction technique. Lines "give the integral form needed for all things", to paraphrase Roberto Longhi.

It was none other than Canella who related the work of the rationalist Milanese architects who gravitated around the figure of Edoardo Persico to the design styles worked on by Longhi, in a lesson on the poetics of Ignazio Gardella (Canella 1999).

Precisely this integrity of form—central to the examples we have mentioned—leads us to prepare a structure not yet compromised by material aspects in our own intervention: i.e. a clean-cut structure delimits the public space from the surrounding buildings. This is an element under which it is possible to linger awhile, discuss, and build relationships.

Arriving in the area of the Farini marshalling yard, during the first on-site visits to the old buildings to be converted, it came to us that it would be necessary to fix its dimension through an ephemeral and reversible element. Precisely in this research on the dimension, the reasons behind the architectural intervention for the settlement of the educational activities lay. The structure testifies to the fact that, in addition, something is happening: beyond that diaphragm, the place needs to be populated. Behind that declaredly ephemeral deck, it is necessary to settle those activities of the Accademia currently deployed among different temporary places.

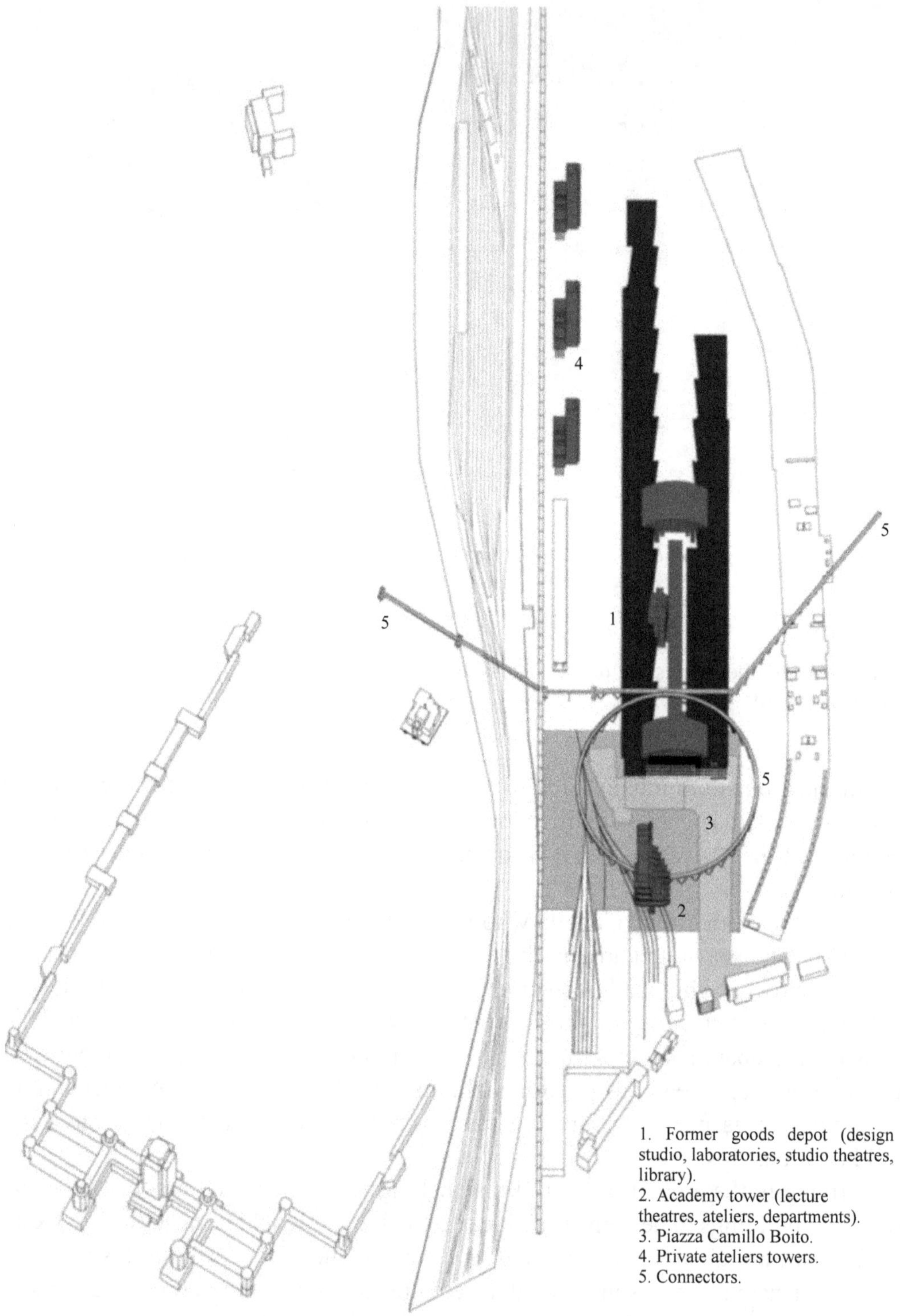

Fig. 3 Research group: L. Monica, M. Acito, L. Bergamaschi, S. Cusatelli, G.L. Ferreri, P. Galbiati, G. Guarisco, M. Nastri, M. Rajabi, Campus for the Arts of Brera. Extension of the Brera Academy at the Farini marshalling yard: General plan

This figure is the minimum construction datum on which to hang, suspend and superimpose themes or even better, to write—always in the form of an allegory—words like civilization, art, taste, aesthetics and morals.

5 Inside the Continuous Monument

The splitting up of the teachings of the Accademia di Brera among multiple locations has increased over the years ever since its evolution toward disciplines which have extended artistic practices in addition to the traditional fundamental ones based on the art of drawing. This is a lengthy process, which certainly began with the new cultural system imposed by Camillo Boito at the Accademia between the end of the nineteenth century and the beginning of the twentieth century, accompanied by the maturing of disciplines related to contemporary industrial arts and to restoration, which might never have found a place in the old Brera building, but arguably more suitable spaces in the oft-cited *capannone* (shed) (see G. Guarisco in this same volume).

Everything began from that Brera building which saw the birth of this institution in 1776, whose distinctive feature from the beginning was the coexistence of multiple activities. A birth that was far from precocious among the academies of art in Italy and in the rest of Europe, but in an absolutely unique and original context, where, in the monastic reality of the original Jesuit College layout, various educational and research institutions, museums, and conservation workshops coexisted.

Starting from the first transformations and additions of spaces from college to academy designed by Giuseppe Piermarini, there were proposals and second thoughts on the most suitable location for each activity. With time, many other variations and internal additions followed, from the construction of the so-called "Hayez Classrooms" (architect G. Voghera, 1852–1857), the pavilion in reinforced concrete to exhibit the plaster casts, conceivably the first in Milan (architect A. Brusconi, 1900), then the didactic library (architect L. Patetta, 1984), up to the apparently incongruous but modern system of mezzanines (Fig. 2).

Today, therefore, it is no longer possible to postpone the reunion of the new seats in a "new monument" that can house spaces more suitable for contemporary teaching methods, divided into two large typological categories, effectively organized in the two grand arcades recovered inside the ex-depot. In fact, these two distribution lines organize both the changed spatial needs and the layout of the traditional teachings (workshops for the various sculpture techniques), and the design laboratory activities (in spaces typologically assimilable to those of schools of architecture, design, etc.).

A third distribution line, that of the free connecting spaces, assumes a further function as a place for individual students to study, share and produce autonomously.

The construction of a common minimum technical–functional denominator could indeed encourage the various forms of operations through simple modular construction elements, the "frames", partly transparent and partly not repeatable and variable with changing needs, and able to maintain one of the typological characteristics which distinguish the former railway depot: the unity of the internal space, also of a certain quality and not devoid of figurative suggestion, the symbol and monument of a different kind of industriousness, apparently equally disordered, but capable of restoring a sense of community in the difference, as in a sort of "continuous monument" (Figs. 4–6).

6 Technique and Construction of Internal Framed Partitions

The study of internal partition systems to fill the two arcades devoted to teaching envisages the design of an independent, self-supporting metal frame with respect to the concrete structure of the building, both for the heavy instrumental laboratories and the light teaching laboratories.

The pattern of modular frames, oriented as the main and secondary load-bearing systems, is related to research and design experimentation as well as contemporary construction, through the definition of:

- stretchers superimposed on the existing apparati, according to a relation of transition with the tectonic, spatial and connective structures in a pervasive form and through fragments, repeated serially in a perspective succession;
- conceptual installations that reveal a functional and visual permeability, as occasions for symbiosis with the existing structure;
- dematerialization of classroom-containers, defined by filters and passages between diversified geometric and dimensional perceptions articulated by a linear design.

The composition of the typological units of the teaching spaces and their structural elements is based on vertical linear frame apparati in metal (steel), comprising:

- normal closure profiles (transparent glass and opaque wooden panels);
- a succession of service modules with shelves;
- realization of the top section as a mezzanine;
- aggregation of flights of stairs;
- a connection with the continuous horizontal plane-route as a suspended walkway, intended to house the study activities of the teachers.

References

Bianchi F, de Lillis L (eds) (2017). Spazio Accademia. Tra espansione e salvaguardia di un'identità culturale. Seminario di studio docenti-studenti. Accademia di Belle Arti di Brera, Milano

Bonini G (2019) Per una grande Accademia. In: Cusatelli S (ed), op. cit., p 8

Bon Valsassina C (2014) Il caso Palazzo Citterio, Skira, Milano

Canella G (1981) Il "Genius loci" della Direttrice Nord-Ovest. In: Hinterland, no 19–20, pp 80–91

Figs. 4–6 Research group: L. Monica, M. Acito, L. Bergamaschi, S. Cusatelli, G.L. Ferreri, P. Galbiati, G. Guarisco, M. Nastri, M. Rajabi, Campus for the Arts of Brera. Extension of the Brera Academy at the Farini marshalling yard: View of the entrance structure; Scale models of the design studio and laboratories

Canella G (1999) Ignazio Gardella: le figure e le città, lezione, 2 giugno 1999, preprint 2000, Politecnico di Milano, now in Architetti italiani nel Novecento, ed by E Bordogna, Christian Marinotti, Milano 2010, p 289 ff

Cassese G (2013) Accademie patrimoni di belle arti, ed by, Gangemi, Roma

Cusatelli S (ed) (2019) Campus delle Arti di Brera. Ampliamento dell'Accademia allo Scalo Farini. Indirizzi di un progetto architettonico, Mimesis, Milano

Fiorese G (ed) (1984) MZ7. Milano Zona sette Bovisa Dergano. ICI-Comue di Milano, Milano

Monica L (2013) Le architetture per l'ampliamento dell'Accademia di Brera. Antecedenti e nuove ipotesi all'Ex Scalo Farini. In: Scarrocchia S (ed) Per Brera Sito Unesco, Accademia di Belle Arti di Brera, Fondazione Cariplo, Edizioni Sestante, Bergamo, pp 99–113

Monica L (2014) Brera. Dove? Nota sulle recenti vicende intorno Palazzo Citterio, le ipotesi per le ex Caserme di via Mascheroni a Milano e la necessità di una istruttoria per l'Accademia di Belle Arti, "Ananke", no 71, pp 41–46

Monica L, Cusatelli S, Ferreri GL, Guarisco G, Galbiati P, Scarrocchia S, Rajabi M et al (2015) Per l'ampliamento dell'Accademia di Brera. Ricerche progettuali. Politecnico di Milano e Accademia di Belle Arti di Brera, Mimesis, Milano

Monica L (2017) Diario breve di una ricerca universitaria per l'ampliamento dell'Accademia di Brera. In: Bianchi F, de Lillis L (eds) op. cit., pp 15–17

Neri R (ed) (2014) La parte elementare della città. Progetti per Scalo Farini a Milano, Lettera Ventidue, Siracusa

OMA and Laboratorio Permanente (2019) Design project "Per la trasformazione e rigenerazione urbana ex scali ferroviari Milano Farini e San Cristoforo". http://www.scalimilano.vision/concorso-scalo-farini/il-team-oma-e-laboratorio-permanente-e-il-vincitore-del-masterplan-concorso-farini/. Accessed 21 June 2019

Persico E (1934) Capocronaca dell'architettura, now in Tutte le opere (1923–1935), vol II, ed by Veronesi G, Comunità, Milano 1964, p 255

Pevsner N (1940) Academies of Art. Past and Present, Italian edition, Le accademie d'arte. Einaudi, Torino 1982

Protasoni S (ed) (2012) Milano scali ferroviari. In: Workshop della Scuola di Architettura Civile. Libraccio, Milano

Veronesi G (1953) Difficoltà politiche dell'architettura in Italia, 1920–1940. Tamburini, Milano

Camillo Boito's "Capannone" for the Accademia di Brera in Milan: Reuse of a Railway Depot

Gabriella Guarisco, Maurizio Acito, Stefano Cusatelli and Mehrnaz Rajabi

Abstract The work concerns the project for the transfer of some departments of the Brera Academy to the area of Scalo Farini located in the centre of Milan. The area is central to the transformations of Milan, and the relocation of the departments of the Academy solves the problem of its development as a school. The transfer occupies the large post office warehouse in the centre of the area, which is the subject of a conservation and evaluation process for the insertion of the new functions.

Keywords Reuse · Preservation · Railway · Depot · Academy of arts

In the heart of Milan, a stone's throw from the Monumental Cemetery, still today we find the physical existence of the two long "wings with a saw-toothed perimeter line" of the former goods depot that unambiguously draw attention to the imprint left by the decommissioned Farini marshalling yard on the city's historical fabric. The building, revamped several times, especially at the end where the subsequent additions are instantly decipherable, constitutes the last physical witness of the history of the transformations that occurred due to the changing routes of the railway system (Aa 1933; Guarisco 2015; Guarisco et al. 2017).

After the early nineteenth-century, in rail transport routes run by private individuals and prevalently for use by travellers, with the establishment of the state railway company, the Ferrovie dello Stato (1 July 1905, Giolitti government), and under the management of the Ministry of Public Works, came the first reorganization of the entire rail network in Milan, and in particular, the fundamental reorganization of the marshalling yard (Rigato 2017–2018). On p. 87: "To expedite the running of the trains, especially in the major stations, it adopted the separation and specialization of transport in passenger and goods, while previously such specialization had never been put into effect" and its buildings. The Central Station became significant and was repositioned towards the edge of the city, while the stations of Lambrate and

G. Guarisco (✉) · M. Acito
Architecture, Built Environment and Construction Engineering—ABC Department, Politecnico di Milano, Milan, Italy
e-mail: gabriella.guarisco@polimi.it

S. Cusatelli · M. Rajabi
Milan, Italy

S. Della Torre et al. (eds.), *Buildings for Education*, Research for Development,
https://doi.org/10.1007/978-3-030-33687-5_23

Fig. 1 Example of the characteristics of the "Hennebique system" structures in a logo for postal envelopes of the Porcheddu company

Porta Vittoria were being built. The points of interchange for goods in use up to that point, in addition to the large Sempione yard (1883–1884), were located near the city gates: Genova (1868–1870), Garibaldi (1873) and Rome (1896). The state railways, as part of a sweeping plan to reorganize the entire railway network of Milan, created the Farini marshalling yard (a transfer of the by-then insufficient one of Garibaldi) and that of Porta Romana (as a result of the decommissioning of Sempione, 1931), and the yards near the railway stations of Porta Vittoria and Porta Genova.

The Farini marshalling yard maintains its original north-west alignment, remaining today among the trajectories of greater accessibility to the centre of Milan. Destined to grow exponentially from the date of its construction, it lies at a crucial point of the city, in the immediate vicinity of one of the historical gates and the passenger station of Porta Garibaldi. Over the last 20 years, the physical aspect of the entire area has changed profoundly: large swathes have been razed, the towers of the Bosco Verticale—"the Vertical Forest"—have risen, Piazza Gae Aulenti was fashioned with new buildings surrounding it that generate a skyline of corporate towers very similar to those of metropolises around the world. Beyond this area, the old marshalling yard has remained (now covering an area of 618,733 m^2) which, in its growth, has extended as far as the Monumental Cemetery (Figs. 1 and 2).

With the *Marshalling Yards Program Agreement* (2017, but the planning process for the abandoned areas had already begun in 2005), it was stipulated *inter alia* that "the marshalling yards [...] can accommodate cultural activities, even of a private nature, tied to music, art, and architecture, by developing the existing buildings wherever possible" (point L, p. 27) with the launch of procedures to regenerate the abandoned areas. It was within this framework that the *Steering Document*[1] appeared,

[1]On 22 December 2017, a letter of intent was signed between the Municipality, FS-Sistemi urbani and the Accademia di Brera, and thereafter (3 May 2018), a more specific convention for the use of part of the large former goods depot as new premises for teaching activities as an expansion of the Academy's historical seat. In the follow-up to research already begun some years earlier (Monica L., Scarrocchia S. 2015), the Academy charged the Polytechnic University of Milan (Head Monica L., consultants Guarisco G., Nastri A. and Acito M.) to prepare a Steering Document that could lay the foundations for the subsequent project phases. On 27 February 2019, in the presence of

Fig. 2 A view of the inside of the depots today

produced in November 2018 with verification of the feasibility to reuse the former depots for some educational and workshop activities of the Accademia di Brera.

Despite relentless consultation of the accessible archives and research on bibliographical bases, up till now it has proved impossible to establish the exact date of construction of the former depots (1910–1914?), even though the research has produced an advancement of knowledge in this regard.

After the state took over the national railway network, the executive committee of the municipality set up a new commission presided over by the Rector of the Polytechnic (the engineer Giuseppe Colombo) which drafted a definitive development plan adopted immediately by the railways. It was in the plan of the municipal engineers Pavia-Masera (1909–1912) that the structure of the Farini marshalling yard, a "new low-speed freight yard" appeared for the first time in all its grandness (Cusatelli 2019). Its mixtilinear profile combines a curve in the northern part (to delimit increases in the number of rails) with a straight line parallel to the tracks which arrives almost as far as the Bovisa gasworks. The works to construct the yard proceeded rapidly: the two wings of the general depots appeared for the first time in the IGM map of 1914 (Figs. 3 and 4).

If the materials available at the Milano-Greco railway (Ministero delle Comunicazioni n.d.; Canella 2010; 6 table Archive of the FF.SS.) archives do not for the moment allow confirmation of the date when the works were completed, the

the authorities, the ceremony to inaugurate the Accademia's academic year took place inside the marshalling yard, in an area made safe especially for the occasion.

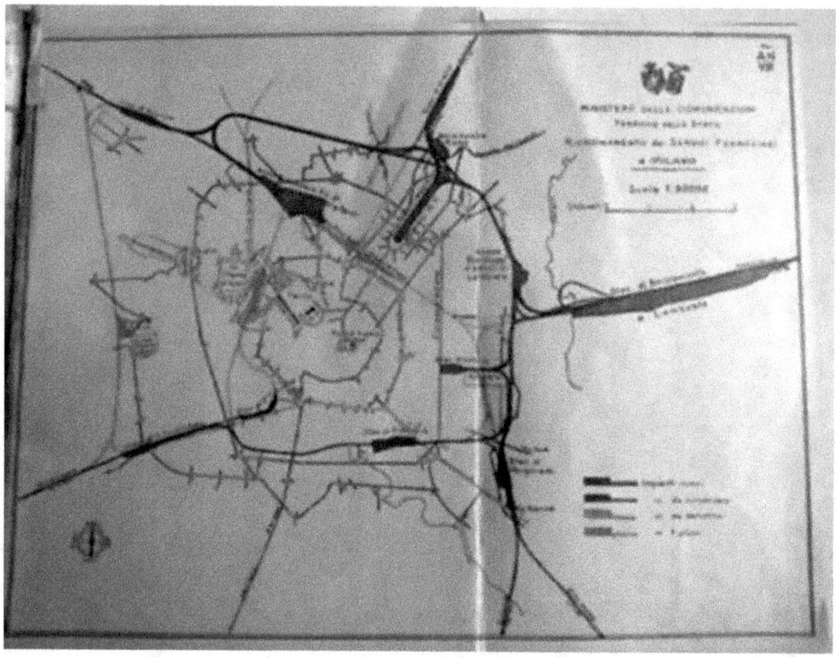

Fig. 3 Ministry of Communications—Ferrovie dello Stato, Riordinamento dei servizi ferroviari a Milano, General Plan (n.d.), Raccolta Bertarelli Milan

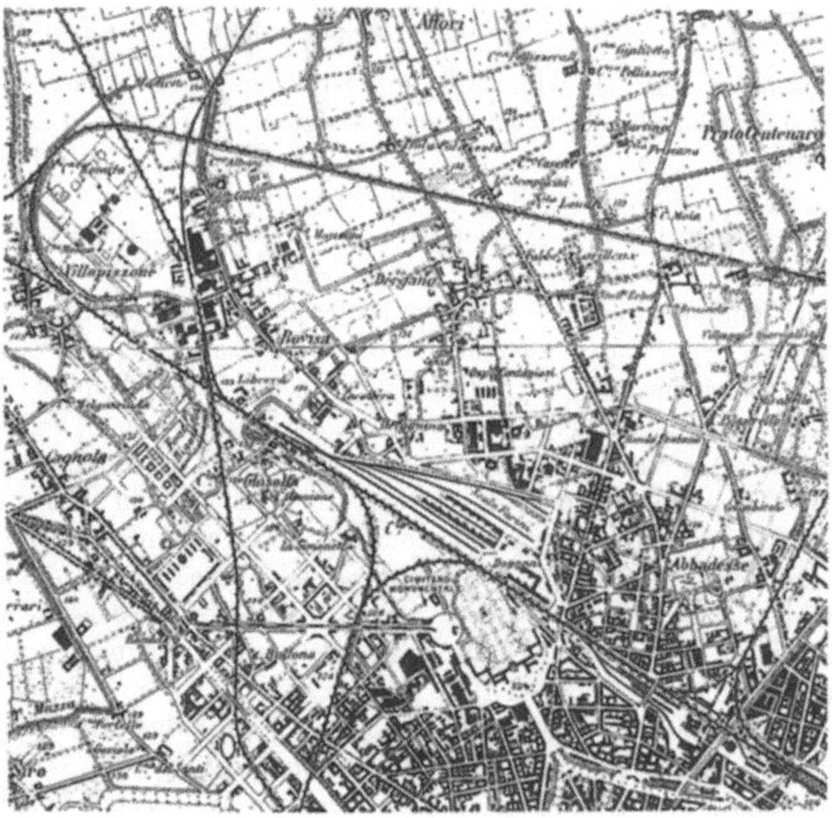

Fig. 4 Istituto Geografico Militare, Panel Bollate and Milan West, 1914, IGM Archive

newspaper *Corriere della Sera* reported the inauguration of the marshalling yard not long after WWI (16 September 1921). The journalist described the depots thus as: "Enormous and ready to accept an upper story that is already likely and for which everything has already been arranged, these depots cover an area of 22,000 m^2, without counting the 6,000 m^2 of open-air loading bays, and are constructed with a saw-toothed perimeter line to make loading and unloading easier on the intermediate and external groups of rails", since the declaration that they were already in use, with every probability, they were constructed at the time of the Great War and "could be relied upon to accommodate many soldiers and machinery" (Elia et al. 2015).[2] Later on, the depots appeared in operation in the photographs of a volume that celebrated the conclusion of the works on the Central Station[3] (1934) (La Stazione Centrale di Milano 1931).

The two long wings are typologically related to industrial sheds and so today can justifiably be considered of interest to industrial archaeology. From the outside they appear as a repeating series of walls with large openings arranged above a high "plinth" (to permit the passage of goods at a height) surmounted by the indispensable canopies, installed (in different periods) to avoid the work being hampered by bad weather. The spaces are equally repetitive inside, rhythmically broken up by the "forest" of pillars surmounted by beams in reinforced concrete. In this respect, it should be noted that this construction system, which has seen widespread use in large buildings that had to be erected quickly, is related to the so-called "*beton armé Système Hennebique*" (Riccardo and Signorelli 1990).

The dissemination and evolution of reinforced concrete techniques in Italy with reference to the Hennebique System, introduced by G.A. Porcheddu, agent and general licensee for north Italy,[4] took place in the years between the end of the nineteenth century and the first decades of the twentieth century. Among the first uses of this technique in Milan, albeit limited to horizontal structures (decking), was the realization (1897–1901, 1898 contract) of the building for the Società Assicurazioni Generali Venezia, which still exists today in Piazza Cordusio and was designed by the architect Luca Beltrami in collaboration with the engineer Luigi Tenenti. This building, with its traditional wall structure, sees the use of floors (originally envisaged as beams and vaulting) made from reinforced concrete according to two structural typologies: the first consisting of slabs with main and secondary ribs (to be used for the upper storeys) for smaller ceilings; the second consisting of flat intrados which required a double slab to be used for the larger ceilings to cover the halls on the ground floor (Figs. 5, 6 and 7).

---

[2]See: Elia M.M., Cantamessa L., Petrucci E. (2015). "Fortunately, the Italian State Railways in the previous two years the war, had begun work to develop the 'strategic' lines and installations with some interventions that dated back as far as 1908 when, as a result of the burgeoning fears of a possible attack by Austria upon expiry of the Triple Alliance, the FS decided of its own accord to scale up the marshalling yard at Mestre".

[3]La Stazione Centrale di Milano—Inaugurata l'anno IX E.F., official illustrated supplement authorized by the Ministry of Communications, Milan 1931, p. 58.

[4]The Porcheddu company's clients included the state railways.

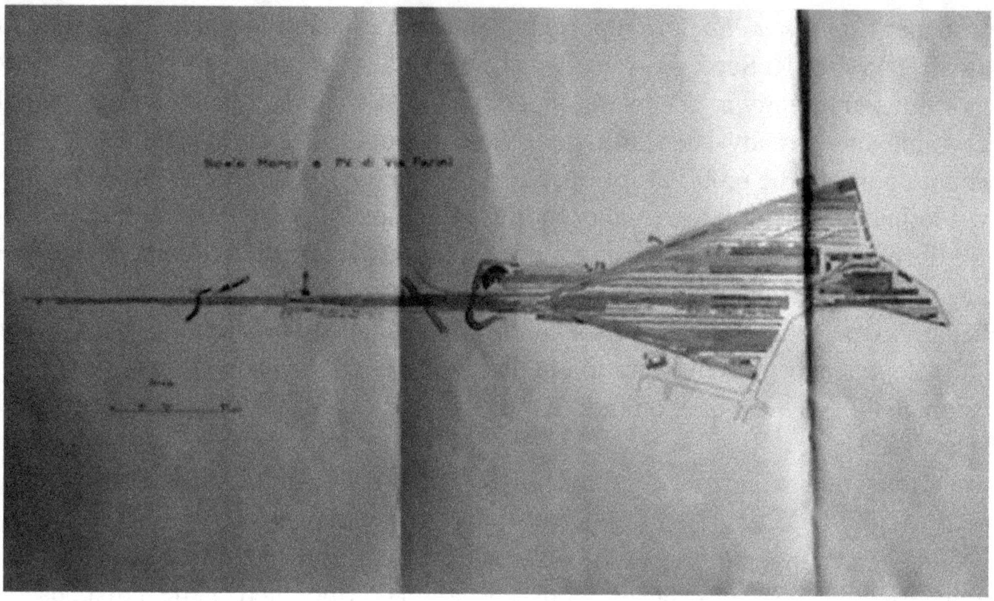

Fig. 5 Ministry of Communications—Ferrovie dello Stato, Farini Railway yard, Project plan, undated, Raccolta Bertarelli Milan

Fig. 6 *Original view of the Warehouse building, undated,* (La Stazione Centrale di Milano 1931)

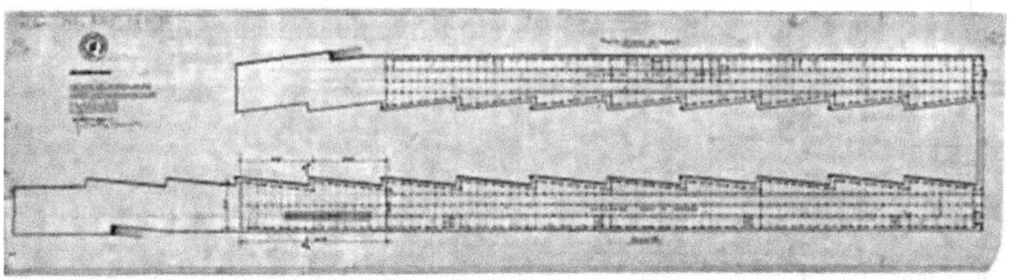

Fig. 7 State Railways Project Office, Plan of the Warehouse building, undated, FF.SS. Archive Milan

Again in Milan, one of the first and most important examples of a structure realized in reinforced concrete with large ceilings is the *Grande Salone*—the Great Hall—(built in 1900) in the courtyard of the Brera building. Measuring 15 × 25 m, it was intended for a classroom that had to temporarily accept works of sculpture submitted for the 1900 Brera Exhibition (Il nuovo grande salone di Brera 1900). The work was designed and completed by the architect Augusto Brusconi[5] of the Regional Technical Office for the Conservation of Monuments.

Despite Hennebique's patent expiring in 1903, the structures of the depots are practically a plastic recreation of the structure represented in the A.G. Porcheddu company logo with which they advertised the Hennebique system, as proof of the strong monopoly that the company had acquired as a patent licensee. In fact, comparing the image of the company logo with an image of the depot structures, we can recognize in both the classic typology of pillars, beams and slabs in reinforced concrete used for industrial buildings, as an evolution that saw for this type of building the passage from constructions of a nineteenth-century type (several aisles with perimeter walls in brick and structures generally in iron) to buildings that adopted reinforced concrete for the horizontal structures (possibly with reinforced concrete pillars in the inner zones) but with façades still in masonry.

The structural elements which were also part of the new architectural language and that characterized the industrial building were the pillar, with characteristic rounded corners, of a generally reduced size (40 × 40 cm, 50 × 50 cm); the main beams depressed by the ceiling slab, with chamfered corners and connected to the pillars via corbels; the secondary beams to stiffen the slabs and fit into the main beams; slabs of reduced thickness and dimensions which could be rectangular or square. Clearly, compared to the previous examples mentioned, in the case of the structures of the two wings, we are in the presence of a further evolution of this type, which also saw industrial buildings freed from load-bearing perimeter walls, almost certainly due to the need to have large openings around the perimeter to facilitate the movement of goods.

Another consideration must be made regarding the use of the Hennebique system by the two Milanese designers Beltrami and Brusconi, the authors of several restoration works, as a matter of common knowledge. The Office (established in 1892 but in operation from 1893 until 1908 when the Superintendencies were set up), is located at the Brera, where two other institutional seats coexist: the administration of the homonymous picture gallery (Corrado Ricci) and the administration of the academy (Camillo Boito). From an examination of the Corrado Ricci archive (Guarisco 1995), a series of private letters to Boito came to light (September 1912) from which it emerges the opinion of both on the members of the Regional Technical Office. Ricci, who was preparing the First Conference of Honorary Inspectors and Superintendents (which took place in Rome in 1912), sought Boito's approval and support for the initiative. But Boito, who saw in the conference the enactment of a

[5] Among other things, the architect Augusto Brusconi would be the leading light of the project "for the general organization of higher education institutes" in Milan, and later on, in the establishment of the new Polytechnic seat of Città Studi.

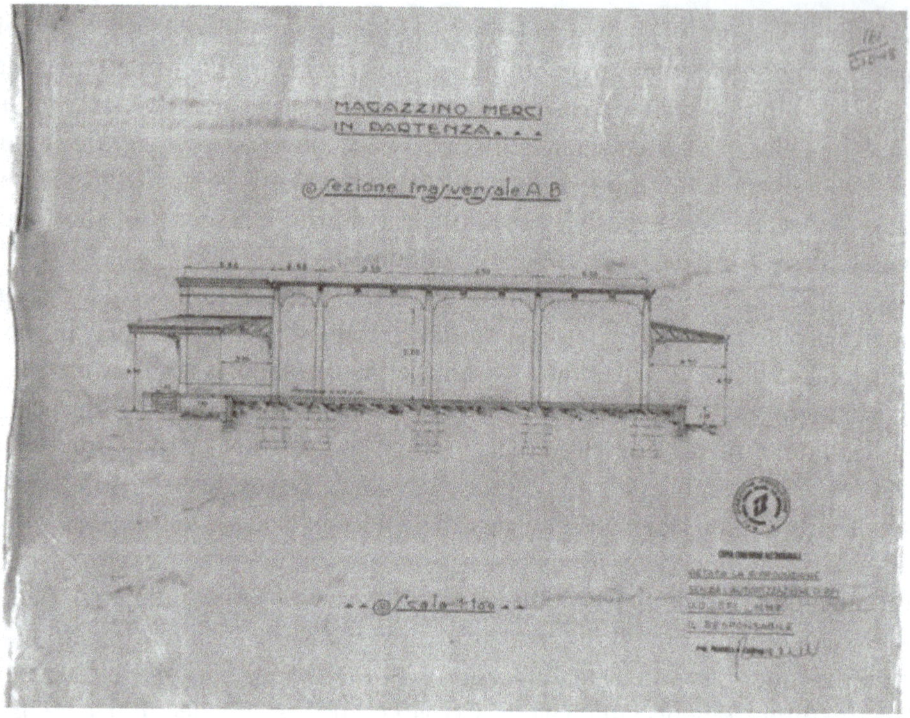

Fig. 8 State Railways Project Office, Cross section, (n.d.), FF.SS. Archive, Milan

"pompous little school" which he considered "unseemly", obliged Ricci to make a pungent and ironic defense through which in the end he would has obtained Boito's blessing. The question would seem of no interest, unless for the fact that Ricci, to bring Boito over to his side, attacked in no uncertain terms first the representatives of the Technical Office and then Boito himself.[6] In short, the atmosphere was not exactly placid. Of course, Boito, in 1912 (he died in 1914), must have already seen (if not directly commissioned?) the project, and the execution of Brusconi's works in the courtyard of the Brera for the construction of that building which already used the Hennebique system, and was supposed to host the 1900 exhibition, and then the *Gipsoteca*—plaster cast gallery—too inconsequential within the picture gallery (which Ricci directed from 1898 to 1903) (Figs. 8 and 9).[7]

At this point, and with this reference framework, also the now famous phrase of Boito becomes clear: "Oh this blessed shed! It would be our anchor of salvation for the Academy and for the exhibition [that of 1900], it would put everything in place for the teachers and the pupils and the artists: I dream of nothing else than the shed,"[8] and

[6]See: Guarisco G. (1995). Ricci to Boito, 27 September 1912: "From the walls of the Palazzo di Brera exude a kind of poisonous humidity that attacks the mood. Beltrami, Brusconi, Moretti, and Modigliani have all come and been touched by it [...]. Reading your ferocious penultimate letter, I said: Sadly, even Boito has become Beltramiated, Brusconiated, Morettatied, and Modiglianiated." The correspondence is kept in the C. Ricci Collection at the Classense Library of Ravenna, under nos. 4010, 4011, 4012, 4013.

[7]See: Pini (2009–2010).

[8]C. Ricci Collection, Correspondence, no. 4041.

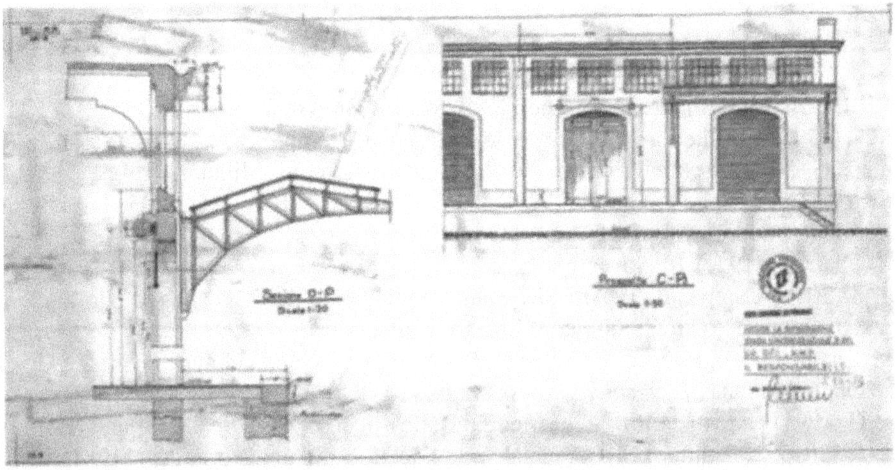

Fig. 9 State Railways Project Office, Front and section (n.d.), FF.SS. Archive, Milan

thus the assumption (which only further extensive research can confirm) that Boito was well aware of the Hennebique system (and this is not something extraordinary, seeing that Brusconi was building it before his eyes) seems evident, but also its use for the construction of large buildings, such as the "shed" referred to. From here to say that the shed wanted by Boito for the academy and its teachings were the Grande Salone or the former Farini depot is still somewhat impulsive since the research does not offer any concrete proof, either regarding the designers or the dates.

It should now be acknowledged that this hypothesis is the result of a close collaboration between contiguous disciplinary areas (restoration, architectural design and structural engineering) that have difficulties seeing eye to eye, but which—when they do—produce unexpected results on the research front.

The first fact-finding investigation of the former Farini depots produced a long series of not negligible particulars in the planning phase to verify the impact for reuse as the seat for some lessons of the Accademia di Brera. Owing to the importance of some of the protagonists of Milanese cultural history in the nineteenth century when it comes to architecture and monuments (Boito, Beltrami, Brusconi, etc.) and due to the importance of the Hennebique construction system used here in a precocious and singular manner, it was already frankly stated in the *Steering Document* that only interventions aiming at the practical conservation of the edifice as it has come down to us would be eligible.

In order to proceed in accordance with the rules laid down in the *Cultural Heritage Code* (2004), an initial phase of cultural valorization would be followed by a physical valorization by reusing the extant remains It is not only to honour Boito's work and that first Restoration Charter (1883) that the conservation of the existing building will be carried out. It is a homage to a school, the Politecnico di Milano (where Boito himself promoted the teaching of restoration much sooner than in the rest of the Country), the continuity of working relationships with the Accademia di Brera (renewed in these studies) and, ultimately, the joint search in the former depots for spaces suitable to teaching activities that are which in both institutions of top quality.

References

6 tables 161/C 10-13 (n.d.) Archive of the FF.SS. Milano Greco Pirelli

Canella G (1979) L'architettura del ferro e del mattone, *Casabella* (451–452), pp 24–28. Reprint in Canella G (2010) Architetti italiani del Novecento. Marinotti, Milan, pp 19–31

Castronovo V, Castagnoli A, Giuntini A, Piccolo S, Ostuni MR (2005) 1905: La nascita delle Ferrovie dello Stato. Hachette Children's Group

Corriere della Sera (1921) 16 November, p 5

Cusatelli S (2019) Campus delle Arti di Brera. Ampliamento dell'Accademia allo Scalo Farini. Indirizzi di un progetto architettonico. Mimesis, Milan

Elia MM, Cantamessa L, Petrucci E (2015) Le Ferrovie Italiane nella Grande Guerra (1915–1918), Fondazione FS italiane, *La tecnica professionale*, (10 October)

Guarisco G (1995) Notizie da Brera: il carteggio Boito-Ricci. In: A-Letheia. Milano restaurata, (6)

Guarisco G (2015) Il Cimitero Monumentale e le linee ferrate: una storia per il riuso dello Scalo Farini. In: Monica L, Scarrocchia S (ed) Per l'ampliamento dell'Accademia di Brera. Ricerche progettuali. Sesto San Giovanni (MI): Mimesis, pp 82–105

Guarisco G, Dezzi Bardeschi M, Fiorese G, Monica L, Pizzi S, Torricelli A (2017) Projects for the new location of the accademia di Brera. In Degli Esposti L (ed) (2017) Milan capital of the modern. Actar Publisher, New York, pp 187–191

Il nuovo grande salone di Brera e la sua copertura in calcestruzzo armato. Sistema Hennebique, L'Edilizia Moderna (1900, September), (IX–IX), pp 87–88

La Stazione Centrale di Milano - Inaugurata l'anno IX E.F. (1931) official illustrated monograph authorised by the Ministry of Communications. Milan, p 58

Ministero delle Comunicazioni - Ferrovie dello Stato (n.d.) Riordinamento dei servizi ferroviari a Milano. General Plan. Raccolta Bertarelli, OP T 58

Pini E (2009–2010), L'attività museografica di Corrado Ricci (1858–1934) e la direzione della pinacoteca di Brera (1898–1903), Specialist Degree Thesis in Science of Cultural Assets and Activities, University of Insubria, supervisor G. Guarisco

Riccardo N, Signorelli B (1990) Avvento ed evoluzione del calcestruzzo armato in Italia: il sistema Hennebique. Associazione italiana tecnico economica del cemento. Edizioni di scienza e tecnica, Milan

Rigato F, (2017-2018) Evoluzione del trasporto ferroviario italiano: dalle origini alla istituzione dell'ente ferrovie dello stato spa, Università degli studi di Padova, Dipartimento di scienze politiche, giuridiche e studi internazionali, Corso di laurea quadriennale in Scienze Politiche, Indirizzo Economico, supervisor Prof. G. Tusset

Vv Aa (1933) Le Ferrovie dello Stato nei primi anni di esercizio 1905–1930. Conferenze tenute dai Capi Compartimento. Istituto Poligrafico dello Stato, Rome

A University Campus for Medical Disciplines in View of the Redevelopment of the Guglielmo da Saliceto Hospital in Piacenza

Piero Poggioli

Abstract The work includes studies and projects for the architectural recovery and completion of the Renaissance period monastery of S. Sepolcro by inserting spaces for a new university of medicine and nursing within the overall regeneration project of the area of the Guglielmo da Saliceto Hospital in Piacenza. The university functions and the other activities compatible with the historical structure envisage a valorization of the monastic complex and contribute on several fronts to supporting the regeneration project of the sector where the hospital function has been continuously performed since 1471. The theme presents an opportunity to question on the relation between the architectural project and the historical buildings and, more generally on the functions and actions compatible with the ancient city. This project requires an extended and in-depth evaluation that also questions the concept of "sustainability", expanded to the "physiology" of the historical buildings and settlements, seeking appropriate solutions that consider the transformations necessary to not disperse the potential accumulated by these areas and to guarantee their active role within the urban organism. The study has been elaborated in the research: "Guidelines and operational solutions for urban redevelopment of the sector corresponding to the Guglielmo da Saliceto Hospital and for the conservation and reuse of the former convent of S. Sepolcro in Piacenza".

Keywords University and city · Historical city · Monasteries regeneration projects · Historical buildings reuse · Historic hospitals · Cloisters · Enclosures · Piacenza

P. Poggioli (✉)
Architecture, Built Environment and Construction Engineering—ABC Department, Politecnico di Milano, Milan, Italy
e-mail: piero.poggioli@polimi.it

S. Della Torre et al. (eds.), *Buildings for Education*, Research for Development,
https://doi.org/10.1007/978-3-030-33687-5_24

1 "Università: Ragione, Contesto, Tipo" (University: Reason, Context, Type)

The above mentioned publication by Canella and D'Angiolini (1975) contains a still eligible operating mode aimed at not separating the definition of large civil functions (of universities in this case) from their insertion within the territory. In the design process, the place—configured through a thorough research of past and present specific characteristics—is put in a position to react with those typological and figurative sedimented paradigms which, by their nature, aim to interpret those "invariants" that can aspire to translate into permanent values, hopefully also valid for unknown future scenarios.

The study[1] which is synthesized in this article aimed to define functional, architectural and conservative hypotheses for the recovery of the Santo Sepolcro Monastery in Piacenza, within a general redevelopment of the historic *Guglielmo da Saliceto* hospital. The university function—public and civil by definition—was the most appropriate not only to give meaning back to the monumental building but also to trigger a series of positive effects for the hospital, its surroundings, the city and the territories that gravitate around it. The Piacenza hospital is a fascinating complex of buildings, including historic ones, within the ancient town center; while boasting a number of specialized excellences, it suffers not only from the limits given by the historicity of the place but also from the "competition" triggered by the health reform that has encouraged phenomena of "health commuting", especially within Lombardy.

Specialist excellence is the main vector of this attraction, and it is therefore clear that the university's presence contributes decisively in qualifying the campuses and increasing their attractiveness, as successfully demonstrated by the nearby hospitals of Parma and Pavia which are structured around highly acknowledged universities. Beyond the ambitions of competitiveness, the Piacenza campus, interpreting the specificity of the territory as a *land of passage*, could be a candidate to constitute a *trait d'union* between regional health systems, and in particular between the Pavia and Parma areas, of which they can also constitute a "detachment".

The relationship between the university and the city is a topic of absolute relevance in the history of Western culture and in the field of urban planning, architecture and architectural typologies. While, on the one hand, Carlo Cattaneo's call to scholars to carry out their work within the city is unquestionable (see Acuto, 1992),[2] on the other hand, there is an extensive literature that reveals, if not a separation, at least an "autonomy" of the university from the city and some of its dynamics.

---

[1] *Linee guida ed operative per la riqualificazione urbana del comparto corrispondente al Presidio Ospedaliero Guglielmo da Saliceto e per la conservazione e il riuso dell'ex convento del Santo Sepolcro a Piacenza.* Research coordinated by P. Poggioli, M. Boriani, M. C. Giambruno, R. Rizzi (Politecnico di Milano, DPA, 2009–2014) for AUSL Piacenza. With F. Cesena, M. Bordin (images and research). All images, excluding Fig. 1, were realized within the research.

[2] Cattaneo C, "La città come principio ideale delle istorie italiane", 1858 "Vogliano gli studiosi compiere questa ricerca delle fonti della scienza sperimentale nel seno delle nostre città", in Acuto A., Università e territorialità: lo Studium Generale a Pavia, "Zodiac", n. 7, April 1992, Abitare Segesta.

From its origins, Harrison notes (Harrison 2008), Plato in The Republic wrote that a philosopher should "shield himself behind a wall" alluding no doubt to the Academy, which was in fact a walled park, much like the immured hunting preserves of the Persian kings, from which comes the word paradise (*paradeisos* in Greek, from the Persian *pairideiza*, or "walled around"). "…Plato's decision to establish his school in a park on the margins of Athens—located far enough to listen to the voice of reason, close enough to stay within earshot of the citizens—set a pattern for the future history of academia in the West".

This hypothesis finds another kind of continuity in a 1927 text by Poëte (1929), where in front of a map of Paris of the sixteenth century, significantly known as the *Three Characters*, he detects the tripartite division in *Ville, Cité* and *Université*. This text, taken up several times by Canella (1968, 1975, 1992), captures next to the religious city of the *Île* and the administrative-mercantile city of the *Rive Gauche*, and the university city of the *Rive Droite*, characterized not only by the university but also by the colleges and convents of the mendicant orders.

The condition of *vagantes*, that is, that of the pilgrim, of the beggar, but also of the student "who went into exile for the love of science", seems to share, at least in the period of the university's origins, the cloistered typology, which appears the most suitable organizational criterion to interpret the needs of concentration and a "considered" isolation from external interference, which is not always favorable.

However, as mentioned by Mumford, "In the university the functions of cultural storage, dissemination and exchange, and creative addition—perhaps the three most important functions of the City—were adequately performed. A cloister and library of the monastery might be called a passive university, so the university might be termed an active cloister" (Mumford 1961).

Despite the good fortune of the theories matured around architectural typologies, the university function has undergone a general disaffection in recent decades. Although enriched by widespread and desirable inventions, it seems in substance to continue to interpret the paradigm of the development of courts or open blocks in various ways, however, generally organized according to the consolidated model of the *campus* which, in its various meanings, seems to still confirm and consolidate the image of an autonomous entity with respect to the city.

G. Canella's "Università: ragione, contesto, tipo" therefore seems to constitute still actual elements of reference, which become indispensable, when the delicate equilibria in urban physiology, the sedimentation of places and types, the relevance of the functions—not separated from their permanence—are faced. Thus, the project synthesis, as in the proposed research, cannot avoid considering them and incorporating them into action, together with the constructive, normative or specifically functional elements.

2 Genesis and Development of the Area in the City

Piacenza is an ancient city; born and structured as a road junction and a military city. Today it expresses a marked vocation for the logistic activity and an extended presence of military areas (over 1,200,000 m^2) located also in the historical center, currently unused/to be divested. It is the capital of a provincial territory, with a markedly eccentric extension, which has about 290,000 inhabitants, while the city population seems to remain just over 100,000. The historic town center, identified by a perimeter of Renaissance walls, in spite of the considerable quality of its construction, made up of monuments and of the considerable liveliness of a number of central axes, has suffered, starting from the 1960s, the outflow of inhabitants and activities, only recently partially compensated by a mainly non-EU turnover, fueling a widespread perception of a "crisis" in the historical center.

The area of the Piacenza hospital, which houses the Santo Sepolcro Monastery, occupies a large area (about 90,000 m^2) at the western end of the historical center and is bordered on the east by the sixteenth-century walls, on the north and south by the Via Campagna and via Taverna streets, a legacy of the two western branches of the Via Francigena (Fig. 1). The hospital complex is a surprising stratification of a history that began in 1471, the year in which the Bishop Campesio—with the approval of Sisto IV and Galeazzo Maria Sforza—founded the Ospedale Grande, which reabsorbed the different *hospitalia* scattered throughout the city and along the main access and transit routes to it (Fig. 2). Over more than five centuries the ancient hospital has continually expanded and has been renewed without ever interrupting its activity. To the original cross-shaped nucleus, still today clearly identifiable despite the various

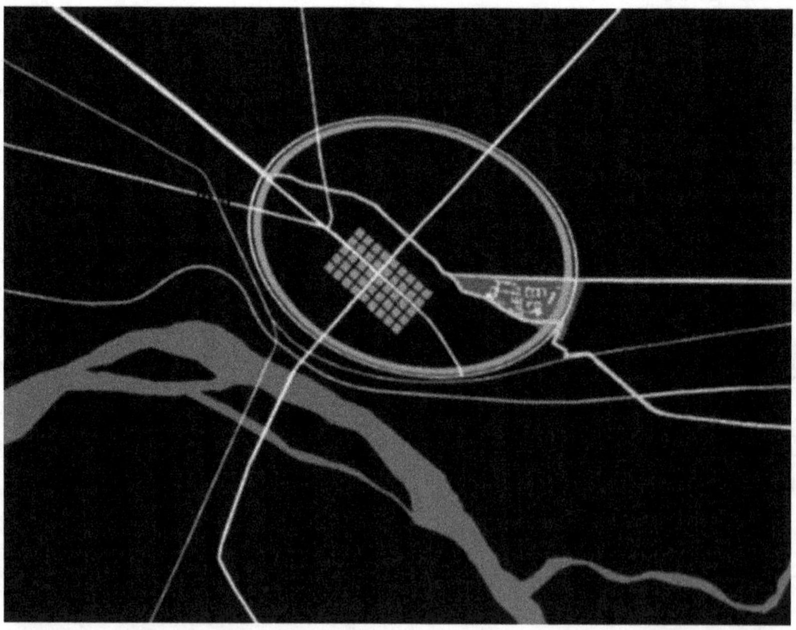

Fig. 1 Synthetic reconstruction of the main urban features and of the study area

Fig. 2 The hospital area in the historic city. Matteo Florimi, view of Piacenza, engraving, late sixteenth and early seventeenth centuries: detail

additions, some neighboring complexes have been added, including the Monastery of the Santo Sepolcro (1498–1516) and of Santa Maria di Campagna (1522–1528) partly designed by Alessio Tramello (Adorni 1998), then acquired and adapted to its current healthcare function. Other structures for specific hospital activities have subsequently been implemented ex-novo, such as the pavilions of the early twentieth century and the Nuovo Polichirurgico building, in the 1980s. The result of this long history is a vast and heterogeneous compound in the historic center, where buildings, open spaces, hospital activity and related activities have a series of complex relationships, from which emerges the relevance of the human presence (patients, staff, daily users, different services, visitors) which directly affects the immediate surroundings as well, the city and the territory. Since the moment of its foundation, the ancient Ospedale Grande has been recognized not only for the service provided to the local community but also for serving the *poor pilgrims*, linked to long-term mobility along the Via Francigena.

3 The Regeneration Project for the Sector

The general project proposal imagines that the health facility, enhanced by the university function, could usefully continue to operate in the historical area by resolving a series of critical issues, among which the most urgent is undoubtedly the limited accessibility and the lack of parking facilities.

The proposed solution, evaluating the availability of adjacent external areas and the possibility of creating an underground parking system, aims to achieve total availability for the free non-built surfaces. Open spaces, recovered for pedestrian use, define—through pavements, furnishings and greenery—the entrances, paths, places for staying and encounters (Fig. 3).

This rediscovered unity of design brings the heterogeneity of the structures built over the centuries into a compositional order, proposing a network of special paths and relationships between the parts and prefiguring a renewed relationship with the city, which finds fulfillment in the functional redevelopment of the Santo Sepolcro Monastery into a university campus (Figs. 4, 5, 6, 7 and 8).

Fig. 3 The general regeneration project for the hospital sector (without volumetric increments)

Fig. 4 Longitudinal section. Restoration project of the system of courts

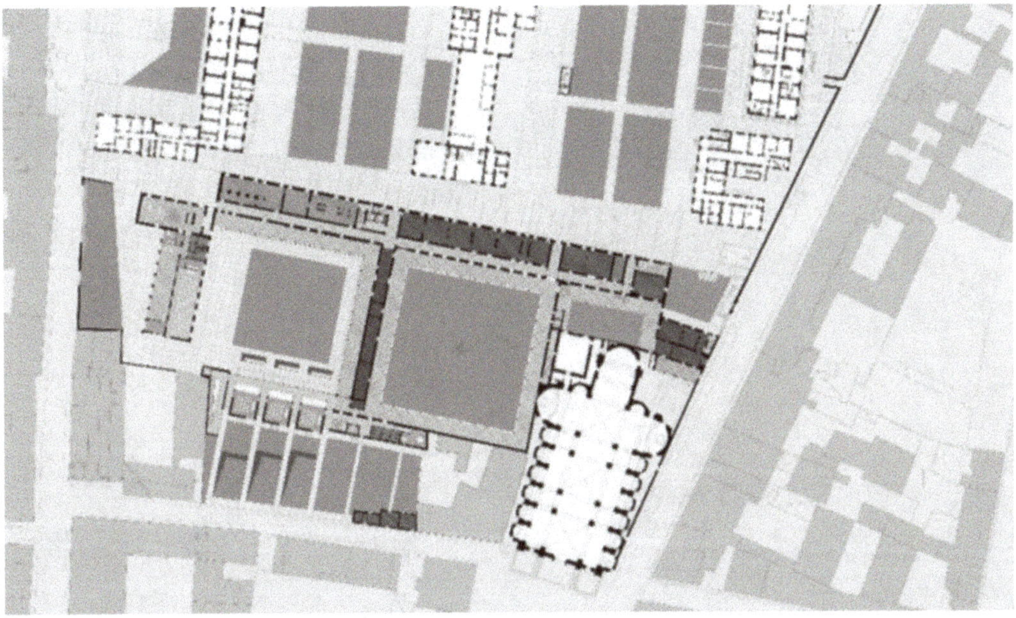

Fig. 5 Ground floor plan. Project of the completion of the cloister with limited volumetric increments

Fig. 6 Exterior elevation. Hypothesis with maximum volumetric increments

4 The Project of a New University Campus in the Ancient Monastery

Given the absolute historical and architectural importance, the complex has been thoroughly studied and surveyed. Although in modest conservation conditions, it is still largely used by the healthcare facility, even if the functions present are not very relevant and are reasonably relocatable.

One of the main features is the presence of a volume with longitudinal development of considerable size, arranged—as if connecting them—between the two urban branches of the Via Francigena. This building with its straight layout of more than 120 m in length, connects, physically or conceptually, a series of spaces, functions and architectural episodes: starting from the north, the Casa dell'Abate (now the AUSL offices), with its small cloisters and gardens, the large church, the main cloister, the important Renaissance library (O'Goorman 1972) and finally the second

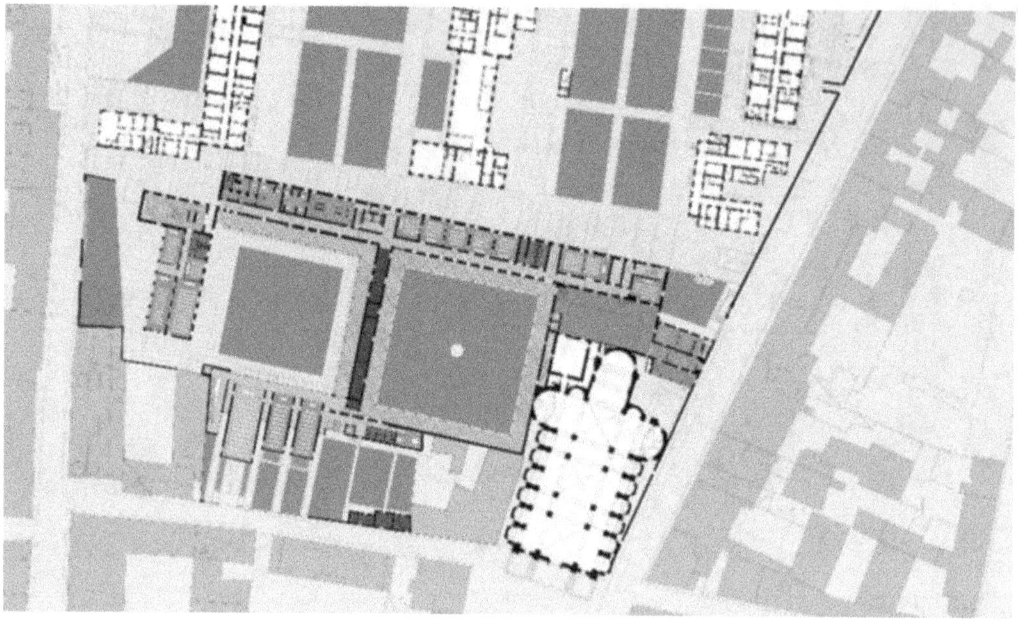

Fig. 7 Ground floor plan. Project of the completion of the cloister with maximum volumetric increment

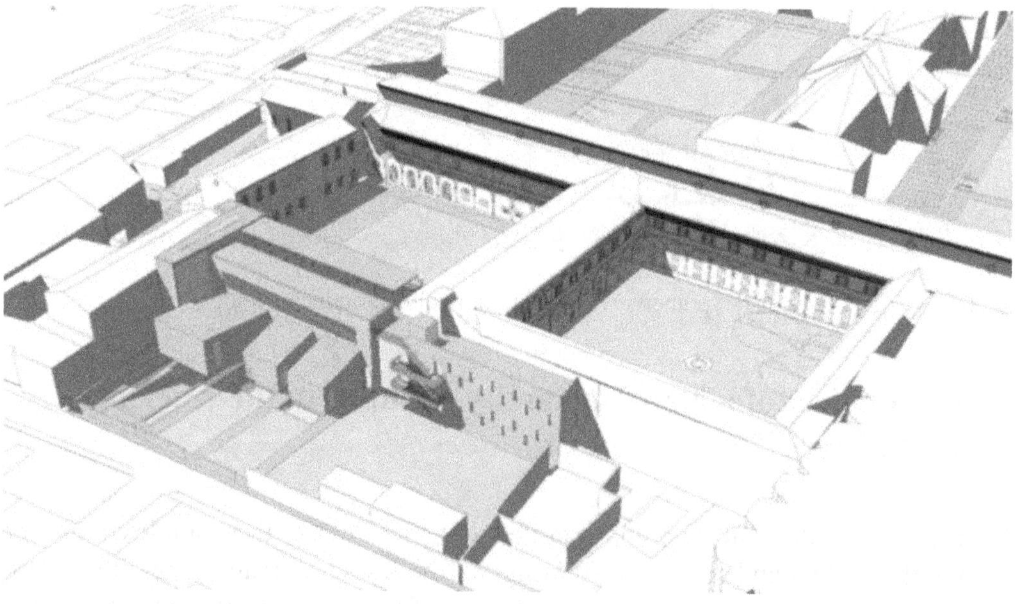

Fig. 8 View of the external volumes of the university classrooms

cloister, which was never completed, but, to a certain extent, subsequently defined by the structures for laundry.

The regeneration project with university functions starts from the consolidated presence of the course in nursing—a three-year degree with approximately 240 students—which does not have spaces inside the complex and must provide an external leased facility. The hypotheses envisaged, in addition to ensuring the spaces for existing courses, expand the function by inserting some courses in medicine, evaluating the possible configurations for the historical structure and its potential extensions. The redevelopment project, in its variations, has placed alongside the university function, facilities for hospitality (hostel/hotel, and related services) and consolidation/promotion of territorial identity (museum of the territory, exhibition halls, spaces for catering and selling local products), with the aim of realizing potential synergies both on the scale of the hospital sector and on the urban/territorial level.

Besides a hypothesis of distinctly conservative intervention to the original construction, a study has been made to evaluate the completion of the cloister according to two different volumetric configurations that allow for accommodating a more substantial and better organized university settlement.

The importance, even semantic, of the *cloistered* typology with respect to the university and *hospitality* functions, finds, together with the lack of space, a latent opportunity in the incompleteness of the second cloister of the monastery, currently an unfinished fragment, above all in regard to the street behind (Cantone San Nazaro) and the buildings of the former Church of San Nazaro and Celso. The definition of the cloister, historically realized only on two sides (north and west), takes place by keeping the complex of the Ex-Laundries (by now historicized) in the south and proposing on the east a newly constructed volume designed in two alternative modes, but with a substantially analogous solution regarding the new section facing the internal courtyard (Fig. 9).

In both cases, the new recomposed cloister, with a central garden marked by seasonal transformations, continues to represent the ideal epicenter that the scientific community has adopted and that in the proposed project becomes the background for multiple possible scenarios of university activity.

The three proposals developed for the university function (no increase in volume, limited volume increment and maximum increment) respectively identifiable with 240, 480 and 930 students, respond to the purpose of defining the minimum

Fig. 9 View of the new cloister

and maximum volumetric, but also expressive, limits to be proposed to the various counterparts (AUSL, Municipality, Superintendence).

4.1 The Project Without Volumetric Increments

The first proposal has an eminently conservative character and foresees the restoration of existing structures with limited demolitions of incongruous additions. The interventions are concentrated in the interior and have a predominantly distributive nature—paths, stairs and elevators, as well as furnishings. In this hypothesis, the monastic complex can still provide the spaces needed for the current university nursing course with its 240 students on the ground floor of the long building and above all in the former laundries. A new design of the pavement and greenery gives an evocative reading of the original project and of the second unfinished cloister.

4.2 Completion Hypothesis with Limited Volumetric Increments

The hypothesis proposes to conclude the missing side of the cloister by adding volumes above the ground. In the basement level there are three large classrooms (70, 82 and 95 m^2), and a series of smaller classrooms/laboratories with related services. To overcome the aero-lighting constraints, zenith lighting systems and light wells have been designed, but above all, on the East side, an excavation frees up the sides of the main classrooms, opening them up to the view and facilitating their access. An inclined plane, mainly treated as a garden, creates a connection between the underground level and the city and provides the classrooms with an optimal view (Fig. 5).

The rather composed facade on the cloister—a large glass wall mediated by a large suspended *brise-soleil*—(Figs. 4 and 9) corresponds to a variety of parts on the back that allude programmatically to the decomposed historical nature of the front that housed disordered additions.

Among these, the largest one is demolished and rebuilt to accommodate adequate services and vertical communication, to which a vigorous external staircase is added which, in addition to safety, allows for independent access to the flat and accessible roof with seats, hanging gardens and canopies.

The east elevation is completed by continuing the theme of the reinvention of the additions and evoking the concept of the "void", disarticulating itself by highlighting the relationship between the glazed and transparent passages and the blocks of the classrooms, or of the stairways that take on a sense of greater solidity.

4.3 *Completion Hypothesis with Maximum Volumetric Increments*

The last proposal tries to define a possible maximum volumetric addition, which is not determined by the current normative and bureaucratic aspects but induced by the discipline of the architectural project. Sharing different characteristic elements of the solution already described (front on the cloister, permeability of routes and views, continuity with the historical structure, accessible flat roofing…), this proposal has a significant increase of the functional spaces and an underground parking area.

Alongside the volumetric increase, the expressiveness is also more accentuated, however limited to the rear front, which, due to its location and history is believed to be able to accept volumes and figures of a certain impact, which was interpreted with the composition tools.

The theme of the reinterpretation of the additions is therefore confirmed and relaunched and more clearly expressed here, highlighting a logic of addition of autonomous volumes to the large brick wall (Fig. 6). Three large volumes faced with metal sheeting and corresponding to the main classrooms (83.6 m^2 for 55 seats) and to the auditorium (167.5 m^2 with 140 seats) are attached to this wall (Fig. 8). The auditorium is emphasized both in terms of volume and shape, also due to a cantilever that extends over the entrance of the underground parking, thus emphasizing rather than concealing its presence.

The pavement design configures the paths and places to stay; the differences in height, the small excavations of the supporting surfaces and the water surfaces—alluding to the presence of the San Sepolcro river—articulate, through lights, shadows and reverberations, the particular and ever-changing perceptions of the metal volumes leaning against the great wall (Figs. 7 and 8).

References

Acuto A (1992) Università e territorialità: lo Studium Generale a Pavia, "Zodiac", new series, n. 7, April

Adorni B (1998) Alessio Tramello. Electa

Canella G (1968), Passé et avvenir de l'antiville universitire, "L'architecture d'Aujourd'hui", n. 137, April-May

Canella G (1992) Università e città, "Zodiac", new series, n. 7, April

Canella G, D'Angiolini LS (1975) Università ragione contesto tipo. Dedalo

Cattaneo C (1858) La città come principio ideale delle istorie italiane. Giulio Einaudi editore

Harrison RP (2008) Giardini. Riflessione sulla condizione umana, (Italian version, 2009). Fazi

Mumford L (1961) La città nella storia, Dal chiostro al Barocco, vol. II, (italian version, 1994). Tascabili Bompiani

O'Goorman JF (1972) The Architecture of the Monastic Library in Italy 1300–1600. New York University Press

Poëte M (1929) Introduzione all'urbanistica- La città antica, (Italian version, 1958). Einaudi

Application of Externally Bonded Inorganic-Matrix Composites to Existing Masonry Structures

Angelo S. Calabrese, Tommaso D'Antino, Carlo Poggi, Pierluigi Colombi, Giulia Fava and Marco A. Pisani

Abstract Fabric-reinforced cementitious matrix (FRCM) and composite-reinforced mortar (CRM) are recently introduced inorganic-matrix composites that have shown promising results as externally bonded reinforcement (EBR) of existing masonry structures. FRCM and CRM comprised high-strength fiber textiles embedded within inorganic matrices. Different fibers and matrices can be used, which lead to a large number of systems characterized by different properties. In this paper, different techniques employed to strengthen the existing masonry structures with EBR. FRCM and CRM composites are presented and discussed.

Keywords Fabric-reinforced cementitious matrix · Mechanical characterization · Flexural strengthening · Shear strengthening

1 Introduction

The numerous seismic events that have struck the Italian territory emphasized the great vulnerability of its built heritage. In the event of an earthquake, strategic constructions such as schools and hospitals have often shown vulnerabilities associated with certain local and global collapse mechanisms.

Over the last decades, the need for improving the behavior of these strategic structures with respect to seismic events led the construction engineering community toward the development of new and innovative strengthening solutions. Among them, composite materials have shown to be an excellent solution for strengthening and retrofitting the existing structures due to their high-strength-to-weight ratio, corrosion resistance, and ease and speed of application. Within this category of materials, those comprised high-strength textiles embedded within cement-based matrices (usually referred to as fabric-reinforced cementitious matrix, FRCM) and those comprising a composite grid embedded within an inorganic-matrix (referred to as composite-reinforced mortar, CRM) represent one of the most recent and promising innovations

A. S. Calabrese (✉) · T. D'Antino · C. Poggi · P. Colombi · G. Fava · M. A. Pisani
Architecture, Built Environment and Construction Engineering—ABC Department, Politecnico di Milano, Milan, Italy
e-mail: angelosavio.calabrese@polimi.it

S. Della Torre et al. (eds.), *Buildings for Education*, Research for Development,
https://doi.org/10.1007/978-3-030-33687-5_25

(Carloni et al. 2016). FRCM has good resistance to high temperatures and excellent compatibility with masonry structures (De Felice et al. 2014). Because of this last feature, as well as the partial reversibility of the strengthening interventions, the use of FRCM reinforcements may be considered a good solution for the strengthening and retrofitting of the existing masonry members (walls, arches, vaults, etc.), which are largely diffused in education buildings in Italy (De Santis 2017).

FRCM composites comprise one or more layers of a high-strength fiber, open-mesh textile embedded within an inorganic mortar. The dimensions of the granules in the inorganic mortar do not allow for a good impregnation of the fiber filaments as is the case for organic resin impregnation. Therefore, to create a mechanical interlock between matrix and textile, the textile bundles/yarns are spaced in such a way as to allow for the penetration of the mortar into the grid spacing. Different types of fiber and inorganic matrices can be paired to achieve different composite properties, suitable for a wide range of applications. Figure 1 shows a selection of textiles commonly used in FRCM composites, even though some hybrid textiles have recently been introduced (Carozzi et al. 2015).

When dealing with seismic retrofitting of existing buildings, one of the key issues is the speed and ease of the execution. This is particularly appealing for educational buildings in order to avoid long interruption of the educational activities. The application of FRCM strengthening is a very effective solution in this regard. Indeed, the application requires a limited number of steps (see Fig. 2): (i) preliminary preparation of the surface that can be mechanically roughened to enhance adhesion, (ii) application of a first layer of mortar, followed by the positioning of the textile by means of a common trowel, and (iii) application of the final layer of mortar on top of it. The impregnation of the fiber textile is of crucial importance for the effectiveness of the application. Thus, the textile shall be gently pressed within the mortar to guarantee its complete impregnation and the absence of voids in the matrix. The curing time is usually of 28 days.

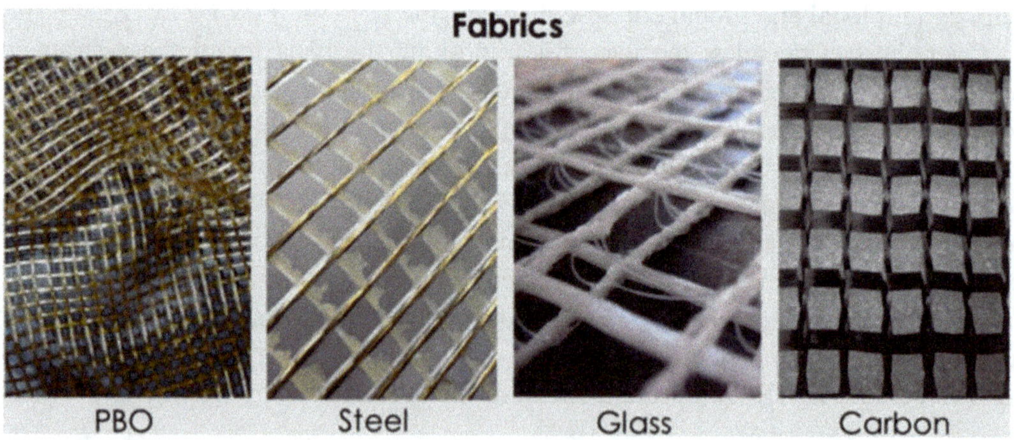

Fig. 1 Main typologies of fiber used in FRCM composites

Fig. 2 Application of a PBO-FRCM strengthening to a masonry substrate

This paper describes the fundamental mechanical properties of FRCM composites, including their tensile and bond behavior. Furthermore, a number of strengthening applications with inorganic-matrix composites, namely the out-of-plane and in-plane strengthening of masonry walls, are described and discussed. The results show that the use of FRCM composites is a valid tool to improve the structural behavior of the existing educational buildings.

2 Constitutive Behavior of FRCM Composites

Several studies have been recently presented regarding the characterization of FRCM composites, aimed at investigating their main mechanical properties, for example, ultimate tensile stress, ultimate strain, and elastic moduli (Carloni et al. 2016). These parameters can be obtained by experimental testing of FRCM coupons in tension using different methodologies.

Tensile coupons can have a rectangular or dumbbell shape and are made by successive layers of inorganic mortar, poured into molds (see Fig. 3) to correctly control the specimen shape. The coupon thickness usually ranges between 10 and 20 mm, depending on the number of fiber (and matrix) layers employed (D'Antino and Papanicolaou 2018).

According to the so-called clamping-grip method (Arboleda et al. 2015), the FRCM coupon ends are gripped directly by the testing machine, which applied the tensile load, under displacement control, up to failure. The specimen ends should be reinforced by applying FRP tabs to promote a more uniform distribution of the clamping pressure and avoid cracking of the matrix at the specimen ends. The load applied to the specimen is measured by a load cell, whereas axial strains may be measured by means of different technologies: extensometers, linear variable displacement transducers, strain gauges, and digital image correlation (DIC) (D'Antino and Papanicolaou 2018). The gripping method has a significant influence on the behavior and failure mode of the FRCM. With the clamping-grip method, possible

Fig. 3 Casting of a rectangular PBO-FRCM tensile coupon using flat molds

debonding of the textile within the matrix is prevented by the pressure applied by the machine grips (Arboleda et al. 2015).

The typical stress–strain curve of a tensile clamping-grip test has a tri-linear behavior (Carozzi and Poggi 2015), as shown in Fig. 4. The first branch is associated with the uncracked state and ends when the mortar attains its tensile strength. At the end of the first branch, cracks occur in the matrix in a number of locations depending on the grid spacing and matrix-fiber bond properties. Owing to matrix cracking (second branch), stresses are transferred from the matrix to the textile and vice versa and the stiffness of the stress–strain curve decreases. When the matrix is fully cracked, a third linear branch develops. This branch is characterized by a slope lower than or equal to that of the stress–strain curve of the dry textile. Finally, sudden failure occurs once the textile tensile strength is reached.

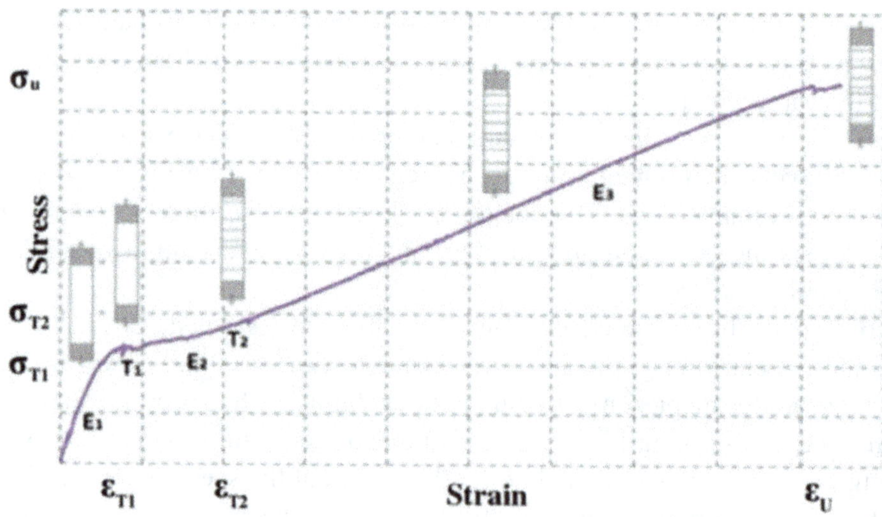

Fig. 4 Typical stress–strain curve (Carloni et al. 2016)

3 Bond Properties of FRCM Composites

The bond between composite and substrate, which is responsible for the stress-transfer mechanism between the composite and the existing substrate, is of fundamental importance in externally bonded strengthening applications.

For FRCM strengthening comprised of only one layer of textile, debonding generally occurs at the matrix-to-fiber interface, with significant slippage of the fiber bundles with respect to the embedding matrix. This phenomenon is in part attributed to the poor impregnation ability of the inorganic mortar, which cannot penetrate within the fiber bundles leading to core filaments slippage with respect to the sleeve filaments (see Fig. 5).

The stress-transfer, from fiber bundles to cement matrix and vice versa, occurs mainly due to shear stress at the fiber-to-matrix interface. The relationship between the interface shear stress and the corresponding slip can be described by a bond slip law (BSL) (Colombi and D'Antino 2019). After the shear stress reaches a limit value (referred to as shear strength), it decreases according to the BSL until attaining a residual shear stress provided by friction/interlocking between the matrix and fiber. When the shear stress attains the residual values, which may be equal to 0 for certain composites, debonding occurs. With a further increase of the slip after the onset of debonding, the load carried by the composites may increase further, provided there is a sufficient bonded length, due to the contribution of matrix-fiber friction. When debonding involves the entire bonded length, the composite becomes completely detached.

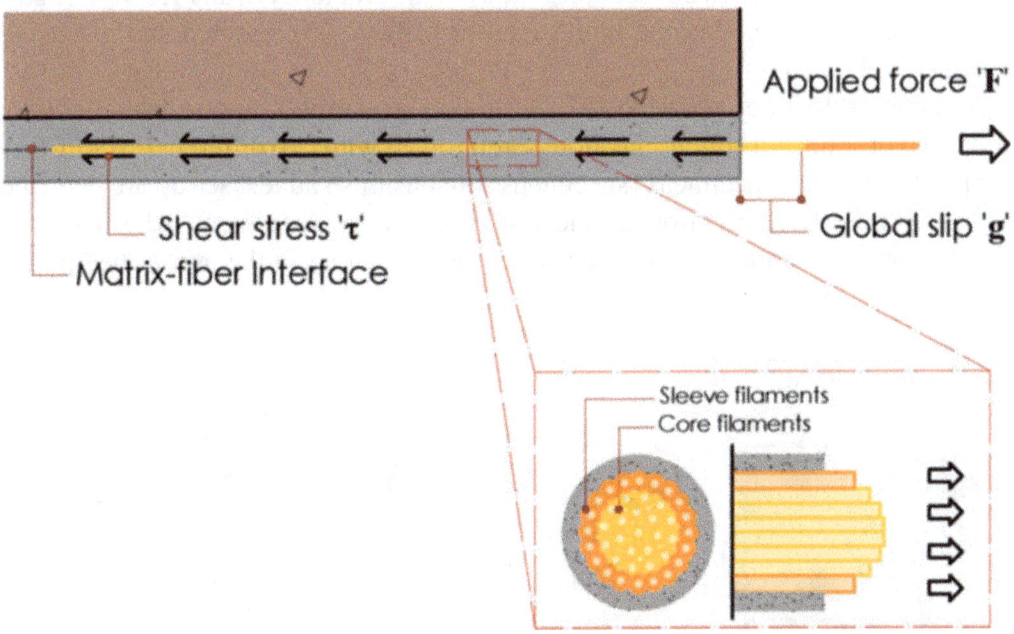

Fig. 5 Debonding at matrix–fiber interface

Several parameters affect the bond behavior of FRCM composites, such as the spacing of the textile, properties of the inorganic matrix, substrate preparation, and level of impregnation of the fibers. In addition, the test set-up may influence the results. Different set-ups are indeed employed to study the debonding process of FRCM composites. The most commonly used are single-lap and double-lap direct-shear tests, where a tensile force is directly applied to the composite. Other set-ups have recently been proposed, such as the hinged and notched beam tests, which indirectly induce a tensile force in the composite as an effect of the presence of a bending moment (Calabrese et al. 2019).

4 Out-of-Plane Strengthening of Masonry Structures

A consistent part of educational building heritage is characterized by a large use of structural and non-structural masonry elements, for example, piers, infill walls, ceilings, and partitions. These elements are extremely vulnerable in the case of seismic events, because the resistance of unreinforced masonry structures to out-of-plane external forces is mainly dependent on its geometry, connection to the adjacent structural elements and on the interlocking between bricks and mortar and on the mortar properties. For instance, the overturning of infill walls in the case of an earthquake is one of the most dangerous events for the safety of the building occupants.

FRCM composites represent an effective solution against these vulnerabilities. They can increase the flexural capacity of bearing walls and prevent the overturning of partition and infill walls. In this section, a study of the efficiency of an FRCM strengthening system, conducted at the Politecnico di Milano, is presented and discussed (Carozzi et al. 2015).

A series of 16 three-point bending tests was performed on masonry elements composed of solid and hollow clay bricks, reinforced with two different PBO-FRCM composites. Certain specimens (specimens' dimensions and test set-up are depicted in Fig. 6) were left unreinforced and tested as control specimens. The load was applied across the entire width of the walls, perpendicular to the mortar bed joints.

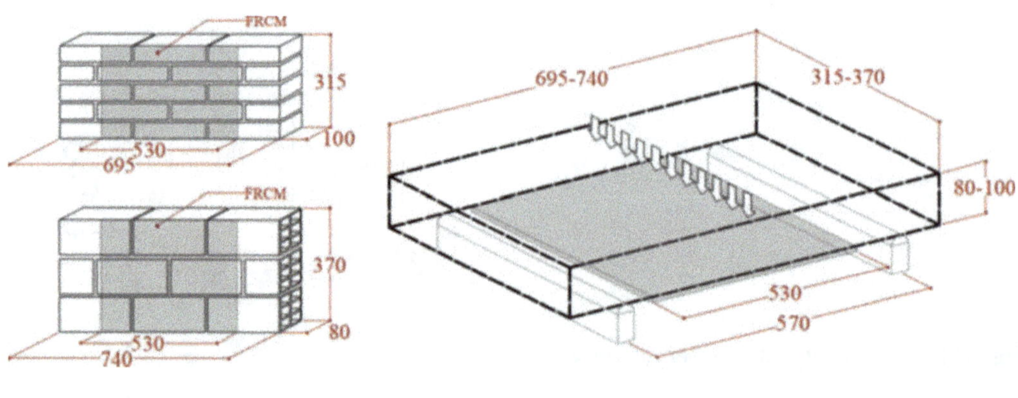

Fig. 6 Specimens' dimensions and test set-up (dimensions in mm)

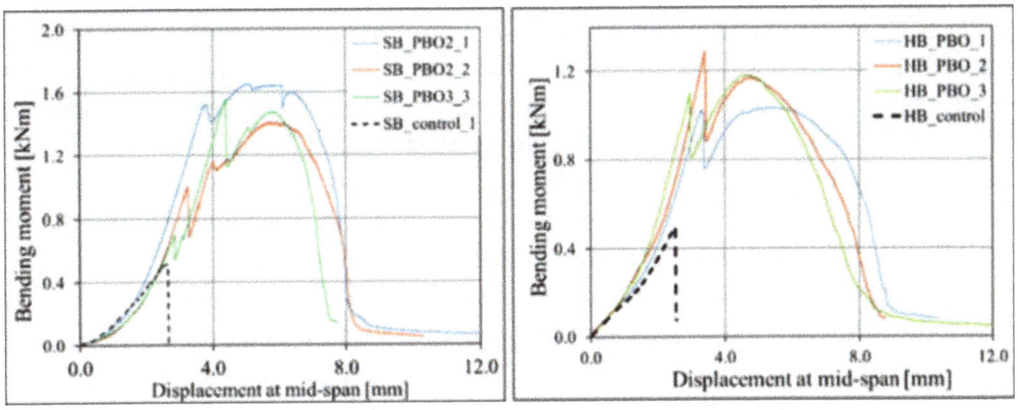

Fig. 7 Flexural responses of solid (left) and hollow (right) brick specimens

The mid-span deflection and the opening of the crack below the point of application of the load were recorded by means of linear variable displacement transducers.

The bending moment versus-mid-span displacement curves are reported in Fig. 7. Each graph in this figure provides the results of solid brick (SB) and hollow brick (HB) specimens strengthened with a specific PBO-FRCM composite analyzed, together with the control specimens.

Five phases could be identified in the strengthened specimen responses. An initial elastic branch, representative of the uncracked state, extended up to the opening of a major crack in the substrate. At this point, the curve shows a vertical drop and a macro-crack appears in the cementitious matrix. After the crack opening, the response may increase (with a low stiffness) due to the presence of friction between the slipping fibers and the embedding matrix. When the mid-span deflection further increases, the propagation of the main crack and the matrix-fiber debonding along the FRCM reinforcement induce a progressive loss of flexural capacity, represented by a descending branch. Finally, the applied load levels off at a low constant value provided by the residual friction between textile and matrix.

Comparing the results of control and strengthened specimens, it is possible to notice a bending moment capacity increase of approximately 170 and 180% for solid brick and hollow brick walls, respectively. Furthermore, a significant increment of the mid-span deflection is observed, equal to 660 and 926% for SB and HB specimens, respectively. This last aspect is of crucial importance because it shows that this reinforcement technique can significantly increase the displacement capacity of the masonry before collapse and guarantee occupants' safety in case of seismic events.

5 In-Plane Strengthening of Masonry Structures

Inorganic-matrix composites may be designed to best fit the substrate characteristics/properties. Recently, new types of inorganic-matrix composites, comprising lime-based and cement-based matrices and high-strength composite grids, were proposed. These composites, which are usually referred to as composite reinforced mortar (CRM), are characterized by the use of reinforcing grids with a yarn spacing higher than 30 mm and by a thickness of the composite higher than 20 mm. In this section, two experimental campaigns involving in-plane strengthening of masonry walls with FRCM and CRM composites are described and discussed.

In both campaigns, historical solid brick masonry walls were strengthened on both sides with FRCM and CRM composites and subjected to diagonal compression up to failure. The comparison between the results obtained from the strengthened members and the control specimens allows for the investigation of the contribution provided by the externally bonded composite systems to the masonry walls' shear strength.

In D'Antino et al. (2019), three historical brick masonry walls (see Fig. 8) of dimension 830 × 830 × 270 mm were strengthened with a CRM system, including a glass composite grid and a lime-based mortar. To anchor the composite grid, four helical inox steel bars were inserted at the corners through the entire thickness and bent against the wall faces. The diagonal compression was applied to the specimens in displacement control, by means of a hydraulic jack with capacity of 1000 kN. Two LVDTs on each side of the specimens were used to measure the vertical and horizontal displacement of the walls.

In Carozzi et al. (2018), two masonry panels (see Fig. 9) measuring 1000 × 1000 × 300 mm, cut from an ancient masonry structure, were strengthened with an FRCM composite comprising a glass fiber textile and a lime-based mortar, anchored with helical steel bars bent against the wall faces. The specimens were tested in situ using two parallel manual jacks equipped with load cells, one on each side of the wall. As in the previous described study, the wall displacement was recorded by means of LVDTs placed diagonally along both specimens' faces.

The typical load response of the tests of both CRM (Fig. 8) and FRCM (Fig. 9) strengthened walls described is characterized by an initial linear ascending branch up to the opening of diffused micro-cracks in the specimens. After the occurrence of micro-cracks, the applied load remains approximately constant with the increasing

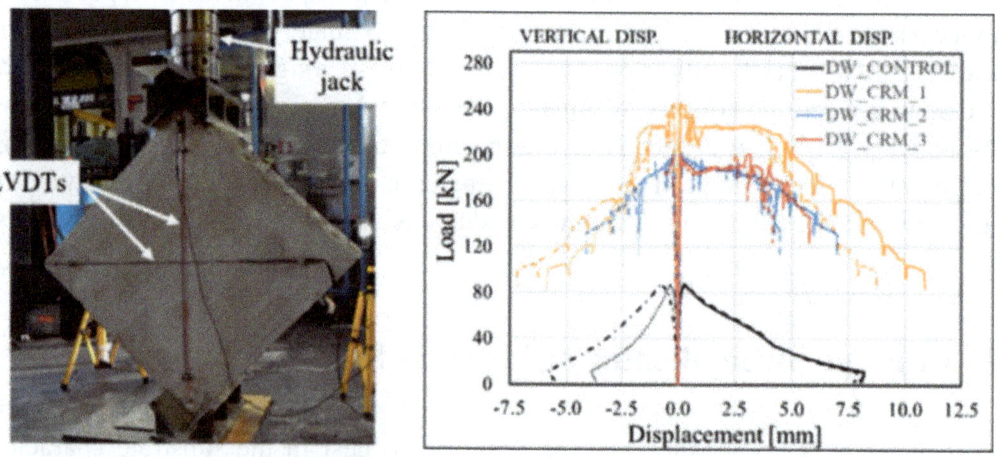

Fig. 8 Specimen set-up and load responses

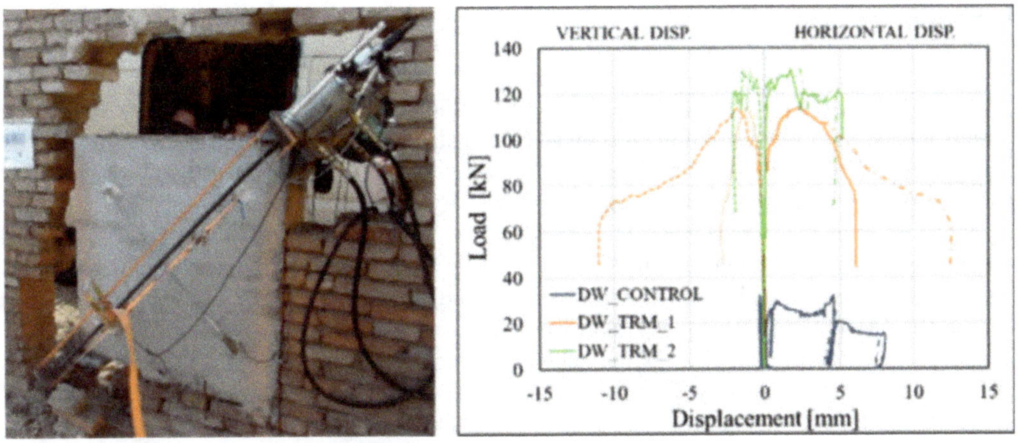

Fig. 9 Specimens set-up and load responses

displacement of the walls, due to the stress redistribution in the bonded composite. This load stage may be considered as a pseudo-ductility stage, which was associated with a slight increase of applied load for certain specimens. When the applied displacement further increased and the stress could no longer be redistributed in the reinforcement, the applied load started decreasing until the specimen eventually collapsed due to the spalling of the external matrix layer, whereas the grid remained anchored to the wall due to the helical bars' presence.

The results of both experimental campaigns show a significant increase in the applied load of the reinforced masonry walls with respect to that of the control specimens. Furthermore, the stress redistribution phenomenon allowed for a relevant pseudo-ductility, which could guarantee the displacement capacity of the masonry with an (approximately) constant load applied.

6 Conclusions

In this paper, some promising applications of fiber-reinforced cementitious matrix composites (FRCM) such as the strengthening of existing masonry structures were discussed. The main mechanical properties of the composite were discussed, with particular attention given to the tensile and bond behavior. An experimental campaign conducted at the Politecnico di Milano, which proved the effectiveness of different PBO FRCM composites for the out-of-plane strengthening of masonry walls, was illustrated. In addition, two experimental campaigns aimed at investigating the contribution of FRCM and CRM materials for the in-plane shear strengthening of masonry walls were described. The results obtained show the effectiveness of the inorganic-matrix composites for strengthening masonry structures. These techniques represent valuable tools to improve the behavior of Italian educational buildings with respect to seismic events in an easy, fast and effective way.

References

Arboleda D, Carozzi FG, Nanni A, Poggi C (2015) Testing procedure for the uniaxial tensile characterization of fabric-reinforced cementitious matrix composites. J Compos Constr

Calabrese AS, Colombi P, D'Antino T (2019) A bending test set-up for the investigation of the bond properties of FRCM strengthenings applied to masonry substrates. Key Eng Mater 817:149–157

Carloni C, Bournas DA, Carozzi FG, D'Antino T, Fava G, Focacci F, et al (2016) Fiber reinforced cementitious (inorganic) matrix. Chapter 9. In: Pellegrino C, Sena-Cruz J (eds) Design procedures for the use of composites in strengthening of reinforced concrete structures, a state of the art report of the RILEM TC 234-DUC, p 501. Springer

Carozzi FG, Colombi P, Poggi C (2015) Fabric reinforced cementitious matrix (FRCM) systems for strengthening of masonry elements subjected to out-of-plane loads. In: Lees J, Keighley S (eds) Proceedings of advanced composites in construction

Carozzi FG, D'Antino T, Poggi C (2018) In-situ experimental tests on masonry panels strengthened with textile reinforced mortar composites. Procedia Struct Integr 11:355–362

Carozzi FG, Poggi C (2015) Mechanical properties and debonding strength of fabric reinforced cementitious matrix (FRCM) systems for masonry strengthening. Compos B Eng 70:215–230

Colombi P, D'Antino T (2019) Analytical assessment of the stress-transfer mechanism in FRCM composites. Compos Struct 220:961–970

D'Antino T, Carozzi FG, Poggi C (2019) Diagonal compression of masonry walls strengthened with composite reinforced mortar. In: Proceeding of mechanics of masonry structures strengthened with composite materials

D'Antino T, Papanicolaou C (2018) Comparison between different tensile test set-ups for the mechanical characterization of inorganic-matrix composites. Constr Build Mater 171:140–151

De Felice G, De Santis S, Garmendia L, Ghiassi B, Larrinaga P, Lourenco PB et al (2014) Mortar-based systems for externally bonded strengthening of masonry. Mater Struct 47:2021–2037

De Santis S (2017) Bond behaviour of steel reinforced grout for the extrados strengthening of masonry vaults. Constr Build Mater 150:367–382

Strengthening of Different Types of Slabs with Composite-Reinforced Mortars (CRM)

Tommaso D'Antino, Angela S. Calabrese, Carlo Poggi, Pierluigi Colombi, Giulia Fava and Massimiliano Bocciarelli

Abstract A great number of buildings built in Europe in the second half of the last century are currently in need of strengthening and retrofitting. One of the more frequent issues is the weakness of the slabs and, in particular, of the intrados covering layer (usually a plaster) and/or of the clay non-structural elements employed to decrease the overall slab weight. The application of composite-reinforced mortar (CRM) systems represents a fast and easy solution to address these weaknesses. Therefore, they are particularly attractive for applications in school buildings, to avoid long interruptions of the educational activities. In this paper, the use of CRM to strengthen different types of slabs is described and discussed on the basis of the results obtained from an experimental campaign conducted at the Politecnico di Milano.

Keywords Composite-reinforced mortars · CRM · Mechanical characterization · Slab strengthening

1 Introduction

Slab intrados deterioration and crumbling is one of the most frequent issues in Italian buildings built in the second half of the last century. The phenomenon is due to the presence of compression stresses that are normal to the joists direction in the slab clay hollow blocks. Compression stresses occur due to several reasons, among which thermal dilation, corrosion of the reinforcing bars, hogging moment due to plate effects, water infiltration, and so on, are the most significant. This scenario is frequent in the case of seismic events, with serious consequences for the building occupants.

Fiber-reinforced composite materials were recently introduced to the construction engineering community and were proven to be an effective solution for the strengthening and retrofitting of existing structures. Fiber-reinforced polymers (FRP) have

T. D'Antino (✉) · A. S. Calabrese · C. Poggi · P. Colombi · G. Fava · M. Bocciarelli
Architecture, Built Environment and Construction Engineering—ABC Department, Politecnico di Milano, Milan, Italy
e-mail: tommaso.dantino@polimi.it

S. Della Torre et al. (eds.), *Buildings for Education*, Research for Development,
https://doi.org/10.1007/978-3-030-33687-5_26

been studied extensively over the last few decades and are largely employed to strengthen existing reinforced concrete (RC) structures (Bakis et al. 2002). However, due to the presence of organic resins, FRP composites present a number of drawbacks (e.g. lack of vapor permeability, low resistance at relatively high temperatures, and difficulties of application on wet substrates) and are poorly compatible with concrete and masonry substrates (de Felice et al. 2014). To overcome these issues, composite materials made by high-strength fibers embedded within inorganic matrices were proposed. Among them, those including high-strength open-mesh textiles with yarn spacing lower than or equal to 30 mm are referred to as fiber (or fabric) reinforced cementitious matrix (FRCM) or textile-reinforced mortar (TRM) composites (ACI 549.4R 2013; CNR-DT 215 2019). FRCM/TRM composites have been gaining increasing attention due to their effective compatibility with substrates, vapor permeability, partial reversibility, and resistance to relatively high temperatures (de Felice et al. 2014).

Recently, new types of inorganic-matrix strengthening materials, comprising a composite grid embedded within an inorganic-matrix and referred to as composite-reinforced mortar (CRM) materials, have been employed to strengthen the existing masonry structures (D'Antino et al. 2019). The grids used in CRMs are made by composite yarns in the warp direction and pultruded elements in the weft one. Usually, the composite yarns are twisted together and around the pultruded elements to realize a stable bi-directional grid (Fig. 1). The spacing between the yarns may vary and is higher than 30 mm. The cross-sectional area of the laminated and pultruded elements is often higher than that of textile-reinforced mortars' fiber textile yarns (Carozzi et al. 2017; D'Antino et al. 2019).

In this paper, a study on the efficiency of three glass composite grids designed as the reinforcement of CRM material, employed for preventing the intrados crumbling hazard, is presented and discussed. The grid contribution to the vertical load induced by the intrados failure (crumbling) was studied by 16 experimental punching tests on different types of specimens simulating horizontal slab, namely a timber slab and a pre-stressed RC slab lightened with hollow clay elements. The glass composite grid was fixed at the slab intrados using anchors. The effect of grid spacing, anchorage types, and grid discontinuity on the system load-bearing capacity is investigated.

Fig. 1 Glass fiber grid (left); strengthening of a slab intrados with a glass CRM (right)

The glass grid studied appears to be an efficient solution to solve intrados crumbling hazards. Owing to the speed of the intervention, it is particularly attractive for applications in public buildings, such as educational buildings, since they cannot remain closed for long periods of time.

2 Mechanical Characterization of the Glass CRM

Three different geometries (layouts) of an impregnated glass fiber composite grid, namely grids with a mesh size of 40 × 40 mm, 40 × 80 mm, and 80 × 80 mm, are analyzed in this study.

Although the grids are all comprised of the same laminated yarns and pultruded elements, the different spacing entails for a different number of twists of the yarns, which in turn may affect the tensile properties of the grid. However, for the sake of brevity, only the mechanical characterization of the 40 × 40 mm grid is reported in this paper. The square grid is made up of rectangular GFRP pultruded weft yarns and two-wire GFRP warp strands (Fig. 1). The cross-sectional area of weft and warp yarns is 5.97 and 5.71 mm^2, respectively.

To measure the grid tensile properties (i.e. tensile strength, elastic modulus, and ultimate strain) in warp and weft direction, tensile tests were performed on warp and weft single-yarn specimens, extracted from the grid, with a length of 700 mm. The tests were displacement-controlled using a servo-hydraulic testing machine equipped with a 100 kN load cell. To improve the specimen gripping by the machine, steel tabs were applied to the specimens ends. An extensometer with a gauge length of 50 mm was used to measure the specimen strain during the test.

The stress–strain curves resulting from the experimental tests are reported in Fig. 2 for weft and warp specimens. The curves showed a linear-elastic behavior up to the brittle specimen failure, which is associated with the glass fiber tensile strength. For some warp specimens, a load drop associated with the failure of a small portion of the twisted bundles was observed close to the peak load. This behavior is caused by

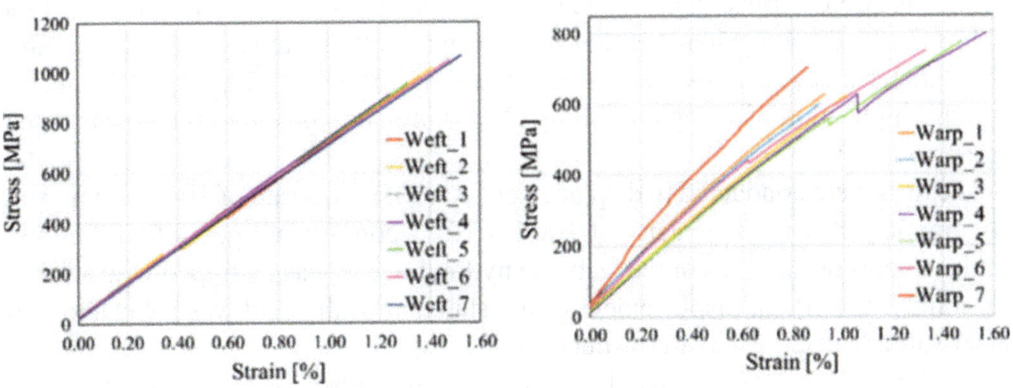

Fig. 2 Stress–strain curves for weft and warp single-yarn specimens

Table 1 Results of tensile tests on glass fiber weft and warp yarns

Sample	F_{max} [kN]	σ_{max} [MPa]	E [GPa]	Sample	F_{max} [kN]	σ_{max} [MPa]	E [GPa]
Weft_1	7.08	1185.9	68.51	Warp_1	5.10	894.1	68.09
Weft_2	6.59	1103.9	68.12	Warp_2	5.06	888.5	65.02
Weft_3	5.17	866.0	72.74	Warp_3	4.27	748.5	62.50
Weft_4	5.42	907.9	72.58	Warp_4	4.57	800.9	61.51
Weft_5	5.70	954.8	68.83	Warp_5	4.41	772.9	60.04
Weft_6	5.94	994.9	70.01	Warp_6	4.30	753.8	67.96
Weft_7	5.94	994.9	68.58	Warp_7	4.91	861.4	79.30
Average	5.98	1001.2	69.91	Average	4.66	817.2	66.3
CoV [%]	11.1	11.1	2.8	CoV [%]	7.7	7.7	9.8

the fiber twisting in warp specimens, which caused stress concentrations where the fibers deviate from linearity. The failure load F_{max} and the corresponding maximum stress σ_{max} obtained by each test, together with the elastic modulus E, are reported in Table 1.

3 Experimental Program

A total of 16 tests were performed on two different specimens simulating horizontal slabs strengthened with the glass composite grid described in Sect. 2. Ten tests were conducted on specimens simulating a one-way timber slab, strengthened with the 40 × 80 mm or the 80 × 80 mm grid, whereas the remaining six tests were conducted on a pre-stressed RC slab including hollow clay elements, which was strengthened with either the 40 × 40 mm, the 40 × 80 mm, or the 80 × 80 mm grid.

Specific steel anchors were employed to fix the composite grid to the slab intrados. These steel anchors consist of one expandable anchor and a 50 mm diameter steel washer and were applied at a distance of 500 mm from one another. The role of the anchors in the specimen load bearing capacity was investigated by varying a number of parameters: (i) washer material (steel or PE); (ii) local strengthening of the anchor with the application of a single mesh of 40 × 40 mm composite grid below the steel washer (Fig. 3).

The tests were conducted in displacement control with a rate of 10 mm/min using a hydraulic jack with a capacity of 100 kN. To accommodate possible differential displacements or rotations of the slab, the hydraulic jack was equipped with a spherical joint. The vertical displacement of the slab and of the grid was measured with linear variable displacement transducers (LVDTs).

The timber slab was simulated by two laminate timber joists supported by rigid steel supports (Fig. 4). The cross-sectional area of the timber joists was 240 (height) × 120 (width) mm and the center-to-center distance between them was 600 mm. The

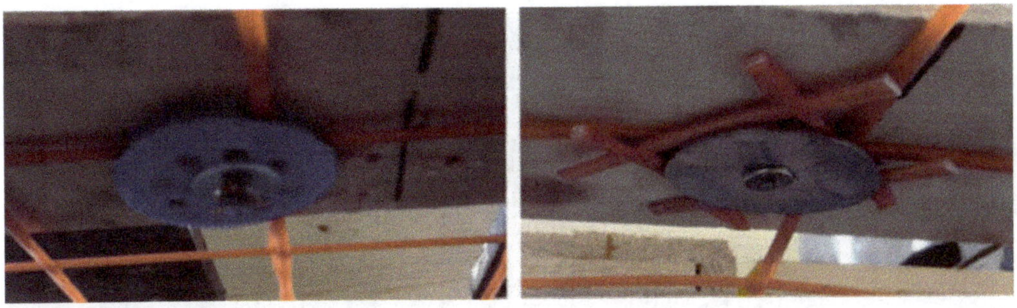

Fig. 3 Anchorage type: anchor with PE washer (left) and anchor with steel washer strengthened using a single mesh of 40 × 40 mm composite grid

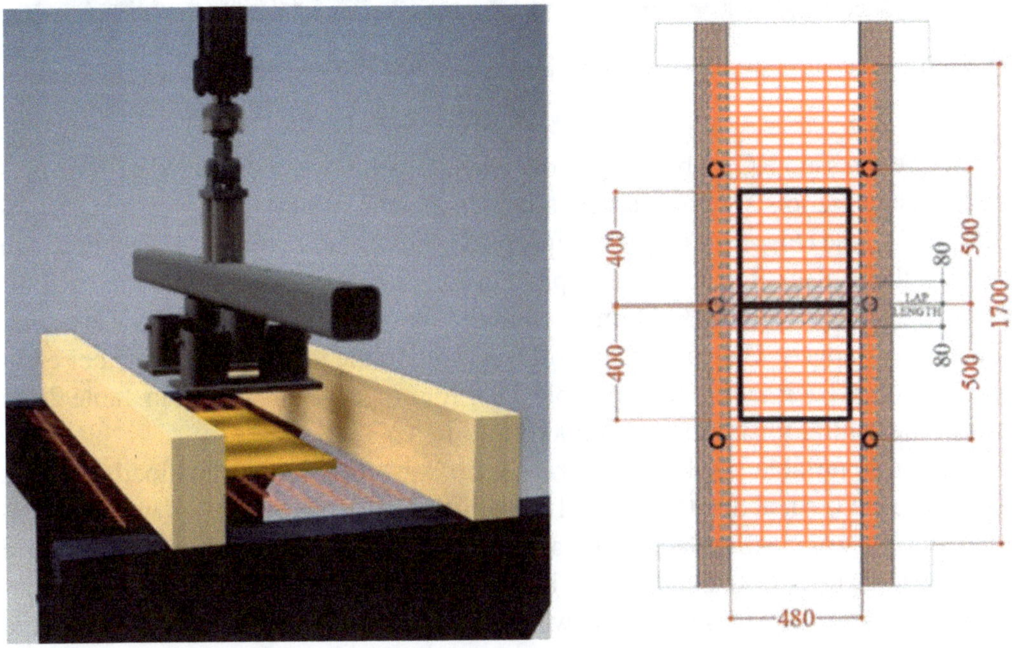

Fig. 4 Tests on strengthened timber slabs (dimensions in mm)

composite grid was fixed to the joist intrados and the load was applied directly onto it by two contiguous timber footprints with areas of 400 × 400 mm (Fig. 4). Two separate load footprints were employed to simulate the contemporary crumbling of two separated hollow clay elements. In these tests, the hydraulic jack was connected to a steel distribution beam. The timber footprints were connected with spherical joints to the steel distribution beam (Fig. 4) and the load was transferred from the jack to the glass composite grid. Four LVDTs were used to measure the vertical displacement at each joist midspan and under the center-point of the footprints. To investigate the effect of grid discontinuity, four tests were conducted on specimens strengthened with two separate grids superimposed for 160 mm at midspan (Fig. 4). Furthermore, two tests were carried out to analyze the behavior of the different anchors on the grid bearing capacity.

Fig. 5 Cross-section of the pre-stressed RC joists and hollow clay elements

The pre-stressed RC slab with hollow clay elements was composed of four pre-stressed concrete joists with a center-to-center spacing of 480 mm fixed to a rectangular steel frame (Fig. 5). The joists supported hollow clay elements placed between the joists themselves. The load was applied to the glass grid with a single timber footprint of dimensions 240 × 350 mm through the space of the central hollow element, which was removed. The composite grid was fixed to the joist soffit with either eight steel anchors (specimens type a in Table 2) or four (specimens type b in Table 2) on each of the central joists, spaced at 500 mm (Fig. 6).

One linear variable displacement transducer was placed under the load footprint to measure the vertical displacement. The vertical deflection of the slab was not measured since it was negligible with regards to grid displacement. To investigate grid capacity, three tests were conducted with the weft yarns aligned with the joists' longitudinal axis and three tests with the warp yarns aligned with the joists' longitudinal axis.

4 Results and Discussion

The results of the experimental tests are reported in Table 2. Specimens were named according to the notation Q_XY_Z_d_A_n, where Q (equal to T = "timber" or C = "pre-stressed RC") indicates the slab type, X and Y are the grid spacing (in cm) in weft an warp direction, respectively, Z is the number of grid layer applied (2 in the case of discontinuous grids), d (equal to a = weft direction aligned with the slab joist or b = warp direction aligned with the slab joist) indicates the grid orientation, A (equal to S = "standard", W = "PE washer", or R = "steel washer") indicates the anchor type, and n is the specimen number.

Table 2 Test results

Name	Max load [kN]	Load footprint [m × m]	Bearing capacity [kN/m^2]	Vertical displacement [mm]
T_48_1_a_S_1	3.40	0.40 × 0.80	10.65	50.86
T_48_1_a_S_2	3.28	0.40 × 0.80	10.27	42.94
Average	3.34	–	10.46	46.90
T_88_1_a_S_1	1.66	0.40 × 0.80	5.19	44.68
T_88_1_a_S_2	1.04	0.40 × 0.80	3.26	55.75
Average	1.35	–	4.22	50.21
T_48_2_a_S_1	4.85	0.40 × 0.80	15.16	64.35
T_48_2_a_S_2	3.93	0.40 × 0.80	12.30	58.56
Average	4.39	–	13.73	61.46
T_88_2_a_S_1	2.23	0.40 × 0.80	6.99	46.76
T_88_2_a_S_2	1.99	0.40 × 0.80	6.22	72.16
Average	2.11	–	6.60	59.46
T_88_1_a_W_1	1.49	0.40 × 0.80	4.66	58.45
T_88_1_a_R_2	1.52	0.40 × 0.80	4.76	53.46
C_44_1_a_S	4.02	0.35 × 0.24	47.92	97.84
C_44_1_b_S	2.76	0.35 × 0.24	32.92	105.26
C_48_1_a_S	1.56	0.35 × 0.24	18.64	65.89
C_48_1_b_S	2.11	0.35 × 0.24	25.14	104.16
C_88_1_a_S	1.32	0.35 × 0.24	15.71	105.36
C_88_1_b_S	1.38	0.35 × 0.24	16.42	81.38

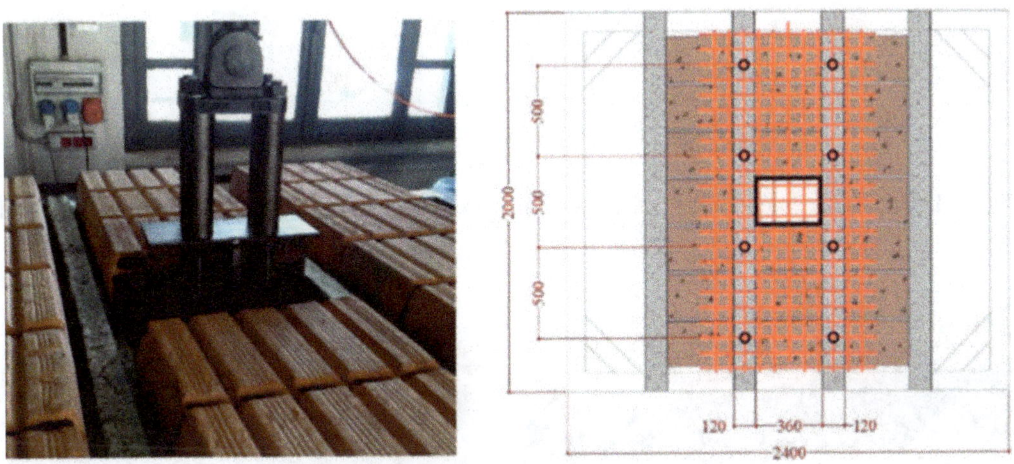

Fig. 6 Tests on strengthened pre-stressed RC slabs (dimensions in mm)

Table 2 reports the maximum load, bearing capacity (i.e. the maximum applied load divided by the area of a single footprint), and the associated vertical displacement. Figure 7 shows the load–displacement curves for timber slabs strengthened with grids with 40 × 80 mm and 80 × 80 mm mesh. All the specimens show an initial approximately linear behavior, which ends with a marked load drop associated with the shear failure of one of the weft or warp yarns close to one of the anchors (Fig. 8). After this load drop, with increasing displacement, the applied load further increased in some cases and remained approximately constant in others due to stress redistribution in the grid bundles.

Specimens strengthened with discontinuous grids showed similar load bearing capacity, although a decrease of the maximum load was observed for some specimens (see Table 2). Also in these tests, the stress redistribution in the grids allowed for an increase in the applied displacement maintaining or even increasing the applied load. This behavior is particularly important in the case of seismic events because it may prevent the sudden collapse of the slab intrados.

All specimens eventually failed due to grid shear rupture close to the anchors due to stress concentrations at the grid warp–weft joints. To decrease the stress concentration at the joints, a single 40 × 40 mm grid mesh was inserted under the

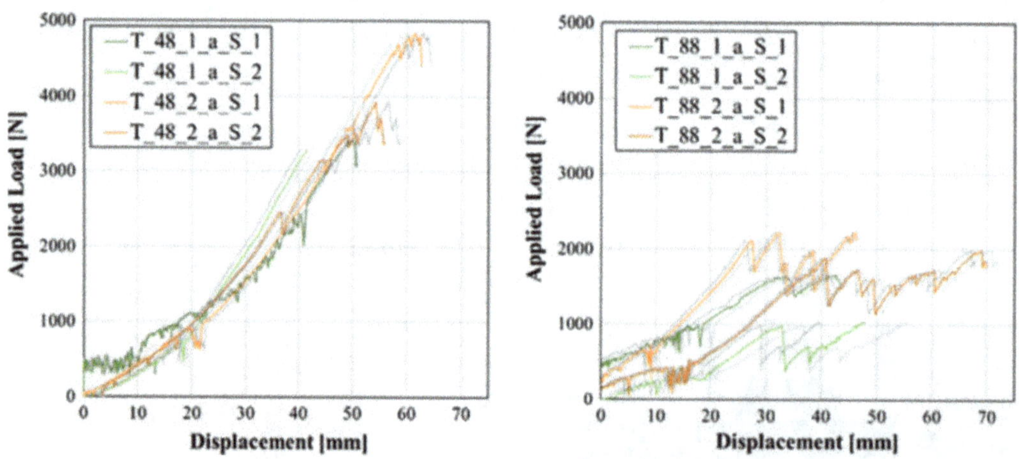

Fig. 7 Load–displacement curves of timber slabs strengthened with a single continuous grid and two discontinuous grids

Fig. 8 Typical grid shear failure at anchor positions

steel washer (Fig. 3). In addition, the steel washer was substituted with a PE washer to investigate the effect of washer stiffness on the response obtained. Figure 9 shows the load responses of timber slabs strengthened with a single continuous grid with 80 × 80 mm mesh with different anchorages (i.e. different washers and the addition of the strengthening single grid mesh). The results show that the different anchorages proposed did not significantly affect the specimen bearing capacity.

The results of the tests conducted on pre-stressed RC slabs are reported in Fig. 10. Specimens strengthened with the 40 × 40 mm grid showed a higher bearing capacity for specimens with warp laminated yarns oriented orthogonally to the slab joist direction with respect to specimens with weft pultruded elements oriented orthogonally to the slab joist direction. As observed for timber slabs, pre-stressed RC slabs strengthened with 40 × 80 mm and 80 × 80 mm grids provided similar results (Fig. 10).

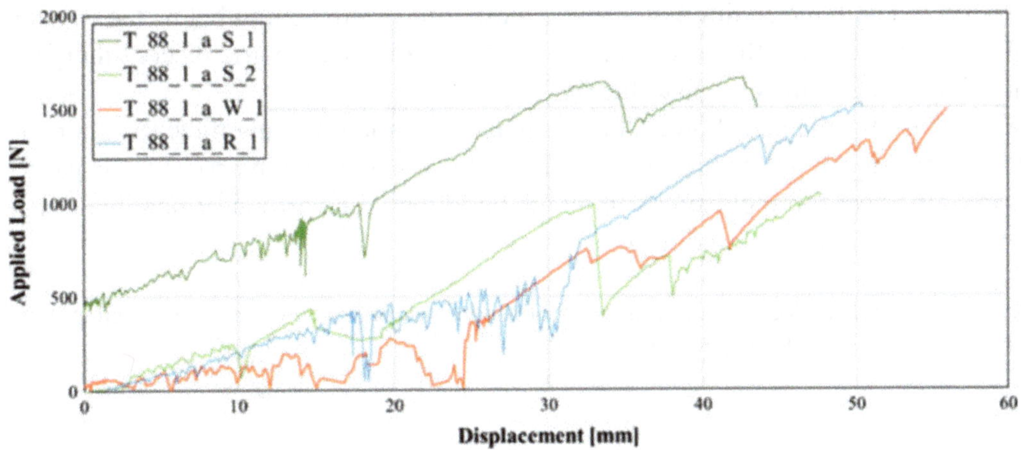

Fig. 9 Load–displacement curves for timber slabs with different anchors

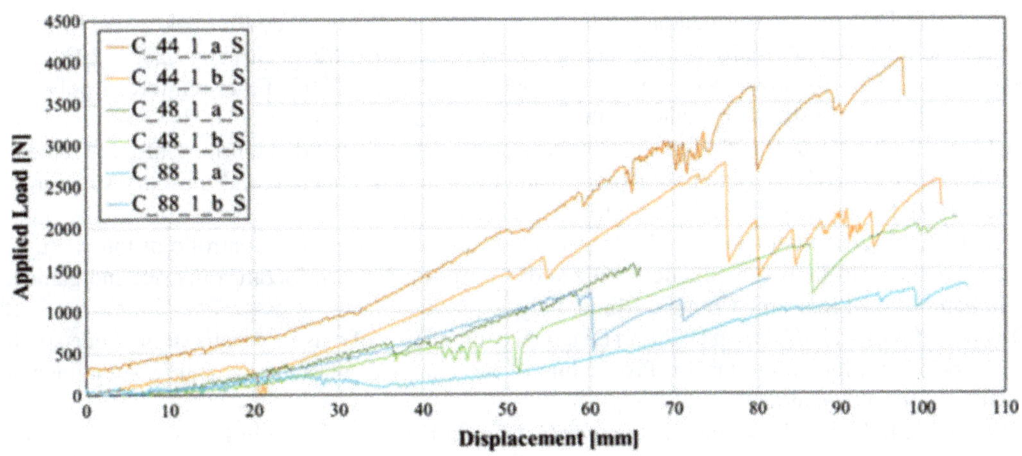

Fig. 10 Load–displacement curves for pre-stressed RC slabs

5 Conclusions

This paper described the results of 16 experimental punching tests on different specimens simulating horizontal slabs, namely timber and pre-stressed RC slabs, strengthened with a glass fiber-reinforced polymer (FRP) grid designed as the reinforcement of a composite-reinforced mortar (CRM). The tests were designed to obtain information on the efficiency of a glass FRP grid in preventing and contrasting the occurrence of slab intrados failure and crumbling. The parameters investigated were the grid spacing, anchorage types, and grid discontinuity. The results obtained showed that a shear failure of a grid bundle close to one anchor always occurred. However, this bundle failure did not determine the specimen failure, because the stress redistribution in the grid allowed for maintaining and in some cases increasing the applied load. All grid spacings (40 × 40 mm, 40 × 80 mm, and 80 × 80 mm) demonstrated similar behavior. The different anchors studied (i.e. steel anchors with steel and PE washer and with the addition of a single mesh grid strengthening below the washer) did not significantly affect the behavior observed. Finally, the grid discontinuity did not significantly affect the load-bearing capacity.

The results obtained showed that the glass grid studied is an efficient solution to solve intrados crumbling hazards and, due to the speed of the intervention, is particularly attractive for applications in public buildings, such as educational buildings, which cannot remain closed for long period.

Acknowledgements TCS s.r.l. is gratefully acknowledged for providing the strengthening materials and for preparing the experimental specimens.

References

de Felice G, De Santis S, Garmendia L, Ghiassi B, Larrinaga P, Lourenco PB et al (2014) Mortar-based systems for externally bonded strengthening of masonry. Mater Struct 47:2021–2037

Bakis CE, Bank LC, Brown VL, Cosenza E, Davalos JF, et al (2002) Fiber-reinforced polymer composites for construction—state-of-the-art review. J Compos Constr 6(2):73–87

ACI 549.4R (2013) Guide to design and construction of externally bonded fabric-reinforced cementitious matrix (FRCM) systems for repair and strengthening concrete and masonry structures. American Concrete Institute, Farmington Hills, MI

CNR-DT 215 (2019) Istruzioni per la Progettazione, l'Esecuzione ed il Controllo di Interventi di Consolidamento Statico mediante l'utilizzo di Compositi Fibrorinforzati a matrice inorganica. Italian National Research Council, Rome, Italy

D'Antino T, Carozzi FG, Poggi C (2019) Diagonal compression of masonry walls strengthened with composite reinforced mortar. In: Proceeding of mechanics of masonry structures strengthened with composite materials

Carozzi FG, Bellini A, D'Antino T, de Felice G, Focacci F, Hojdys L, Laghi L et al (2017) Experimental investigation of tensile and bond properties of carbon-FRCM composites for strengthening masonry elements. Compos B Eng 128:100–119

Energy Retrofit Potential Evaluation: The Regione Lombardia School Building Asset

Fulvio Re Cecconi, Lavinia Chiara Tagliabue, Nicola Moretti, Enrico De Angelis, Andrea Giovanni Mainini and Sebastiano Maltese

Abstract This chapter summarizes a long list of research activities aimed at defining a method to assess the retrofit potential of school buildings, based on maintenance needs, energy-saving potential, and the life cycle cost of the retrofitted building. New concepts are introduced as the gained comfort cost (GCC) as well as new methods are suggested as a probabilistic approach to describe users' behavior. Moreover, innovative methods as artificial neural networks have been employed to predict school buildings' energy performances. The GCC is a new key performance indicator employed to compare different retrofit strategies, focusing on a single classroom. Furthermore, the retrofit potential is evaluated also for the whole school building, exploiting building information modelling (BIM) to collect and transfer information to the building energy model (BEM). This method to analyze energy savings associated with the retrofit of a school building is combined with a method to manage and forecast the running costs of building stocks. The cost forecasting method has been validated through 11 case studies. Eventually, the scale is widened to all the school buildings in Regione Lombardia and the potential energy savings are computed by artificial neural networks (ANN) and Geographical Information Systems (GIS). These methods allow to evaluate energy retrofit potential of school buildings and their life cycle costs at different scales of intervention, from the single classroom to all the buildings in a region, allowing the public decision-maker to choose the best policy for retrofitting his school building stock.

Keywords Energy retrofit · Data-driven process · Artificial Neural Networks (ANN) · Geographical Information System (GIS)

F. Re Cecconi (✉) · N. Moretti · E. De Angelis · A. G. Mainini
Architecture, Built Environment and Construction Engineering—ABC Department, Politecnico di Milano, Milan, Italy
e-mail: fulvio.rececconi@polimi.it

L. C. Tagliabue
Department of Civil, Environmental, Architectural Engineering and Mathematics—DICATAM, University of Brescia, Brescia, Italy

S. Maltese
Institute for Applied Sustainability to the Built Environment, University of Applied Sciences and Arts of Southern Switzerland—SUPSI, Canobbio, Switzerland

S. Della Torre et al. (eds.), *Buildings for Education*, Research for Development,
https://doi.org/10.1007/978-3-030-33687-5_27

1 Introduction

Although educational buildings account for about more than 4% of the European built stock in terms of net floor area (even less, about 3%, in terms of energy consumption, BPIE 2011, and little more than 1 MToe, following Citterio and Fasano 2009), they represent a critical asset: because of their age and maintenance needs; even more, because of their influence in the learning performance of their students.

Comfort, safety and security conditions, have a strong influence on the learning process. The learning performances (Chatzidiakou et al. 2014; BB90 2014; BB93 2014; BB101 2006) report a possible upgrade from 16 (average) to 50% in learning rate if adequate air quality and natural lighting are provided. European buildings are old (ENEA FIRE 2012) and, although slightly less aged than the average built asset, educational buildings are even less "updated" than others. This is particularly true in Italy, where more than 40,000 buildings hosting the more than 52,000 Italian public educational institutions (Source: MIUR Open Data: an older estimation, Citterio and Fasano 2009, accounted about 43,000 buildings, from other sources) have been poorly retrofitted and maintained, in the last 50 years and scarcely adapted to innovative teaching models. Analyzing the recent open dataset provided by the Italian Ministry of Education (about 2017/2018 teaching session), we can report that more than 17% of them is considered fully outdated ("vetusto"); more than 20% are lacking any basic analysis of their risks and safety issues and, most important, among those located in a high (Zone I) or medium/high (Zone II) seismicity area, about 80% of them have not been analyzed even only from a seismic point of view.

In a less recent document about possible actions on the Italian educational building stock (ENEA 2012), the percentage of the number of buildings highly needing a retrofit was estimated as higher than 35%. In any case, 75% of actual school building dates back before any Italian energy laws (1976).

Very few of them (less than 10%) have been retrofitted to empower their acoustical performances (but a clear analysis of the acoustical needs is widely lacking, also in highly disturbed areas); only 5% of actual school buildings have been fully (envelope and plants) retrofitted to enhance their energy efficiency (mainly to achieve a basic *C-level* energy label[1]) and no more than 25% have been only partly refurbished, with scarce impact on energy needs.

The following paper reports the main results of some recent studies, performed by the authors, about the effects of school building retrofitting: how to evaluate

[1]A *C-level* building, in the context of the European EPBD European Directive 2010/31 (Energy Performance of Buildings Directive-EPBD) and 2018/844 practice, is a building, in a specific climatic condition, that consumes as the "average building", i.e. the benchmark of all the building realized as prescribed by the national/regional standards before EPBD. An A-level building is a better—more efficient—one, that consumes less than 50% of the *C-level* benchmark.

them, their feasibility and potential, and the data sources we can access and organize to enforce the evidence of our understanding, either for a more general, strategic decision, or for the optimization of a specific case study.

2 How to Assess Comfort and Real Energy Needs

The evaluation of the real energy needs of a building is a fundamental step to reliably assess the potential annual savings and the payback time of investment of a retrofit strategy. The real energy needs are highly influenced by the real building use, which is the main issue when an extended time of use (beyond main teaching activity) is promoted. The occupancy profiles and users' habits can help to predict the variability of the energy performance of the building (Tagliabue et al. 2016). Moreover, the occupants' awareness about energy use combined with low-cost strategies has an estimated 20% effectiveness on energy reduction (ENEA 2012), especially in case of total replacement or integration of thermal plants and without smart control devices. Italian school buildings are mainly equipped with heating systems for winter, avoiding cooling systems for summer, however, climate changes and the extended use of the buildings entail the need of mitigation measures for overheating in the middle and summer seasons. In this paper, refurbishment strategies referred to an adaptive comfort approach are mainly proposed, considering the building envelope as a passive control system of the indoor conditions. Moreover, since 2009 (DPR 59/09) national regulations introduced dynamic thermal properties to be assessed for building envelope in order to reduce and effectively control the heat gains (Decreto Interministeriale 2015). In any case, the bioclimatic approach encompasses evident advantages such as a lean and cost-effective implementation in addition to its affordability. Measurement for commissioning and dynamic simulations is crucial to define quantitative advantages of bioclimatic design strategies; nevertheless, they are complex and time-consuming due to the amount of hourly data that are managed and finally the passive behavior of a building is not effortlessly synthesized. A comparison of hourly consumption can be used for air-conditioned buildings; meanwhile, buildings with no active thermal control in summer need more sophisticated statistical analyses to account for the thermal inertia effect (Di Perna et al. 2011).

3 Average Building School Conditions and Related Performances

The retrofit of the envelope and the thermal plants of a building can heavily reduce the energy consumption and the associated running costs of a building. The investment costs, therefore, may be compensated by the running cost reduction for refurbishment

Table 1 Frequently adopted envelope typologies for the Italian school building stock

Opaque envelope component	U_{factor} [W/m^2 K]	Y_{ie} [W/m^2 K]
ROOF: Flat with reinforced brick-concrete slab, low insulation	1.01	0.19
WALL: Hollow brick masonry, low insulation (25 cm)	0.80	0.19
WALL: Hollow brick masonry, low insulation (40 cm)	0.76	0.06
FLOOR: with reinforced brick-concrete slab, low insulation	0.98	0.19
FLOOR: Concrete floor on soil, low insulation	1.24	0.11
Transparent envelope component	U_{factor} [W/M^2 K]	SHGC [W/M^2 K]
Double glass, air-filled, wood frame	2.8	0.75
Double glass, air-filled, metal frame without thermal break	3.7	0.75

interventions, that are more cost competitive when associated with envelope and plant maintenance due to their physical or technical obsolescence.

70% of the school buildings have reinforced concrete frame structure, brick infill walls and are equipped with gas boiler systems for heating (average efficiency $\eta \leq 0.9$). In any case, for buildings realized after the L. 373/76 was established, a thin insulation layer in the opaque envelope can be expected. Focusing only on the schools built from 1976 to 1990, the average and most frequently adopted envelope typologies in Italian school buildings are presented in Table 1 to define the framework in which the envelope technologies and thermal properties of the simulation baseline scenario are limited. The main reported parameters are: U-factor is the thermal transmittance, Y_{ie} represents the periodic thermal transmittance value and SHGC is the solar heat gain coefficient.

It is worthy to note that, in addition to thermal transmittance for both transparent and opaque envelope, and solar heat gains control strategies, a suitable level of thermal inertia is crucial to improve comfort conditions and energy savings in particular when adaptive thermal comfort models are assumed. Depending on the calculation methodology, the building type and use (Aste 2009; Karlsson 2013), the influence of the inertia in the thermal behavior of a building can vary from 30 to 80%.

In old school buildings where the transparent/opaque envelope surface ratio is low, the effect of thermal inertia decreases while air change rate and permeable coverings interact more efficiently with a time constant and energy saving (Di Perna et al. 2011). Nevertheless, thresholds of suitable internal areal heat capacity related to periodic thermal transmittance (Y_{ie}) have also been defined for school buildings envelopes ranging between 50 kJ/m^2 K for $Y_{\text{ie}} \leq 0.04$; 70 kJ/m^2 K for $0.04 \leq Y_{\text{ie}} \leq 0.08$ and 90 kJ/m^2 K for $0.08 \leq Y_{\text{ie}} \leq 0.12$.

4 Assessing the Gained Comfort Cost

The methodology adopted in the present study focuses on the assessment of the thermal indoor conditions into a representative unit or classroom of a school building in northern Italy, equipped with traditional envelope (Table 1) and compared to improved scenarios including refurbishment strategies.

The base case has been compared with five improved alternatives with different energy retrofit strategies for enhancing energy performances and improving indoor thermal comfort (Table 2).

The results are based on four levels of intervention, from micro (single class in a school building) to macro (school building stock in the Lombardy region). The analyses have been carried out according to different methodologies ranging from the simple dynamic analysis of energy performance to statistical evaluation of the result to forecasting methods based on neural networks.

The simulation test cell is a single classroom in the south oriented with three identical windows (1.25 × 2.5 m) on the only wall-facing outdoors. The energy saving retrofit strategies (Table 2) and the variation in associated costs GCC are represented by the segments in the graph in Fig. 1 for winter (blue lines) and summer (yellow lines).

The assessment is based on the hourly indoor air temperature as a comfort parameter under free-floating conditions. The LCC has been calculated for six cases to show how the cost categories (i.e. construction, operation, maintenance) influence the total cost. The installation cost is proportional to the maintenance cost, while operational energy costs are related to the performance of the component. The energy cost has a predominant role because installation and maintenance costs are strictly related to the envelope (opaque and transparent). In case of installation and maintenance, if the costs of different components (e.g. finishing, floor, partitions, and systems) would be included, the ratio would change. The preventive maintenance strategy has been chosen, as more convenient than the corrective strategy.

Figure 1 shows the cost for achieving one unit of comfort. This is done comparing the LCC and the comfort (in winter and in summer) for each alternative with the base case. According to the LCC (*Y*-axis), the best solutions are the positive ones, with an LCC lower than the base case. According to the comfort (*X*-axis), the best solutions

Table 2 Tested combinations of energy-saving retrofit strategies

No.	Evaluated test case	U_{av} W/m^2 K	U_w W/m^2 K	SHGC [–]
1	Base case	0.96	2.8	0.75
2	Improved U_w	0.96	1.0	0.50
3	Improved U_w and SHGC	0.96	1.0	0.35
4	U_{av} reduced	0.29	2.8	0.75
5	U_{av} and U_w reduced	0.29	1.0	0.50
6	Best case	0.29	1.0	0.35

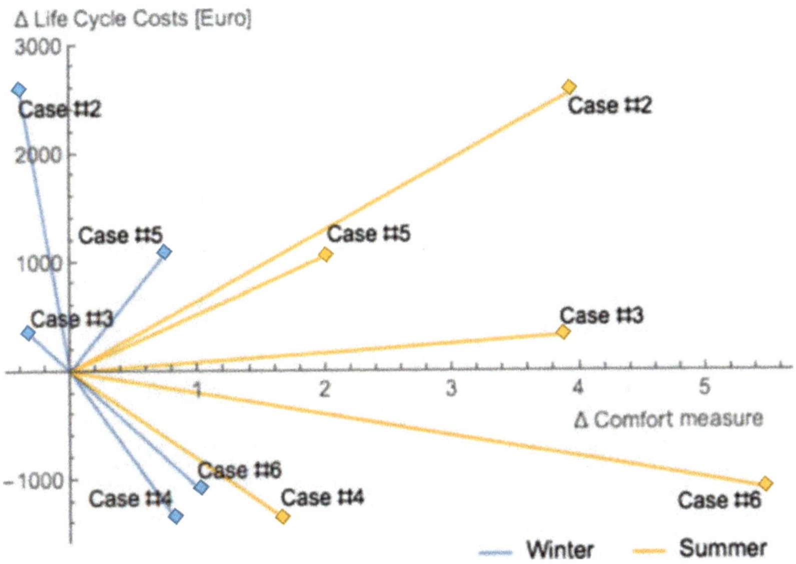

Fig. 1 Gained comfort cost variation graph (Tagliabue 2017)

are the most distant from the origin; a negative value means that the alternative has a lower comfort than the base case. According to what was mentioned above, case 6 (best case) is the most suitable in terms of comfort, while case 2 is the most suitable in terms of LCC.

5 Retrofit Potential of a School Building

At building-level we considered, as a case study, a primary school located in northern Italy, in Nerviano, Milan province, built in the 1950s and not refurbished except for the ordinary maintenance operations. The occupants, both students and staff are estimated at around 400 people. The floor area is around 3000 m^2 counting classrooms spaces, toilets, multi-task rooms, offices and canteen. A BIM model has been created on the documental data and information derived by a survey. Each building envelope component identified during the survey has been modeled in terms of thickness and materials (Table 2). The internal height has been assumed based on external measures and standard height for the built environment (~3 m). A BIM model was used to define retrofit intervention related to pathologies and costs have been deduced. Furthermore, the BIM model could be connected to Facility Management (FM) software. The EnergyPlus simulations result has been used to calculate a cost simulation for a retrofit intervention of Exterior Insulation and Finishing System (EIFS).

Owing to the need to refurbish the facade, one or more energy retrofitting alternatives can be evaluated on the same components:

(a) Replacement of windows and doors with a more thermally resistant model;

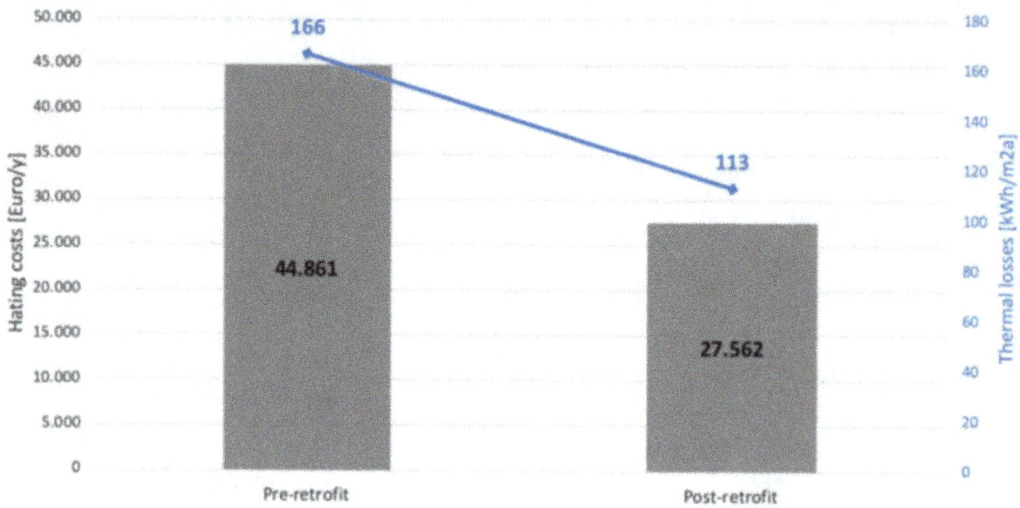

Fig. 2 Economic results comparison: Thermal losses and cost (Tagliabue 2017/2)

(b) Replacement of the external layer with an EIFS to be applied on an existing wall.

The thermal loss decrease has been calculated as −38%, calculated on the opaque envelope (Fig. 2). Therefore, the main results are listed as follows:

(c) Envelope refurbishment cost: 108,284 €
(d) Total retrofit cost: 222,925 € (~75 €/m^2)
(e) Annual saving: 17,299 €/year (~− 38%)
(f) Payback time: ~5 years

6 Managing Running Costs Data of a Portfolio of School Buildings

The analysis focused on how expenses for schools were managed in the Municipality of Seregno pointed out the main lack: there was no management system. Expenses were forecast without exact knowledge of how much was spent the previous year, no records were taken on who did what and how much was paid for maintenance operation except for paper documents that were collected, stored and never used anymore. Seregno's 11 school buildings expenses bills are stored in paper documents archived in folders in different offices of the Municipality.

All the data acquired and elaborated had been used to implement predictive models for future expenses (heating and gas, electricity, maintenance) to solve one of the main municipality's problem—the reliability of costs forecast. The following figure shows, for example, the linear model to be used to predict heating and gas expenses for school "Rodari". Models like this one are available to the municipality for each school and for all the three types of expenses (Fig. 3).

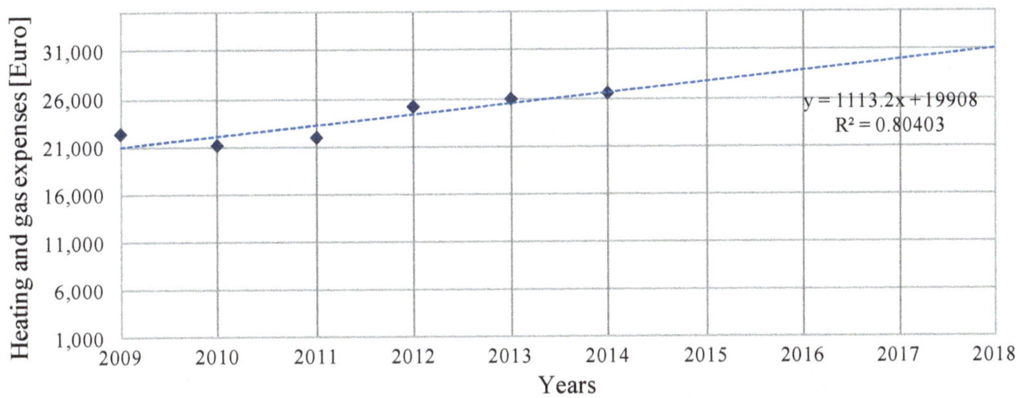

Fig. 3 Example of a predictive model for energy costs (school "Rodari")—(Re Cecconi et al. 2017)

7 Evaluate Retrofit Potential in Regione Lombardia's School Buildings

The last results are included within a regional level framework. The implementation of the set of methods proposed through this research starts from a specific dataset: the CENED DB provided by Regione Lombardia. This DB includes all the energy labels provided by the accredited professional in the Lombardy region. CENED open data have been used to train a set of multi-layer feed-forward ANN that proved to be reliable instruments to forecast energy performance of school buildings. School buildings have been analyzed and classified according to their age (in an overall time span of more than one century) and performance of their envelope. This process gave as an outcome the definition of homogeneous classes of comparable school buildings. For each class, specific retrofit strategies, suitable for their characteristics, have been defined and the potential energy savings have been computed through the trained ANN. The output data have been imported in a GIS environment, through which it has been possible to carry out a spatial analysis for the whole Lombardy region territory. Although there is a gap between actual and ANN-computed performance, the proposed process balances the reliability of energy savings forecasts, with the necessity of saving time and resources to carry out the estimation. A more precise energy demand estimation method would be certainly too expensive to be applied on an extended school buildings stock. Moreover, the proposed method allows to easily spot the most convenient retrofit strategy for the whole portfolio, even in the very early stage of the decision-making process (Fig. 4).

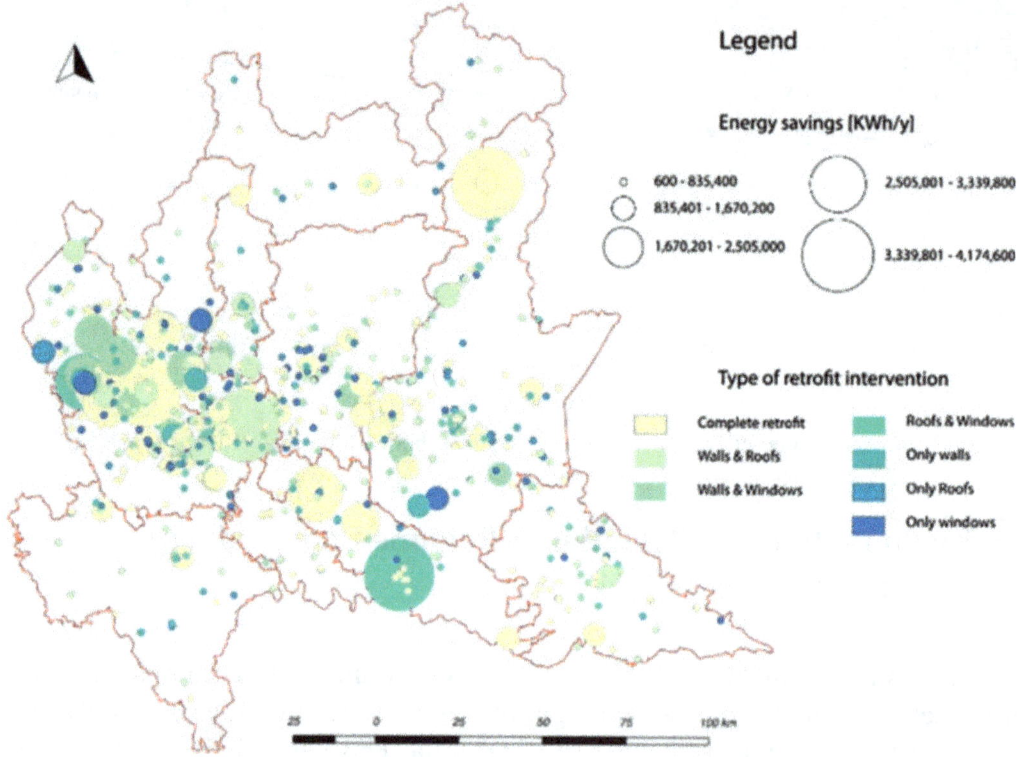

Fig. 4 ANN computed energy savings by type of intervention in the Lombardy region. (Re Cecconi et al. 2019)

8 Conclusions

The main barriers to the effective recovery of existing school buildings are the lack of a coherent and comprehensive methodology to address the process and an ever-limited intervention budget. To overcome this situation, it is necessary to structure the decision-making process, as well as prioritize the choice of interventions according to a multi-stage matrix that is able to guide interventions and satisfy the different stakeholders involved during the process. Likewise, it will be equally important to support the creation of new financial instruments and sources of funding. It is evident that it is not possible to evaluate only the aspect of costs as a driver for decision-making. During the last few years, the improvement of the building's energy efficiency has been a very important lever not only to raise funds but also to stimulate the renovation of existing school buildings. The reason for this choice lies in the immediate perception of the economic return due to the sudden reduction in consumption, which translates into lower maintenance and operating costs for the municipal administration. In some cases, some interventions were the occasion to focus also on the improvement of the building seismic performance, which sometimes is underestimated but that in particular areas can strongly affect the safety of the people work in and studying in these spaces. On the other hand, new teaching and learning methods require a modified use of the school spaces, able to accommodate of daily changes, reaching the maximum flexibility for teaching and learning activities. At the same time, this need of flexibility should relate to added building envelope and systems capabilities, for example, by the use of natural ventilation and daylighting or the use of low-temperature heating systems, increasing the need of monitoring indoor quantities and fostering a continuous commissioning approach.

Further development of the research will focus on new functional requirements, which combine energy-related as well as functional, and maintenance measures.

References

BPIE (2011) aa.vv. Europe's buildings under the microscope, 2011, here

Building Bulletin BB90 (2014) Lighting design for schools. Education Founding Agency, 11 March 2014. https://www.gov.uk/government/publications/building-bulletin-90-lighting-design-for-schools

Building Bulletin BB93 (2014) Acoustic design of schools—performance standards. Education Founding Agency, 19 December 2014. https://www.gov.uk/government/publications/acoustics-lighting-and-ventilation-in-schools/acoustics-lighting-and-ventilation-in-schools

Building Bulletin BB101 (2006) Ventilation of school buildings, version 1.4—5th July 2006. https://www.gov.uk/government/publications/acoustics-lighting-and-ventilation-in-schools/acoustics-lighting-and-ventilation-in-schools

Chatzidiakou L, Mumovic D, Dockrell J (2014) The effects of thermal conditions and indoor air quality on health, comfort and cognitive performance of students. The Bartlett, UCL Institute for Environmental Design and Engineering London, here

Citterio M, Fasano G (2009) Indagine sui consumi degli edifici pubblici (direzionale e scuole) e potenzialità degli interventi di efficienza energetica, ENEA, Report RSE/2009/165

Decreto Interministeriale 26 giugno (2015) Applicazione delle metodologie di calcolo delle prestazioni energetiche e definizione delle prescrizioni e dei requisiti minini degli edifici, Supplemento Ordinario alla Gazzetta ufficiale n.162 del 15 luglio 2015

Di Perna C, Stazi F, Ursini Casalena A, D'Orazio M (2011) Influence of the internal inertia of the building envelope on summertime comfort in buildings with high internal heat loads. Energy Build 43(1):200–206

Directive 2010/31/EU of The European Parliament and of the Council of 19 May 2010 on the energy performance of buildings (recast), Off J Eur Union L 153/13

Directive (EU) 2018/844 of the European Parliament and of the Council of 30 May 2018 amending Directive 2010/31/EU on the energy performance of buildings and Directive 2012/27/EU on energy efficiency. Off J Eur Union

DPR 59/2009 Regolamento di attuazione dell'articolo 4, comma 1, lettere a) e b), del DLgs 192/05 concernente attuazione della direttiva 2002/91/CE sul rendimento energetico in edilizia

ENEA-FIRE (2012) Guida per il contenimento della spesa energetica nelle scuole

Legge 30 aprile 1976, n. 373 Norme per il contenimento del consumo energetico per usi termici negli edifici

Re Cecconi F, Tagliabue LC, Sebastiano M, Ciribini ALC, De Angelis E (2017) A performance based management system for cost prediction suitable for school building stock, ISTeA 2017—Re-shaping the construction industry

Re Cecconi F, Moretti N, Tagliabue LC (2019) Application of artificial neutral network and geographic information system to evaluate retrofit potential in public school buildings. Renew Sustain Energy Rev 110:266–277

Tagliabue LC, Manfren M, Ciribini ALC, De Angelis E (2016) Probabilistic behavioural modelling in building performance simulation—the Brescia eLUX lab. Energy Build 128:119–131

Tagliabue LC, Mainini AG, Re Cecconi F, Sebastiano M, De Angelis E, Zani A (2017) Thermal and economic efficiency of progressive retrofit strategies for school buildings by a statistical analysis based tool (2017), IBPSA 2017—building simulation 2017 conference

Tagliabue LC, Sebastiano M, Re Cecconi F, De Angelis E, Ciribini ALC (2017) Integrated design and modelling-based smart school concept to renovate the existing school building sector (2017) ISTeA 201—Re-shaping the construction industry

Energy and Environmental Retrofit of Existing School Buildings: Potentials and Limits in the Large-Scale Planning

Giuliano Dall'O' and Luca Sarto

Abstract This chapter summarizes the results of research activities promoted by a group of researchers working in ABC Department—Politecnico di Milano aimed at energy and environmental requalification of school buildings located in the Lombardy region (Italy). The buildings subject to energy audits have been selected considering various factors, including the type of user (e.g. kindergartens, elementary schools, middle schools, etc.), construction period, construction technology and degradation. The methodological approach considers energy retrofit scenarios with different energy performance targets and required investments. The results of the research, which is concerned with a substantial and diversified existing building stock, provide public administrators decision-making tools and indicators supporting the energy and environmental retrofit actions for the existing schools. Although the potential for energy savings and reduced environmental impact is important in all scenarios, the achievement of very high energy performance targets is not always economically convenient and is sometimes technically impossible to reach. Large-scale energy planning, therefore, always requires in-depth energy audits that allow defining the optimal energy performance targets. The research activities demonstrate that it is convenient, when the energy performance of a building is improved, to consider also the environmental aspects. For some sample school buildings simulation analyses were carried out in accordance with the LEED® protocol, and the higher cost due to environmental enhancement (e.g. the choice of ecological materials, the recycling of demolition materials or the use of renewable energy sources) is absolutely acceptable in the intervention economy.

Keywords Energy efficiency in school building · School buildings retrofit · Sustainability of school buildings · Green energy audit of buildings · LEED® protocol

---

G. Dall'O' (✉)
Architecture, Built Environment and Construction Engineering—ABC Department, Politecnico di Milano, Milan, Italy
e-mail: giuldal@polimi.it

L. Sarto
Milan, Italy

S. Della Torre et al. (eds.), *Buildings for Education*, Research for Development,
https://doi.org/10.1007/978-3-030-33687-5_28

1 Introduction

Improving the energy and environmental performance of public buildings, particularly as regards schools, is important for the promotion of a culture of energy efficiency among the local population. Indeed, the European Union has developed strategies particularly for this sector using specific legislation. Article 5 of Directive 2012/27/EU of 25 October 2012 on energy efficiency requires Member States to ensure that, as from 1 January 2014, 3% of the total floor area of central government-owned and -occupied, heated or cooled buildings is renovated each year to meet the minimum energy performance requirements that each Member State has set in application of Article 4 of the Energy Performance of Buildings Directive (2010/31/EU).

Nowadays, there are over 40,000 buildings in Italy for exclusive or prevalent school use—of which one-third is concentrated in ten provinces—with thermal consumption of 9.5 TWh/year and electricity consumption of 3.66 TWh/year. At the School Building Registry it appears that in 58% of school buildings, measures have already been implemented to save energy, installing photovoltaic panels, double glazing and double windows or isolating the external walls and roof (Dall'O' and Sarto 2013).

In this chapter the results of three researches promoted within the ABC Department of the Politecnico di Milano are reported. The first concerns a study of 49 school building complexes in the Lombardy region: through detailed energy audits, three different energy retrofit scenarios were evaluated.

The second research concerns a research extended to 49 high schools owned by the province of Milan (now Metropolitan City of Milan). The energy consumptions calculated with the actual energy consumption and savings estimates are made on three energetic retrofit scenarios.

The third research proposes and discusses a study which provides a considerable improvement in the environmental quality of 14 school buildings (pre-schools, primary and secondary) located in two municipalities in Milan Province, northern Italy.

2 Energy Retrofit of Existing School Buildings: A Case Study for Schools up to Lower Secondary Schools

For public authorities, improving the energy efficiency of public buildings is an important goal. Data from CESTEC, the energy register of the Lombardy region, concerning energy certificates in the Lombardy region show that 49.4% of school buildings are in class G (the worst efficiency class according to the classification scale); 13.4% are in class F; 10.8% are in class E; 11.6% are in class D; 9.7% are in class C; 3.3% are in class B; and only 1.9% are in class A or A+ (www.cestec.it 2019).

On the other hand, to improve the energy performance of public buildings, and in particular school buildings, large investments are required by the Public Administration (PA). The economy over the next several years and the Stability Pact, which is now mandatory for the PA of Italy to reduce the public debt, will most likely limit direct investments.

As regard the performance quality, the energy retrofit of school buildings aims at high energy performance comparable to that of new buildings. The improvement in energy performance, however, has a specific cost, which increases exponentially the closer we get to the high energy classes. In order to make investments on energy retrofit cost-effective, it is useful to understand to what extent it is convenient to upgrade existing buildings. In other words, is it always economically convenient to push energy performance up to the highest level? The aim of this study was precisely this: to outline different scenarios and evaluate the economic convenience limits in energy retrofit investments (Dall'O' and Sarto 2013).

The study is based on data collected from an energy audit campaign. The energy analysis concerns school building complexes owned by 16 municipalities. The school building stock (49 school building complexes comprising 77 buildings) includes a large variety of building types (pre-schools 33%, primary schools 18%, secondary schools 12% and mixed schools 37%). The energy performance of the buildings varies widely because of different building features related to the various construction periods.

The year of construction of the school building complexes ranges over a wide period, from 1920 to 2009: 9 complexes up to 1960, 7 complexes between 1961 and 1970, 25 complexes between 1971 and 1980, and 8 complexes built after 1981. The distribution of the construction years is related to the social needs, in terms of the number of children of school age in the period in question.

Figure 1 shows the comparison between the actual and calculated primary thermal energy consumption due to space heating of the school building stock. The dashed line represents the perfect match between the two values, while the regression line represents the average situation of the entire school building stock. The two regression lines are comparable, the measured energy consumption of all the buildings is 22% lower than the predicted consumption and this can be considered a good match. Thus the heating plant is switched off during night and this could achieve a reduction in energy consumption of more than 30% as stated in EN 13790 Standard (ISO 2008) as a function of inertia and other parameters which cannot be evaluated with the available data.

As regards to the energy retrofit actions, three scenarios were considered:

- In the *standard scenario*, the objective is to provide a technological upgrade of the heating plants with minimal investment;
- In the *cost-effective scenario*, the objective is to significantly increase the energy performance of the building envelope and the heating plants;
- In the *high-performance scenario*, the objective is to greatly increase the energy performance of the school building complexes up to the standards required by near-zero energy buildings (Art. 9 Directive 31/2010).

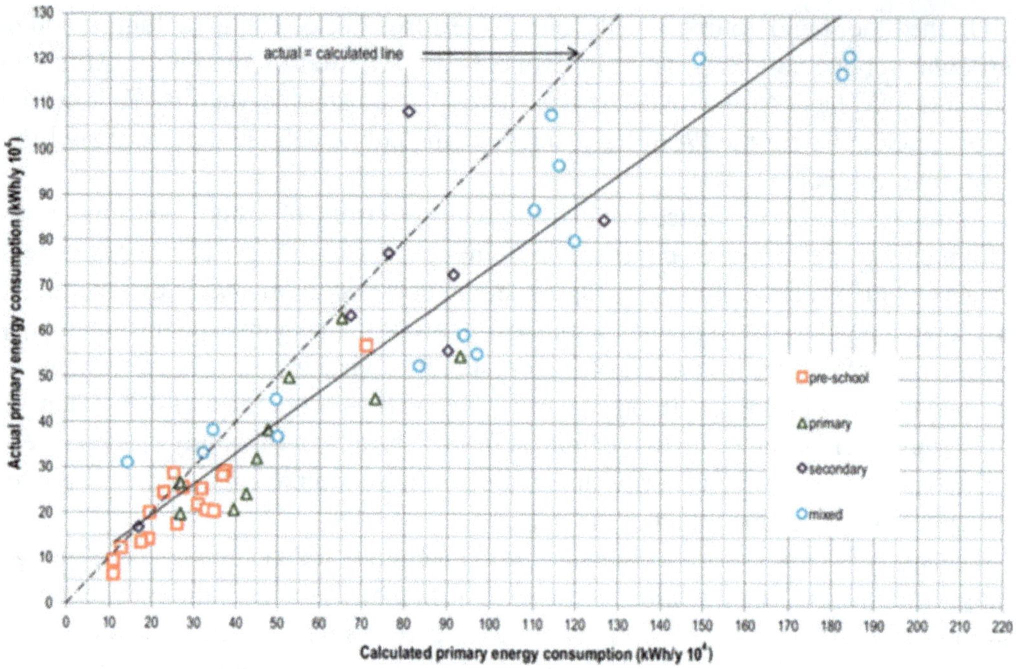

Fig. 1 Comparison between the actual and calculated primary thermal energy consumption due to space heating of the school buildings stock

In the *standard scenario*, the achievable energy savings are 15%, with an investment, referred to the net floor area of the buildings, of 14.0 €/m^2 and a simple payback (SPB) of 5.7 years. This scenario represents a typical situation of low-profile maintenance actions.

In the *cost-effective scenario*, the energy retrofit measures are applied to all the school building complexes. Considering the entire building stock, the achievable energy savings are 67%, with an investment required of 121.9 €/m^2 with a SPB of 13.4 years.

In the *high-performance scenario*, the energy retrofit measures are applied to all of the school building complexes with the objective of obtaining the maximum energy performance. This scenario does not consider cost effectiveness (i.e. a limitation of the SPB), but rather the technical and physical constraints due to the fact that we are acting on existing buildings, some of which are historical. For this reason it is not always possible to obtain the maximum projected energy performance and some building complexes do not reach the class A standard but a lower standard (e.g. class B or class C) according to the energy classification scheme of that time (2013).

Considering the entire building stock, the achievable energy saving is 81%, with an investment required of 479.4 €/m^2 with a SPB of 42.4 years. Figure 2 shows the cost distribution of high-performance scenario.

This study demonstrates that reaching high levels of energy performance to comply with the EPBD recast could be very difficult or not cost-effective in many cases. Sometimes the cost of energy rehabilitation for the increasing of heating performance is comparable with the cost of a new building.

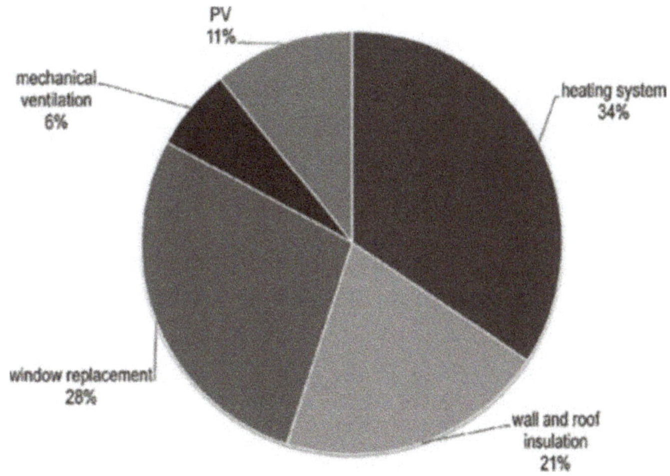

Fig. 2 Cost distribution in scenario 3

If the target is less ambitious (effective scenario), it is possible to reduce consistently energy consumption for space heating with a reasonable simple pay-back time, thereby reducing the environmental impact of school buildings.

3 Energy Retrofit of Existing School Buildings: A Case Study for High Schools

The case study presented in this section concerns a research which was carried out on an existing school building stock located in the Lombardy region in 2014. Unlike the case discussed in Sect. 3, in this case the school buildings, owned by the Province of Milan, concern upper secondary schools. It constists of 59 large school complexes with volumes ranging between 4,600 m^3 and 164,860 m^3 (average volume 45,545 m^3).

For each school complex, a detailed energy audit was made in order:

- to define digital models of the buildings according to the ISO 13790 standard (ISO 2008);
- to compare the theoretical energy consumption with the actual energy consumption normalized with the standard day degrees;
- to calibrate the digital models;
- to define possible retrofit actions based on three scenarios.

Figure 3 shows the comparison between real normalized energy consumption and calculated theoretical consumption. From the graph it can be observed how the correlation is not high ($R^2 = 0.2715$). Considering the set of cases, the analytical calculation overestimates real consumption by 53%.

The differences that energize in individual cases, shown in Fig. 3, are due to several factors:

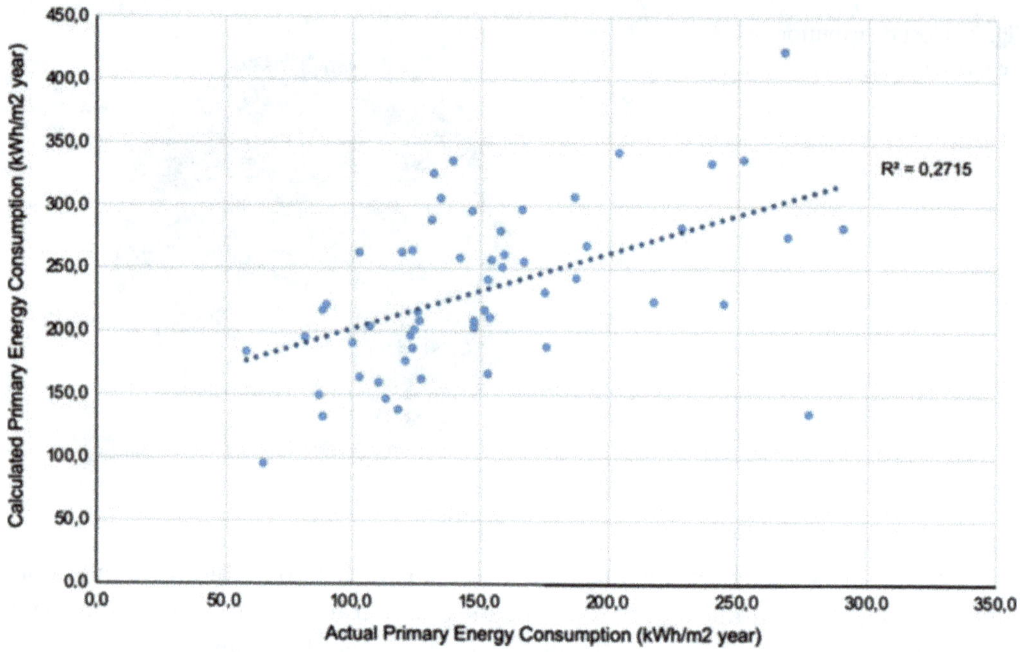

Fig. 3 Comparison between calculated and normalized actual primary energy consumption

- in the calculation we consider a standard internal temperature (20 °C), often different from the real one;
- the periods of use of the spaces may be different from those declared;
- there are inefficiencies in the heating system regulation system.

Regarding the energy retrofit scenarios, the following choices were made:

- *Scenario 1*: minimum interventions aimed at restoring the complete operation of the plants;
- *Scenario 2*: interventions both on plants, with replacement of components and systems that are no longer efficient, and easy and low-cost interventions on the building envelope;
- *Scenario 3*: Heavy requalification of the building envelope (e.g. ETICS—External Thermal Insulation Composite System and window replacement) and requalification of the plants.

In some cases not all the three scenarios were simulated because interventions were not needed or not applicable for architectural constrains.

Possible uses of renewable energy sources have not been taken into consideration.

Considering the entire building stock, starting from the specific consumption indicator weighted on the surfaces of 144 kWh/m^2 per year (baseline), with Scenario 1 it reduces to 131 kWh/m^2 per year; with Scenario 2 it reduces to 115 kWh/m^2 per year; and with Scenario 3 it is reduced to 75 kWh/m^2 per year.

The cost of Scenario 3, roughly equal to 1000 €/m^2 makes the application of energy retrofit actions not acceptable.

4 Increase in Environmental Sustainability in School Buildings: Case Studies

This paragraph discusses a feasibility study which provides a considerable improvement in the environmental quality of 14 school buildings (pre-schools, primary and secondary) located in Cesano Boscone and Trezzano sul Naviglio, two municipalities in Milan province of northern Italy. The objective is to ensure the requirements for LEED® certification according to V2 version of the protocol (USGBC 2009).

For the school buildings the "Green Energy Audit" (GEA) procedure described in (Dall'O' et al. 2012; Dall'O' 2013) was applied, in order to verify the possible improvement of energy efficiency and environmental quality, in accordance with the LEED® for schools rating system. The objective of the study was to ensure at least the minimum requirements for obtaining LEED® certification (Dall'O' et al. 2013).

The aim of GEA is to evaluate the degree of improvement in sustainability of the building as a whole that can be obtained through the proposed choices; such choices do not necessarily generate an advantage in terms of energy saving, but they can generate many advantages as regard to sustainability. If the standard of comparison is the LEED® Protocol (USGBC 2009), then the problem is in understanding how the application of a certain retrofit action can help to meet the credits. GEA, therefore, integrates two strategic elements, energy saving and environmental impact reduction, by mixing the Energy Audit and LEED® methodologies.

This synergy strengthens the role of the classic energy audit by providing a method that not only optimizes the energy performance of existing buildings but also achieves a green retrofit of buildings.

Table 1 summarizes the main technical characteristics of the school buildings considered in the study, while Table 2 shows also the green house gases (GHG) reduction, assessed as total savings of each building, resulting from the implementation of all retrofit measures, and the energy saving for each building school calculated according to the prerequisite 2 of the LEED Protocol.

To obtain LEED® certification, the applicant projects must satisfy all the prerequisites and should be qualified for a number of points to attain the minimum established project ratings equal to 40 points (red line in Fig. 4).

Having satisfied the basic prerequisites of the program, the applicant projects are then rated according to their degree of compliance within the rating system: eight buildings fall within the level of Certified with an average score equal to 46.1, while the remaining six reach the Silver with an average score equal to 50.7. So our objective to achieve LEED® certification for all buildings while maximizing energy performance has been achieved. The study shows that there is a technical feasibility: the credits are between 42 and 54 (see Fig. 4).

The economic evaluation was conducted considering the costs of retrofits (hard cost) but also soft costs and the cost of Green Building Certification Institute (0.4%).

Cost items considered in the economic evaluation concern: Building envelope retrofit cost, heating systems retrofit cost, ventilation systems cost, solar PV cost (for the installation of a polycrystalline PV system), Green Building Certification

Table 1 Data of some characteristics of the buildings

Bldg.	Type[a]	Year	Occupants	Net surf. (m^2)	Volume (m^3)	Site area (m^2)	Bldg. footprint (m^2)
#1	PS	1965–1966	260	2345	9920	5770	1521
#2	SS	1980	352	5190	31345	8144	2060
#3	PS	1976	303	3300	21504	18259	2696
#4	PS	1972	238	2805	11634	6339	1980
#5	NS	1974	180	1144	5468	5143	1266
#6	PS–SS	1974	617	6019	28808	12210	3302
#7	NS	1973	132	688	3045	14974	837
#8	NS	1976	185	1124	4248	4491	1190
#9	NS	1974–1984	131	714	2598	5132	841
#10	NS	1973	137	1144	5468	4787	1265
#11	PS	1962	253	1833	8120	45608	13799
#12	NS	1968	146	876	3478	31246	11300
#13	NS	1933–1974	137	773	2978	29555	9535
#14	PS	1966–1984	242	2882	12809	63322	19187

[a]PS primary school, SS secondary school, NS nursery school

Table 2 Data for primary energy demand and percentage of primary energy savings

Bldg.	Primary energy for heating and ventilation (kWh/m^3 year)	Primary energy for domestic hot water (kWh/m^3 year)	Primary energy for lighting (kWh/m^3 year)	Primary energy for process energy (kWh/m^3 year)	Primary energy for renewable energy (kWh/m^3 year)	Emissions savings (tCO_2)	Percentage of primary energy savings (%)
#1	5.65	0.27	2.58	8.32	3.36	52.64	67.6
#2	5.71	0.12	1.80	6.35	3.32	140.89	66.4
#3	6.77	0.15	1.67	6.44	3.62	79.51	64.6
#4	6.73	0.21	2.63	9.19	5.87	67.51	72.0
#5	7.92	0.34	2.28	8.66	2.90	28.90	62.4
#6	5.81	0.55	2.63	8.88	3.11	163.29	66.8
#7	9.57	0.45	2.46	8.54	6.73	14.20	66.5
#8	9.76	0.45	2.88	10.48	5.67	26.16	65.8
#9	12.96	0.52	3.00	11.71	11.47	16.85	71.4
#10	9.91	0.26	2.28	7.20	6.81	19.50	64.3
#11	10.65	0.23	3.60	10.61	4.54	48.22	61.3
#12	10.68	0.44	1.95	10.94	10.13	22.73	74.6
#13	10.97	0.48	3.29	11.55	7.27	19,89	67.1
#14	8.38	0.20	2.30	9.62	6.33	75.10	70.5

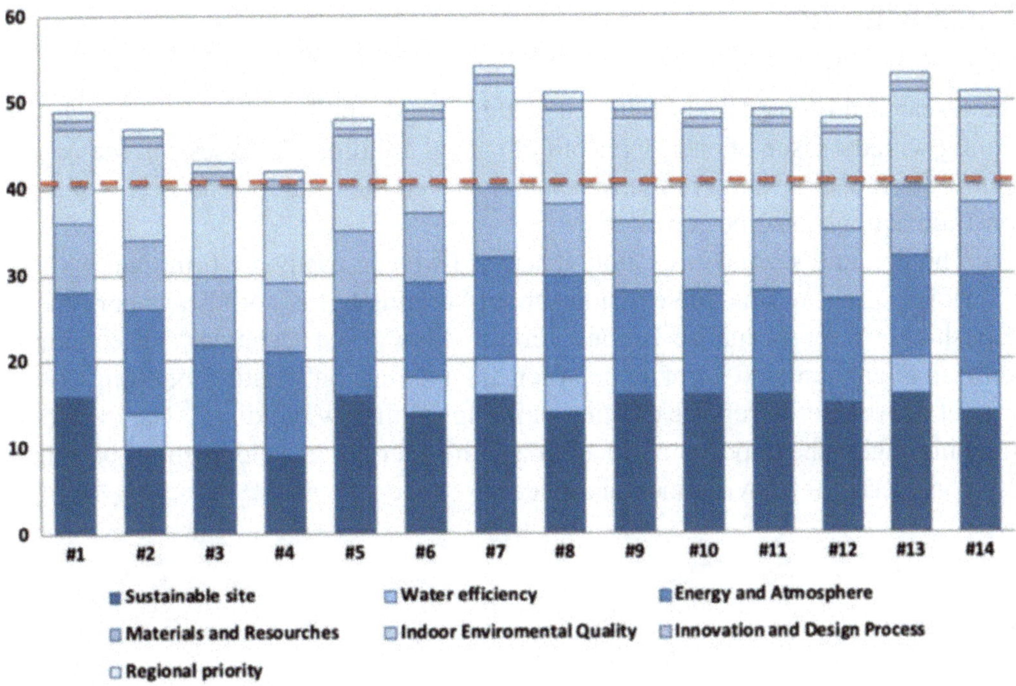

Fig. 4 Potential to improve sustainability of school buildings

Institute cost (related to LEED® certification), soft cost (related to building design that meets LEED® standards), increased renovation cost (related to higher cost of renovation to satisfy LEED® standards), water efficiency cost (related to installing water flow reducers and double flow toilets).

The cost of building envelope retrofit is the highest cost item with 53.2% of the total cost; heating systems retrofit is the second largest cost item with 29.7% of the total cost (the cost of building envelope and heating systems retrofit is therefore 82.9% of the total cost).

The economic issue remains, however, and is even greater when operating inside the public market, which is made up of public buildings such as schools.

The following question arises: should in the sector of public building retrofit strategies should be limited to an improvement of the energy performance or should aim to improve the sustainability as well? The purpose of this study was also to give a response to this question. Considering the feedback emerging from our research, which is based on concrete examples of school buildings subjected to green energy audit, we can state that it is a more appropriate aim to improve the sustainability.

Given that the increased spending is due to the portion of energy retrofits, when a building is under redevelopment we should look beyond. It is time to orientate strategies toward sustainability targets. This choice is particularly important for the school buildings for a better comfort and with a higher indoor air quality contributing to improve the conditions for learning.

5 Conclusions

The research presented and discussed in this chapter highlights a great interest in dealing with the issue of energy retrofit of school buildings. In assessing the opportunities for reducing energy requirements, however, the economic aspects that often constrain actions must be considered.

In the first and second case study discussed we can easily confirm that, while it is very important to upgrade existing school buildings, it is not always convenient to push energy performance beyond certain values. The technical and economic constraints encountered in practice when intervening on existing buildings often make it convenient to replace existing buildings with new buildings. The third case study highlights the opportunity to approach the energy redevelopment of buildings also considering the environmental aspects.

References

Dall'O' G, Speccher A, Bruni E (2012) The green energy audit, a new procedure for the sustainable auditing of existing buildings integrated with the LEED Protocols. Sustain Cities Soc 2012(3):54–65

Dall'O G (2013) Green energy audit versus LEED® protocols. In: Green energy audit of buildings, a guide for sustainable energy audit of buildings, 1st edn, Chapter 9, pp. 213–240. Springer, London

Dall'O' G, Bruni E, Panza A (2013) Improvement of the sustainability of existing school buildings according to the leadership in energy and environmental design (LEED)® Protocol: a case study in Italy. Energies 6:6487–6507. https://doi.org/10.3390/en6126487

Dall'O' G, Sarto L (2013) Potential and limits to improve energy efficiency in space heating in existing school buildings in northern Italy. Energy Build 2013(67):298–308

International Organization for Standardization (ISO) (2008) ISO 13790: 2008— Energy performance of buildings—calculation of energy use for space heating and cooling

United States Green Building Council (USGBC) (2009) LEED reference guide for green building design and construction. USGBC, Washington, WA, USA, ISBN: 9781932444346

www.ingramcontent.com/pod-product-compliance
Lightning Source LLC
LaVergne TN
LVHW082117200926
848484LV00008B/1964
* 9 7 8 1 0 1 3 2 7 2 5 1 6 *